·City A–Z·

Edited by
Steve Pile and Nigel Thrift

London and New York

First published 2000
by Routledge
11 New Fetter Lane, London EC4P 4EE

Simultaneously published in the USA and Canada
by Routledge
29 West 35th Street, New York, NY 10001

Routledge is an imprint of the Taylor & Francis Group

Designed and typeset in Sabon
by Keystroke, Jacaranda Lodge, Wolverhampton
Printed and bound in Great Britain
by TJ International Ltd, Padstow, Cornwall

British Library Cataloguing in Publication Data
A catalogue record for this book is available from the British Library

Library of Congress Cataloging in Publication Data
City A–Z / edited by Steve Pile and Nigel Thrift.
p. cm.
Includes bibliographical references and index.
1. Cities and towns—Miscellanea. 2. Sociology, Urban—Miscellanea.
3. Urban geography—Miscellanea. I. Pile, Steve II. Thrift, N. J.
HT153 .C557 2000 99-053842
307.76—dc21

ISBN 0-415-20727-4 (hbk)
ISBN 0-415-20728-2 (pbk)

Contents

Contributors

John Allen is Head of Geography at the Open University, where he has edited and co-authored such volumes as *A Shrinking World? Global Unevenness and Inequality* (1995), *Rethinking the Region: Spaces of Neoliberalism* (1998) and *Human Geography Today* (1999). He has published widely in economic geography and is currently researching Berlin's changing fortunes.

Ash Amin is Professor of Geography at Durham University. He is currently working on the social economy and on the democratic city. His latest book is *Beyond Market and Hierarchy: Interactive Governance and Social Complexity*, co-edited with Jerzy Hausner. He co-edits the journal *Review of International Political Economy*.

Marc Augé is Director of Studies at the École des Hautes Études en Sciences Sociales, Paris. He has published widely on the anthropology of the contemporary world. Of his many books, the most recent are *Non-Places: Introduction to an Anthropology of Supermodernity* (1995), *A Sense of the Other* (1998), *An Anthropology of Contemporaneous Worlds* (1999) and *The War of Dreams* (1999).

Anette Baldauf, cultural critic, studies and teaches at the New School for Social Research in New York and at the University of Vienna, is co-editor of the anthology *Lips. Tits. Hits. Power? Feminismus und Pop* (in press) and the co-author of *Women's Studies im Internationalen Vergleich* (in press). She is currently working on an extensive research project entitled 'Shopping: a comparative analysis of two shopping streets: Broadway (NY) and Mariahilferstraße (Vienna)'.

Andrew Barry teaches sociology at Goldsmiths College, University of London. He has previously worked as a laboratory technician in London and Stockholm and as a political researcher in Brussels and Luxembourg. He has written extensively on the history and sociology of technology.

David Bell teaches cultural studies at Staffordshire University. His publications, as editor or author, include *Mapping Desire* (1995), *Consuming Geographies* (1997), *Hard Choices* (2000), *City Visions* (2000) and *The Cybercultures Reader* (2000).

Walter Benjamin appears with the kind permission of Verso Press – he has a ghostly presence throughout this book.

Gargi Bhattacharyya is employed by the Department of Cultural Studies and Sociology, University of Birmingham. She is the author of *Tales of Dark-Skinned Women* (1998), among other dubious fictions. Long-time teller of tall tales and shameless fibber, she is working hard to reintroduce an occasional giggle into the tortuous process of education. Who knows with what success?

Alastair Bonnett is Lecturer in Human Geography at the University of Newcastle upon Tyne. He edits an increasingly irregular magazine, *Transgressions: A Journal of Urban Exploration*, and is the author of *Anti-racism* (1999) and *White Identities: Historical and International Perspectives* (1999).

Iain Borden is Director of Architectural History and Theory, and Reader in Architecture and Urban Culture, at The Bartlett, University College London. A founding member of Strangely Familiar, he is co-editor of a number of books, including *Architecture and the Sites of History* (1995), *Strangely Familiar* (1996), *The Unknown City* (2000), *InterSections* (2000), *Gender Space Architecture* (1999) and *The City Cultures Reader* (2000).

Stephen Cairns teaches architectural design and theory at the University of Melbourne. He has published on the intersections of architectural theory and postcolonial criticism, and is the Arts Editor of *Postcolonial Studies*.

Iain Chambers teaches at the Istituto Universitario Orientale, Naples. Author of *Migrancy, Culture, Identity* (1994), and editor (with Lidia Curti) of *The Post-colonial Question. Common Skies, Divided Horizons* (1996), he has written extensively on modernity, music and metropolitan cultures and is presently working on architecture, geography and the cultural implications of dwelling.

Ivan da Costa Marques is trying to explore the partial connections between enchantment and impoverishment of the Third World by modern technology. He has performed as an electronic development engineer, a policy maker and a businessman in the 1970s and 1980s. He is now teaching and doing 'science studies' at the Federal University of Rio de Janeiro.

James Donald is Professor of Media at Curtin University of Technology in Perth, Western Australia. He is the author of *Imagining the Modern City*, and *Sentimental Education: Schooling, Popular Culture, and the Regulation of Liberty*, and editor of a dozen books on aspects of the media (especially cinema), education, and cultural theory.

Arturo Escobar was raised in Cali, Colombia, where he studied chemical engineering, after which his interests shifted to the culture and politics of development. He currently teaches in the anthropology department at the University of North Carolina, Chapel Hill. His recent work is on rainforest political ecology. In the summers of 1996 and 1997 he began preliminary research at the City Planning Office in Cali for a future year-long project on urban political ecology in this city.

Jill Forbes is Professor of French at Queen Mary and Westfield, University of London. Her research focuses on postwar French society and cultural studies, and her most recent books are *French Cultural Studies* (with Michael Kelly, 1995) and *Les Enfants du Paradis* (1997). She is currently writing a book on European cinema.

David Gilbert is Senior Lecturer in Geography at Royal Holloway, University of London, and lives in West London. His recent work has concerned London's role as capital of empire, and he is the editor with Felix Driver of *Imperial Cities: Landscape, Display and Identity* (1999). He is now developing a broader project exploring the cultural geographies of twentieth-century London.

Stephen Graham is Reader at the Centre for Urban Technology (CUT) in Newcastle University's School of Architecture, Planning and Landscape. His research addresses the relations between new information technologies, urban development, urban theory and urban policy. He is the joint author, with Simon Marvin, of *Telecommunications and the City: Electronic Spaces, Urban Places* (1995).

Dolores Hayden's poetry has appeared in

Witness, *Poetry Northwest*, *Hellas* and *The Formalist*. Her chapbook, *Playing House*, has just come out (1998). An architect and urban historian, her most recent non-fiction book is *The Power of Place: Urban Landscapes as Public History* (1995). She lives in Guilford, CT, and teaches at Yale.

Steve Hinchliffe is Lecturer in Geography in the Faculty of Social Sciences, Open University. His research interests focus on science, technology and nature and he has written on a diverse range of subjects from energy debates, to outdoor management training, to city-natures and ontologies of disease.

Jane M. Jacobs teaches geography at The University of Melbourne. Her books include *Edge of Empire: Postcolonialism and the City* (1996); *Cities of Difference* (1998) and *Uncanny Australia: Sacredness and Identity in a Postcolonial Nation* (1998).

Joe Kerr is Senior Tutor in Humanities at the Royal College of Art, London. A founding member of Strangely Familiar, he is co-editor with Iain Borden, Jane Rendell and Alicia Pivaro of *Strangely Familiar: Narratives of Architecture in the City* (1996) and *The Unknown City: Contesting Architecture and Social Space* (2000). He is currently writing, with Murray Fraser, a book entitled *The Special Relationship: the Influence of America on British Architecture since 1945*.

Anthony D. King flies between the UK and the US where, as Professor of Art History and of Sociology, State University of New York, Binghamton, he parks on one of 500 earth-sealed slots provided for 12,000 faculty, students and staff. With Tom Markus, he co-edits Routledge's Archi*text* series on architecture and social/cultural theory.

Mary King grew up in the town of Elmwood, Connecticut. She graduated from the Massachusetts College of Art in Boston and is a graduate student at the University of Massachusetts. As an anthropology major she combines a love of art, story and poetry with her work among economically disadvantaged youth.

Ross King is Dean of the Faculty of Architecture Building and Planning at the University of Melbourne; he has practised as an architect, planner and policy analyst, and taught at the universities of Sydney and Melbourne. He has most recently published *Emancipating Space: Geography, Architecture, and Urban Design* (New York: Guilford, 1996).

John Law is Professor of Sociology at Lancaster University. He has written extensively on the sociology of technology, organisational sociology, and on actor-network theory, and is currently working on disabilities, medical technologies, materialities and subjectivities. His recent publications include *Machines, Agency and Desire* (co-edited with Brita Brenna and Ingunn Moser) (1998), *Organizing Modernity* (1994) and *A Sociology of Monsters* (editor) (1991).

Deborah Levy is a novelist, playwright and critic. She has published five novels, including *Swallowing Geography* (1993), *The Unloved* (1995), *Diary of a Steak* (1997) and most recently *Billy and Girl* (1999).

Peter Marcuse, a lawyer and urban planner, is Professor of Urban Planning at Columbia University. Now at work on a history of the New York City Housing Authority, he has written widely in the field, including articles on housing policy, racial segregation, urban divisions, New York City's planning history, property rights and privatization, and the history of housing.

Satomi Matoba is a Japanese artist living in London and Hiroshima. Her work explores the sense of existing between two cultures, not truly at home in either; needing constantly to renegotiate her space between conflicting systems. In the series *Map of Utopia*, which includes 'Pearl Harbor/Hiroshima', she dissolves familiar maps in order to reconstruct the world, suggesting an open-ended potential in the act of shifting boundaries.

Eugene McLaughlin teaches criminology at the Open University. His primary research interest is in urban policing and his recent publications include *Community, Policing and Accountability* (1994) and *The Problem of Crime* (1996). He is currently writing a book on the theme of policing 'after' postmodernism, and is researching London's policing following the Stephen Lawrence inquiry.

Nicola Miller teaches sociology at Goldsmiths College, University of London. She is currently writing up her Ph.D. thesis on the state funding of dance.

Harvey Molotch is Professor of Sociology at New York Uinversity and the University of California, Santa Barbara where he works in the areas of urban political economy, industrial design and environmental studies. He has been Visiting Centennial Professor at the London School of Economics as well as Visiting Professor at Lund, Essex, Stony Brook (New York) and Northwestern Universities; besides whatever insights into cities his travels may have brought, they have earned him frequent flyer Premier status with United Airlines.

Miles Ogborn is Lecturer in Geography at Queen Mary and Westfield College, University of London. He has written extensively on power, the production of space and the cultural politics of the changing city. He is author of *Spaces of Modernity: London's Geographies, 1680–1780* (1998), and lives in Hackney.

Adrian Passmore considers the canon of urbanisation to include Berlin, Paris, and Bodmin. If he became Miss World he too would crave global peace; however, since he is male he wants a gun, an atlas or a motorbike for Christmas. He suffers from an intellectual version of Tourette's Syndrome.

Steve Pile teaches in the Faculty of Social Sciences at The Open University. His recent books include *The Body and the City* (1996), *City Worlds* (1999, co-edited with Doreen Massey and John Allen) and *Unruly Cities?* (1999, co-edited with Chris Brook and Gerry Mooney). Now, he is looking forward to getting down to some serious dreaming.

Roy Porter is Professor of the Social History of Medicine at the Wellcome Institute for the History of Medicine. He is interested in eighteenth century medicine, the history of psychiatry and the history of quackery. Recent books include *London: A Social History* (1994), and *'The Greatest Benefit to Mankind': A Medical History of Humanity* (1997). His general history of the Enlightenment in Britain is forthcoming.

Allan Pred is Professor of Geography at the University of California, Berkeley. Writing on the historical geography of the present, he has sought a congruence between textual strategy and theoretical and empirical contents. Recent books include Recognizing *European Modernities* (1995), *Reworking Modernity: Capitalisms and Symbolic Discontent* (1992, with Michael Watts) and *Even in Sweden* (forthcoming).

Jane Rendell is Lecturer in Architecture at the University of Nottingham. She is an architect and architectural historian, interested in gender and architecture, theory and practice. She is co-editor of *Strangely Familiar: Narratives of Architecture in the City* (1995); *The Unknown City: Contesting Architecture and Social Space* (2000); *Gender Space Architecture: An Interdisciplinary Introduction* (2000); and editor of 'A Place Between', *Public Art Journal* (1999).

Jenny Robinson is Lecturer in the Faculty of Social Sciences at the Open University. She is author of *The Power of Apartheid* (1996), and is currently working on spaces of interaction and transformation in South African cities. Living in London, she heads off to South Africa and Durban's beaches whenever possible.

Saskia Sassen is Professor of Sociology at the University of Chicago and Centennial Visiting Professor at LSE. Her most recent books are *Globalization and its Discontents* (1998) and *Losing Control? Sovereignty in an Age of Globalization* (1996). She directs the project 'Cities and their Crossborder Networks' sponsored by the United Nations University and continues work on her project 'Governance and Accountability in a Global Economy'.

Gail Satler is an Associate Professor of Sociology at New College of Hofstra University. Her book, *Frank Lloyd Wright's Living Spaces*, was out this year. She is currently writing a book on the Chicago School's architecture and sociology and is also engaged in research projects on New York City restaurants and Fourteenth Street, Manhattan.

Helen Scalway is an artist working in London, currently a Research Associate at Camberwell College of Arts. She has a particular interest in the representation of the experience of urban space. The drawings by London Underground passengers are taken from her artist's book, *Travelling Blind* (1996).

Rob Shields is an editor of *Space and Culture* and author of *Places on the Margin: Alternative Geographies of Modernity* (1992), and *Henri Lefebvre: Love and Struggle* (1999). He has edited several books on aspects of spatiality, including *Cultures of Internet* (1996) and *Lifestyle Shopping* (1992). His work includes comparative studies of urban cultures in Korea and Canada, and ethnographies of innovation and building construction. He is Associate Professor of Sociology and Anthropology at Carleton University, Ottawa.

David Sibley teaches cultural geography at the University of Hull. He is particularly interested in psychoanalytical interpretations of place and landscape and in the placing of nomads, children and animals. He likes cycling round cities, and envies middle-aged Italian pedallers who can dress like racing cyclists without looking stupid.

Susan J. Smith is a geographer concerned about social inequality in all its guises. She has worked on: victimization and fear of crime, the politics of race, gender inequality in housing, and health inequalities in Britain. She dabbles in the world of music and, recognising that this is as much about the exercise of power as it is about the enjoyment of aesthetics, worries about uneven access to cultural resources and to the political spaces they occupy.

Helen Thomas is Reader and Head of Sociology at Goldsmiths College, London. She has published widely on issues around dance and society. She is editor of *Dance in the City* (1997) and *Dance, Gender and Culture* (1993), and author of *Dance, Modernity and Culture* (1995). She is co-editor of a new series for Macmillan on *Dance and Culture*.

Nigel Thrift is Professor of Geography in the School of Geographical Sciences, University of Bristol. His chief interests are in the spaces of technology, the growth of soft capitalism, and theories of weak subjectivity. He is currently writing a social history of time consciousness with Paul Glennie. His most recent books include *Spatial Formations* (1996), and *Money/Space* (1997, with Andrew Leyshon).

Fran Tonkiss is Lecturer in Sociology at Goldsmiths College, University of London. She teaches and writes in the areas of urban studies and economic sociology, with a particular interest in changing forms of economic and urban governance. She is the author (with Don Slater) of *Markets, Modernity and Social Theory* (in press).

Katharina Weingartner, cultural critic, writes and produces radio features for German, Swiss and Austrian NPR out of New York, and is co-editor of the anthology *Lips. Tits. Hits. Power? Feminismus und Pop* (in press). She is currently participating in a research project, 'Shopping', with Anette Baldauf and working on her first film documentary, *Prison Blues*.

Sharon Zukin is Broeklundian Professor of Sociology at Brooklyn College and the City University Graduate School in New York City. She has written *The Cultures of Cities* (1995), *Landscapes of Power: from Detroit to Disney World* (1991), and *Loft Living* (1982) and is currently working on a book on shopping.

PREFACE

This book can be seen as a contribution to a wave of experimentation which is concerned with writing the city. This wave is gaining momentum, washing up against the problems encountered by previous generations of urban experimentalists and then splashing on against a few more for good measure. And, as the wave breaks, so the city's cultural sightlines are being shifted once more. New losts and new founds. New smells and new sounds.

At least the problems are clear enough. To begin with, there is the issue of writing the city. Granted that writing the city is never simply an act of reportage, but can we go to the other extreme and think of the city as a text without source? Even without a documentary source, however, 'writing the city is not the same as writing about the city' (Wolfreys, 1995, p. 8). Then, following hard on the heels of the first issue, there is the question of what can in any case be brought back to print. Perhaps the city is such a complex jumble of practices that it cannot be reported? Or perhaps somewhere there is a transcendental map which, though it consistently shifts registers and locations, can act as a sort of compass, for however brief a flash of time: the city's dream of itself? Then there is the question of what has to be left out of any account of the city. Can the pulse of bodies and the push of objects ever be captured in language? Even many poets have their doubts (Perloff, 1996). Then, there is one more issue: how to find a rhetoric that balances the serious, the patient, and the oppressive side of cities with their spontaneous vitality and capacity to induce play? Cities are too often written as either monster machines or improvisatory frontierlands, rarely as both at once (Lefebvre, 1995).

What is also clear is the sheer amount of work currently underway in an attempt to restate and sometimes even answer these questions. Take only the example of London. The sheer amount of experimental and quasi-experimental writing on London currently transcribing and translating every highway and byway of the city is almost beyond comprehension by itself. There are the various collections of new writing (e.g. Granta, 1999), there are Peter Ackroyd's often mystical rounds of the city, Michael Moorcock's magical histories, and Iain Sinclair's psycho-geographical excursions. There are numerous collections of poetry. And this is to ignore all the other media, like film, art, music, theatre, performance poetry, and the various forms of events that try to temporarily transform and re-map the city, such as the series of performances arranged by Artangel which have taken in walks across London with digital camera and DAT recorder, legislative theatre, and the refurbishing of a disused underground station as a kind of prehistoric cave.

Most of this kind of work at one point or another plays with the idea of reinvesting traditional forms of urban address (the walking tour, the guidebook, telephone yellow pages, the armoury of signposts, the slide lecture, the pre-recorded video guide, the night out (Schlör, 1998), and so on).

This book is no different. By way of the conceit of an A to Z guide, one of the most important means of detailing the city's presence, our various authors attempt to rewrite the commonplaces which inhabit the city as something both general and specific: as a set of maps, but in other words.

In this way, we hope to provide not so much an overview as an 'underview' of the confusion and profusion of the city, and an understanding of the city's complexity which is correspondingly modest. The goal is similar to that of a recent book, by Latour and Hermant (1998), on Paris, in which, reflecting on the difficulties of seeing Paris as a whole, these authors play on the double meaning of 'overlooking' as simultaneously seeing everything in the city and inevitably missing most of it (see also Réda, 1996). This volume contains the same muted panoramic ambition (which is not to say it has no ambition at all!). Hopefully, what the volume can do is provide a means of demonstrating that city maps and guides are made up of points of departure (Moretti, 1998). They are territories in which every place is always a junction on the way to somewhere else: the roundabout heres of constant circulation.

A volume like this inevitably depends on the contributions of many intermediaries to produce a smooth passage. Of these, we would like to thank in particular, Sarah Lloyd and Sarah Carty – and in the earlier stages of the project, Tristan Palmer – all at Routledge. We have benefited from assiduous help from within our institutions, including Simon Godden (who drew the city letterheads), John Hunt (who redrew many of the diagrams), and Karen Ho, Michèle Marsh, Kit Kelly and Margaret Charters (who typed some of the entries). We have been able to reprint 'entries' by Walter Benjamin with the kind permission of Verso Press. Others have been equally willing to give permission to reprint various items; they are credited in the body of the text.

Most of all we want to thank our many authors for their enthusiasm, imagination – and patience.

Nigel Thrift
Steve Pile

References

Granta (1999) 'London: the lives of the city', *Granta: the magazine of new writing*, No.65.

Latour, B. and Hermant, E. (1998) *Paris. Ville Invisible*, Paris: Institut Sythelabo/La Decouverte.

Lefebvre, H. (1995) *Introduction to Modernity*, London: Verso.

Moretti, F. (1998) *Atlas of the European Novel, 1800–1900*, London: Verso.

Perloff, M. (1996) *Wittgenstein's Ladder. Poetic Language and the Strangeness of the Ordinary*, Chicago: University of Chicago Press.

Réda, J. (1996) *The Ruins of Paris*, London: Reaktion.

Schlör, J. (1998) *Nights in the Big City. Paris, Berlin, London. 1840–1930*, London: Reaktion.

Wolfreys, P. (1995) *Writing London. The Trace of the Urban Text from Balzac to Dickens*, London: Macmillan.

Cover artwork

The images on the cover of this 'A to Z' are taken from an exhibition of the work of Satomi Matoba and Helen Scalway, entitled *Remappings*, held at the Cable Street Gallery in London (15 January to 7 February 1999). Their project to 'reconfigure familiar maps to produce unfamiliar insights' has strong resonances with the aims of this book. And more so because they are seeking to reconfigure familiar maps of cities. The book cover is a combination of two such works – Satomi's *Map of Utopia* and Helen's *Travelling Blind* – but let them explain them themselves:

'Map of Utopia'
by Satomi Matoba

I started the series 'Map of Utopia' by transplanting my hometown Hiroshima into Western Europe. Traditional Japanese conformity is a hierarchy of sexual and class discrimination, supported by filial piety, interdependence and self denial, particularly by women. I was dreaming of how this could shift if the underprivileged have a chance to see another world where they could have a better life and find that their oppressive system is not unchangeable. On the other hand, I was thinking of how European people might find relief in the security of belonging to a supportive community, in comparison with the rather isolated hard life of Western individualism. My interest is in the process of the interchange of people, which may become inevitable when their territory overlaps and they have chosen to cohabit.

In 'Pearl Harbor/Hiroshima', I am asking myself: if these two places had been closely located, could such brutal crimes have been committed? I wonder how we can live with others and Others, preferably more happily, how we can resist our own hostility and apathy against outsiders; how we can be more friendly to strangers who may not share our code, who because of that appear potentially dangerous; how we can expose ourselves and protect ourselves at the same time, what makes strangers into enemies, how individuals become excluded from communities, how we can leaven the rule of old-fashioned community with a sense of justice, and so on.

Contemporary lives both in the West and the East are quite different from what they used to be, but we are still controlled by the ghosts of obsolete power, as they are invisible from inside. For example in Japan, the traditional farming community depended upon the subordination of the needs and desires of the individual to the expectations and well-being of the community as a whole. Whilst these structures are nominally obsolete, their influence clearly moulds contemporary social expectations, to the extent that the rejection by an individual of conventions and values is liable to invite exclusion. Consequently, Japanese often hesitate to speak out, and this tendency prevents Japanese contemporary society from abolishing discriminations. The borderline between the Japanese principle of 'community in harmony' and totalitarianism is very easy to cross, as is the border between Western individualism and egoism. I believe that the idea of democratic community should not be given up, no matter how utopian it sounds; I want to find hope in the subtle changes of our mentality and attitude which we can achieve through everyday personal experience.

Nowadays, life-styles and mentalities in urban areas, even in different countries, may be more similar than life-styles existing within the same country's urban and rural areas. Moreover the generation gap within a family could be more significant than anything else. Cultural gaps exist less and less geographically and we are losing the safe distance between ourselves and the 'alien' world more and more. The conflict becomes inevitable. We can't ignore each other when we interfere with each other, but we can't kill each other when we need each other. When we think of our own brutality which enables us to do anything against people with whom we have nothing in common, we find the importance of interdependence and mutual reliance.

A typical example of this dilemma can be found in the mixed-race marriage and its hybrid offspring, or in the confusion of a rapidly changing society like Japan where education given after the Second World War is critically incompatible with that given before. In such situations we may not be able to understand each other, but we can still acknowledge each other, especially our difference, which may be the most important thing.

Another possibility: could we open ourselves and make the distinction between insider and outsider invalid? In order to do that, it may be important to stay in the transitional state, not to cleave to the safety of convention, because we are living in a world in transition. An unsettled feeling of living in 'no man's land' often becomes unbearable; nevertheless, it may be the right way to give everyone a loving and secure place, in other words, utopia.

For those who experience alienation and exclusion from their unhomely 'home', the liminal space left for them is the frontier to establish their new 'home.' The ambiguous advantage of liminality can be to achieve a flexible, open, and independent state of mind. I feel that to retain ambiguity is to retain the possibility of living within an open potential, and such an attitude is important. It keeps us from prejudice as well as protecting us. Moreover the recognition that we are all strangers to each other protects us from becoming the oppressors of others. Utopia is not a paradise to dream of, but the home we need to recover, from which we can see the world in order. For me it is a desire to find my place among others.

The 'Map of Utopia' is a metaphorical landscape of the postmodern global world of pluralism, or the innerscape of a cosmopolitan. It is an invitation to a 'no man's land' where convention is not reliable; so that we have to be vigilant and think independently, in order to gain better understanding of ourselves and to acquire the skill to navigate in the ever shifting world. It is an invitation to be a 'stranger'.

'Travelling Blind'
by Helen Scalway

Drawings collected from passengers on the London Underground

Fineliner on paper 1996.

These drawings are a small sample from a much larger number of mental maps which I collected from passengers waiting for trains on London's Underground.

How do people in cities mentally construct the urban spaces they move through in order to inhabit them and to navigate them? I was curious to bring to light the personal geographies of Tube travellers whose private copings with the city's space might mingle strangely with the authoritative suggestions of official maps.

In London the iconic Tube Map, based on Harry Beck's original diagram, achieves its remarkable elegance by distorting or compressing time, space and direction. For who would wish to see in the darkness of the city under the city?

So I sat about on Underground station benches and asked in turn anyone who unwittingly sat down next to me, over a hundred total strangers waiting for their trains: 'Please draw your London Underground network.'

The drawings were gathered from just two stations in the first instance: Fulham Broadway in south-west London, and Leyton Tube Station in north-east London. The locations were chosen to give some sense of the polarity of the city. In a second series, I started collecting drawings at the rate of five an hour from the first train at 5.07 a.m. from a suburban tube station, moving inwards, to Piccadilly, the centre, on a Friday at midnight. At five in the morning, no one wanted to draw. Midnight at Piccadilly, everyone did. Taken by surprise, trains due in one or two minutes, no one had time to embellish or pretend.

As travellers drew 'their' Undergrounds, their hands re-enacted their different experiences of the space. There are drawings that register in the uncertain pressure of the pencil, how the mind crept uncertainly from point to point. Other hands dashed or swept confidently around. Many hands hovered tentatively above the paper, and then sketched doubt or question in the air before descending to make the mark. Elderly hands trembled the pen across the paper. In every case the line traced a thought, despite differing levels of ease with the concepts of 'drawing' and 'network'.

As different individuals drew 'their' underground networks, unfamiliar tube maps briefly, hesitatingly, triangulated themselves in an unlocatable territory between actual routes, imagined routes and the routes officially mapped.

But do these drawings offer anything more than the poignant spectacle of mass misconception and the multiplicity of human isolations?

I think something else emerges from these pictograms. It is that the very spareness of the official Tube Map has freed our conceptions of our London Underground journeys from all constraints of geographical accuracy; and in doing so has opened up a different kind of space for imagination. This is a space collectively imagined, dreamed forth, where each fragment contributes to a specifically London variant of the image found so often in the literature of the modern city: a shape-shifting labyrinth, shared and yet not shared. Largely shaped and mis-shaped by the London Underground Map, this labyrinth of the communal imagination can be seen as a space simultaneously physical, imaginary and metaphoric; a richly charged territory.

And what did passengers say, when asked: Please draw your London Underground network? Their responses went like this:

> That's easy that's impossible I can do that easily I'm really bad at that sort of thing
>
> I can't draw I can't draw that should be quite simple I really couldn't no no I couldn't oh all right then
>
> There's no picture in my head I ignore it I don't look it's vague it's clear I just don't know
>
> Oh my god oh my god oh my god oh lordy lordy oh the shame of it
>
> I haven't a clue I haven't a clue not a clue not a clue
>
> Something like this something like that somewhere over here somewhere over there somewhere
>
> I don't know where it goes I don't know I don't know of course I know I thought I knew

Are you making fun of me you are making fun is this a joke this is really cruel

This is really interesting I want to see if I can do it that was my train never mind I really want to finish this I want to see if I can do it thankyou thankyou that was so interesting

I hate this drawing by hand stuff it makes me seem tremulous I am not a tremulous person I wish I could do this on my computer then I could use straight lines I'm a straight line person I hate seeing myself like this the readymade characters in my computer would be much better

Don't ask him lady ask me he's useless I can do it better no you can't yes I can c'mon Todd c'mon Wurzel let's show the lady god this is difficult actually its impossible look here's our train thank god I told you you'd be useless you were worse I wasn't you were I wasn't bye, lady, bye

You want me to draw? – I will try . . . I'm not much use . . . I'm so sorry, so sorry, so sorry

I failed every exam at school this is a sort of 'nother test yes yes I see it as a test I really have to do this there's a train gone it doesn't matter I have to do this I have to finish it there's another one gone I'm going to do this there's another and another I've missed five trains doing this I don't care

I go there . . . then I go there . . . and then there . . . then round there . . . and up here . . . and round there and along here and then here . . . mmm . . . then I come back like this . . . and like this and like this . . . mmm . . . mmm . . . here I go, here . . . mmmmm . . .

I'm stoned . . . you want what? me to draw the London Unground? . . . who knows? . . . who knows? . . .

Guide to readers

by Steve Pile and Nigel Thrift

Stop! Close your eyes. Imagine a city. Think of all the things you can see. (You can open your eyes now!) Now try listing them. The inventory quickly grows . . . endless, isn't it? But there is more to the city than simply an inventory of the visible. There are some things that can't quite be seen: feelings, perhaps, or maybe a sense of the city's social relations. Eyes open or eyes closed, there are so many things to keep in mind. This doesn't mean that we should give up on thinking about cities, whether as a whole, or in the fine threads that link them together, or in their tiniest details. Far from it.

This book is born of a curiosity about cities. If we know one thing about cities, then perhaps it is that they are too vast to be imagined in their entirety: that is, as *a* city. More often, commentators have attempted to grasp cities by identifying one thing as symptomatic of the whole – the walled city; the Imperial city; the city of capital; the capital city – yet none has quite succeeded in capturing the whole of the city. Nor can cities be understood as if they were only the sum of their parts. They are not the sum of roads + buildings + money + people + . . . plus whatever. For cities are more than the inventory of things to be found in them, precisely because they are also about the social relations that constitute them. Cities are never static, or complete, ready to be held frozen in some timeless encyclopaedia of facts and fictions.

So, what are we left with? On the one hand, we have attempts to grasp the whole of cities by selecting a part or parts of the city. Even the 'whole' of the city, in this sense, becomes one of its parts. On the other hand, there are the parts of cities that never add up to the city. (As when 'the city' refers to the financial district.) Between these two options, there must be other approaches. Approaches that demand and foster other interpretations and analyses of the city, of cities, of city life. In some ways, this book is an attempt to create a city-like space for interpreting and analysing cities. To this extent, at least, it is a conceptual book. (For those of you who wish to explore further the more theoretical aspects of this book, there is a 'technical note' at the end of the book.) Here, however, we are more interested in how this book might be used to develop an understanding of cities and city life. Mostly, we suspect, this understanding will emerge from the way in which readers gradually build up a picture of cities from the various writings contained in this book. Of course, it will matter how this book is read – and how the picture is built up. For us, one of the most intriguing spin-offs of the book has been that it develops the (intellectual?) skill of connecting up disparate aspects of urbanity and, by extension, of thinking in more abstract ways about the various social relations that link cities and/or parts of cities. In a modest way, reading this book is an act of theorising – in part because different thoughts about, and associations between, the A to Z 'entries' will happen at different moments . . . depending. How like a city?

There are many ways in which this volume might be read to develop alternative readings of the city. Let us outline some possible tactics, all (or none!) of which you might wish to try.

First, you might wish to read this book from cover to cover, as it were, from A to Z. As you will see, the 'entries' have been listed alphabetically. In part, this is to echo the way in which the city brings unlikely elements into close proximity. As you read A to Z, you might find that the arbitrary juxtaposition of entries allows you to think something new or different about the city that you might not otherwise have realised. Maybe you will see common threads, say, between 'Celebration' and 'Cinema' or between 'Invisibility' and 'Job Centres'. In this way, the book might surprise – and become more than the sum of its parts. Just like the city itself.

Second, you might simply dip into the book. Allow the book to fall open and just read the entry there. Skipping around the book will enable you to focus on the disparate elements in the city, linked by no apparent underlying logic, simply there. In this way, you will be able to concentrate on the specific arguments being offered in each entry. Or perhaps you might make links between entries that seem otherwise unconnected. One might notice, say, how ideas of aesthetics or of social inequality crop up in disparate pieces. As likely, you will get a sense of how different the pieces are! Let us spend a little time, then, on the kinds of entries included. To begin with, you will not find a consistent 'voice' between entries. The book echoes the cacophony of voices found in the city, some angry, some ecstatic, some ironic, some playful, some obscure, some angry – but all expressive of experiences of the city. The entries deal with cities around the world, allowing the differences between them to sit side by side, seemingly without privileging one set of experiences over any other. In this way, the book seeks to accommodate and evoke the differences within and between cities, among and between people living in cities.

Third, it is possible to use the index to this book to identify places where common 'addresses' are used. We have emboldened 'entries' in the index that refer to 'entries' in the book to help you navigate through the A to Z.

Fourth, you might do what many visitors to a new city do: look at a map, try and figure out where you are and where you want to go, and follow the route that gets you there (though many get lost and end up going to unexpected places). At the back of the book is a map that outlines routes through this 'book of cities'. Like any map, it is partial. (And not altogether satisfactory.) However, you might want to follow specific routes, or to jump from one route to another, perhaps by using the interchange points. Importantly, cities are not entirely arbitrary, not simply a collection of fragments. So, through tracing the connections between fragments, you can begin to glimpse the ways in which the city is ordered. Or disordered. The map we have provided is suggestive of how you might wish to make these connections. Each line is thematic, and the 'stations' in the line are linked by this theme. However, themes like 'traffic' and 'nature' cover a host of sins. And some entries work more and less well under their thematic headings. Nevertheless, there is an important point being made here. It isn't so much that we think this map shows the essential connections between entries, or that this map is (or is not) comprehensive – there were, for example, many themes we did not use (to name a few: 'inequality', 'violence' and 'fantasy') and many connections between entries that we could not show (for example, we thought 'networks' were better networked!). What is being demonstrated is the capacity to make (y)our own maps of the city – and thus to follow routes that only you know about. Again, the city is comprised of many paths and passages – some are highways, some byways, but perhaps this book has most in keeping with those that are called 'rat runs'. And you may wish to use the map to scribble down new meeting points, sites of interest and lines of connection as you go along. Indeed, create your own map!

In some ways, however, it doesn't matter which tactic you use to read this book, for each has its advantages and disadvantages – something that might be worth thinking about. What is important is that this book becomes a resource for thinking the city differently. Again, think about the city and its contents. Does it feel the same now that you are collating the things you imagine are there?

There is more to be gained than simply a 'fragmentation' or a 're-integration' of the city, however. For what we have begun to ask is *how* the city is put together. These entries are not included on the grounds of whether or not the fragment can be found in the city (for what would be excluded?), but rather on their exploration of what is urban about them. That is, on the basis of asking, what is the 'cityness' or 'urbanness' of this part of the city? This is, perhaps, the most fundamental question, since it relates to how cities come to be the way they are. Of course, there is no simple or singular answer. But, in answering this question, we – and you – can begin to piece together how the city becomes a city. And in piecing together the city, you might come to understand the city anew. And even to consider new ways of intervening in, and changing, city life.

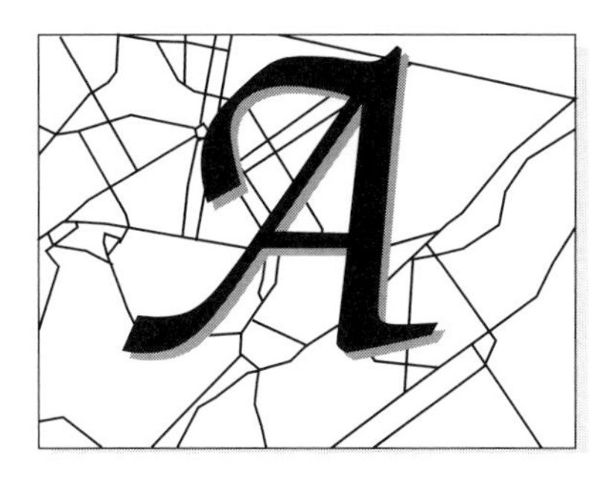

•A TO Z•

by Fran Tonkiss

> As an end product, a representation of reality, a map may be judged either by its artistic qualities – the fineness of line, harmony of colour and lettering and balanced layout or design, or by its usefulness. Whether a work of art or of science, such a map furnishes a new view of reality, albeit a subjective one. (Board, 1967, p. 712)

When urban theorists describe how people find their way around cities, they sometimes use a term borrowed from experimental psychology. As rats navigate a maze to reach a morsel of food, so people use 'cognitive maps' to orient themselves within urban spaces. It's a good metaphor, and one especially useful to the planners whose job it is to make cities legible. But it misses something in the translation: people are not always as purposeful as rats.

Part of the art, or science, of making maps in the city is the possibility of losing one's way. Sometimes it is as good to go nowhere as it is to get somewhere. Looking for oddments along canal byways, kicking over ideas or arguments on football pitches and in bandstands. Setting out as if for work each morning, before riding all day on the London Underground. Being around and about in local streets in the afternoon, as television sets pass out of the backs of vans into the waiting arms of men on the footpath. Walking in the city with eyes downcast, looking for something you might have lost once coming this way at night, or for coins someone else might have dropped. Are you really making a map if you're not *going* anywhere?

Cities are more fluid than mazes. It isn't always necessary to follow the prescribed routes. It isn't always sensible. Desire paths – those unofficial routes made by users rather than planners – simply seem more rational than do the rationalities of urban design. Beating the city at its own game. It's more logical to cut through the park than to skirt its edges, keeping off the grass. In a fit of misplaced municipal energy, the local council is paving the desire path that countless walkers have beaten diagonally across my local park. This might seem like a democratic thing. But the council's paved walkway is a means of organising desire, and a new path is taking shape at an oblique angle to the old one – a small subversion on the way to the tube station.

According to the notion of cognitive mapping, one knows the city in this way: we try to work out a pattern, to find a position from overhead or orient ourselves to some monumental or personal landmark. To essay a map between two points and to say, 'then at this street, turn', or, 'on your left you will see a shop with a blue awning'. In unplanned cities, it can be tricky to figure out the complicated secrets of the streets. The rationally planned city in which I grew up took much of the drama out of map-making. One mile square, edged by parklands, it had a grid of wide streets, the usual monuments and statuary of a colonial city, and a profusion of clocks. Social theorists have been fond of noting how the building of railways made necessary the standardisation of time; they might also notice how the attendant station clocks provided the best spots for meeting people – making time and place.

A well-made city might lend itself to the mapmaker's art. But paths through the city are as much stories as maps (de Certeau, 1984). There are different ways of making the streets tell. Architecture, the dull weight of history sedimented in buildings, is only the most literal. The inert form of the city houses a multitude of little spatial histories told by bodies moving within it. Movement gives us the city in real time, a geography of memory as much felt in the body as seen in built forms. To walk in the city, de Certeau writes, 'is to lack a place. It is the indefinite process of being absent . . .' The movement of individuals composes the city as a 'universe of rented spaces', occupied for a moment, passed through (de Certeau, 1984, p. 103). These many paths through the city cannot be charted. Here is the uncertainty principle of mapping the city – the more you know about someone's movement, the less clearly you can say where they are standing. A moving target is harder to hit.

The easiest way to map the city is on foot or bicycle. Machines are less nimble. Cars have to stick to the proper A to Z, and planners use 'traffic calming' measures to stop them following the rat-run lines of their desire. The map of a city given over to cars is limited, it can be gridlocked. During the 1980s an anarchist project to 'Stop the City' sought to block a number of key intersections around the City of London, with the aim of halting the traffic and ultimately of bringing down the Stock Exchange (similarly, in the 1990s barricades at Bishopsgate formed a 'ring of steel' to protect the City against terrorist attack). More recently, a campaign to 'Reclaim the Streets' of the city as a realm of freedom for walkers, cyclists and children has taken to occupying major junctions, trunk roads, bits of flyover (under the motorway, the beach and so on). There aren't many stories you can tell with a car, and they're all going to end badly.

There is something else which the notion of a cognitive map misses in its translation from rats to people. The 'cognitive' is defined in distinction from the emotional, the wilful, the mystic or embodied. It is about how one thinks rather than feels the city. Its documentary quality resembles that of the photograph, 'a work of art or of science' that – like the map – has certain claims to record the real. It shows us what is evidently there. The New York photographer Duane Michals accompanies one of his pictures – of a scene in a bar – with a text telling what else was there, but is imperceptible from the image. How he was hot, and sweat trickled under his shirt; how he felt menaced by one of the drinkers; how he drank a beer. He titles it, 'There are things here not seen in this photograph'.

Just so, for every sense-making map there are story-telling maps. They might tell, for example, about walking in big cities. About how dirty you get. About how being alone in a well-peopled city can be less fearful than you might think. On every page of the A to Z you could write: there are things here not seen on this map.

References

Board, C. (1967) 'Maps as models', in R. J. Chorley and P. Haggett (eds), *Physical and Information Models in Geography*, London: Methuen, pp. 671–725.

de Certeau, M. (1984) *The Practice of Everyday Life*, London: University of California Press.

•Air (1)•

by Andrew Barry

Bad air has always been a problem for city dwellers. The city is a place of smells and fumes. The smells of kitchens and animals; the fumes of motor transport, and smokers and factories; the odour of other people. There has been an extraordinary effort to regulate, manage and displace these smells. Whether through the law (governing the emissions from factories, regulating the space of smoking and dog fouling), or physical barriers (car windows, face masks, automatic doors), ventilation systems, or perfume. But to some extent the smell of the city is unavoidable. For while

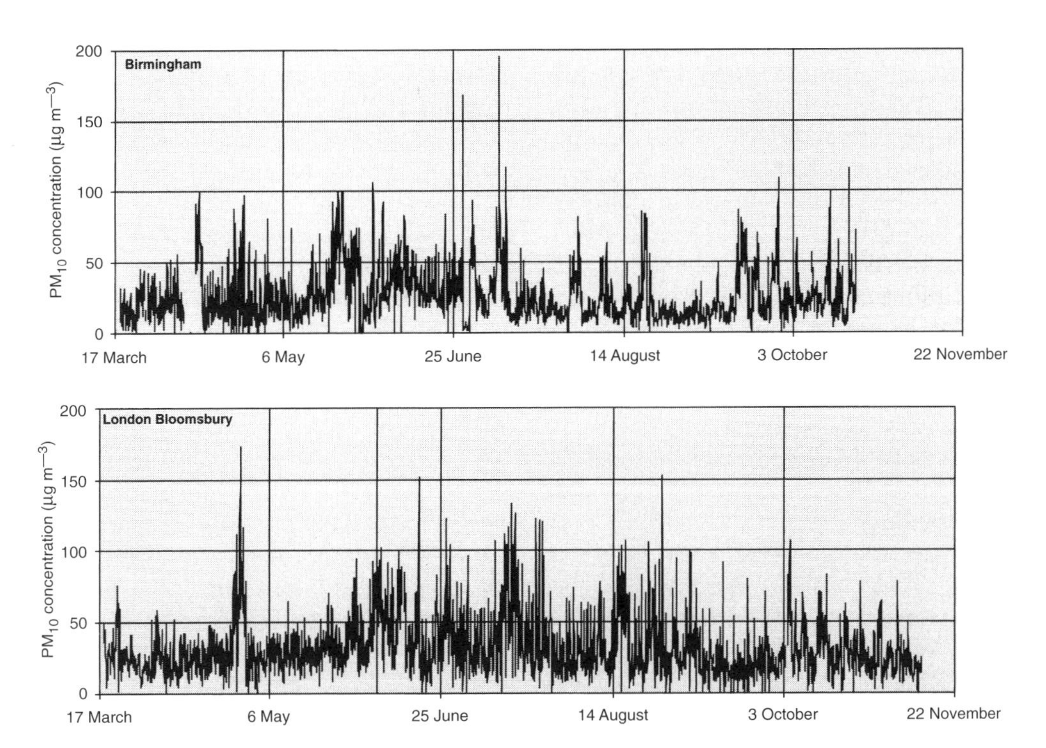

Figure 1 Hourly Mean PM_{10} concentrations from the London and Birmingham enhanced urban network sites (March to November 1992). Reprinted with kind permission

local sources of smell can be disguised, and fumes dispersed by technical means, the air of the city does smell. To filter out all sources of smell would be extremely costly. The idea of bottled air is a joke. The odour of the city is there to be endured, but also enjoyed.

The problem of bad air today, however, is said to be not just a matter of smell, or even of the hidden contagions that unpleasant smells might reveal. It is also a matter of the chemical composition of air, and of the complex and uncertain relation between this composition and the bad health of the urban population; not least the growing incidence of asthma and lung cancer in the advanced industrial countries. It is common enough for health professionals to talk about the serious consequences resulting from taking a *cocktail* of drugs, or drugs and alcohol. But today, the air is also understood by experts as a cocktail, in which various chemicals and particles interact in uncertain ways, with potentially harmful consequences for those who breathe them in. We abuse the air; and the air abuses us in return. The air is an alarmingly active agent.

But how can this invisible pollution be made visible? How can the urban population be convinced of its significance and importance? Consider the case of a small experiment which took place in the South London Borough of Southwark in 1994 on the Old Kent Road, a congested main road which leads visitors from continental Europe into the West End of the capital. The aim of the experiment was to measure the pollution emitting from individual vehicles. It used two American devices, FEAT and SMOG DOG™, which had already been used extensively in US cities to detect vehicles with faulty catalytic converters. In the experiment, cars and lorries were stopped by a policeman when measurements exceeded a certain figure.

In the early modern anatomy lecture theatre the demonstrator was a person who literally pointed out the object of which the lecturer spoke. In this way, words were connected to a visible object, and the students' gaze aligned with that of the lecturer. The experiment on the Old Kent Road was, likewise, in the technical sense a *demonstration*. It was not just a scientific experiment. It was intended to point out the existence of invisible pollutants to the wider audience. In this case, the students of the demonstration were the motorists of South London.

But if the SMOG DOG™ experiment was a technical demonstration, it was also in some sense a political one. For although all the drivers of 'polluting' vehicles were not fined (the legal basis did not exist) they were encouraged to become 'exhaust aware'. They were told how bad the exhaust fumes from their vehicles were. They would be hailed by the scientists and police and, at the same time, as Althusser would say, interpellated as environmentally conscious citizens. Environmental information would make them more environmentally conscious. And this in turn, it was hoped, would lead to a change in behaviour.

The experiment on the Old Kent Road is one manifestation of the importance of technical devices to the conduct of contemporary urban government. But it also plays a part in a story about the relation between the local and the global. For in South London SMOG DOG™ and FEAT came to have a particular and novel political purpose. They were partially funded by the European Commission in Brussels. And one of the effects of the Commission money was to draw the local authority, for the first time, into Europe. Both professionally (officials began to talk to their counterparts in Copenhagen and Paris), but also technically. FEAT and SMOG DOG™, along with a more permanent network of monitoring devices, made it possible to measure the quality of air in the capital against European and world standards. And the composition of the air in the world's cities could be compared from day to day. In this way, a small part of London became drawn into an emerging regime of transnational government.

Elsewhere in London, in the mid-1990s, other demonstrations occurred. People started to occupy and block roads, and to 'reclaim the streets', using bodies, bicycles and abandoned vehicles. These were technical demonstrations too, but of a different order to those conducted by the city authorities. For what was made visible was not the poor quality of the air, but the physical occupation of the city by the car. These were political demonstrations against an occupying force. Bad air is just a symptom.

•AIR (2)•

Overview of air quality in twenty of the world's largest cities

City	*Sulphur dioxide*	*Suspended particles*	*Airborne lead*	*Carbon monoxide*	*Nitrogen dioxide*	*Ozone*
Bangkok	Low	Serious	Above guideline	Low	Low	Low
Beijing	Serious	Serious	Low	(no data)	Low	Above guideline
Bombay	Low	Serious	Low	Low	Low	(no data)
Buenos Aires	(no data)	Above guideline	Low	(no data)	(no data)	(no data)
Cairo	(no data)	Serious	Serious	Above guideline	(no data)	(no data)
Calcutta	Low	Serious	Low	(no data)	Low	(no data)
Delhi	Low	Serious	Low	Low	Low	(no data)
Jakarta	Low	Serious	Above guideline	Above guideline	Low	Above guideline
Karachi	Low	Serious	Serious	(no data)	(no data)	(no data)
London	Low	Low	Low	Above guideline	Low	Low
Los Angeles	Low	Above guideline	Low	Above guideline	Above guideline	Serious
Manila	Low	Serious	Above guideline	(no data)	(no data)	(no data)
Mexico City	Serious	Serious	Above guideline	Serious	Above guideline	Serious
Moscow	(no data)	Above guideline	Low	Above guideline	Above guideline	(no data)
New York	Low	Low	Low	Above guideline	Low	Above guideline
Rio de Janeiro	Above guideline	Above guideline	Low	Low	(no data)	(no data)

continued . . .

Overview of air quality in twenty of the world's largest cities (continued)

City	*Sulphur dioxide*	*Suspended particles*	*Airborne lead*	*Carbon monoxide*	*Nitrogen dioxide*	*Ozone*
São Paulo	Low	Above guideline	Low	Above guideline	Above guideline	Serious
Seoul	Serious	Serious	Low	Low	Low	Low
Shanghai	Above guideline	Serious	(no data)	(no data)	(no data)	(no data)
Tokyo	Low	Low	(no data)	Low	Low	Serious

Source: UNEP/WHO, 1992, *Urban Air Pollution in Megacities* (Oxford, Basil Blackwell). Cited in United Nations Centre for Human Settlements (HABITAT), 1996, *An Urbanizing World: Global Report on Human Settlements 1996* (Oxford, Oxford University Press), p. 145.

Note: These are based on a subjective assessment of monitoring data and emission inventories.

•AIRPLANES•

by Anthony D. King

In Newark airport, New Jersey, a plaque commemorates Lindbergh's flight from there across the Atlantic. It also states that in the previous year (half a dozen before the birth of Clint Eastwood, Sean Connery – and myself) all the airlines in the USA had a total of twenty-eight aircraft. Had all taken off together with every seat occupied, 112 passengers would have been airborne.

In the 1990s, over forty-five million passengers passed through Newark every year. The number of passenger miles flown annually worldwide resembles the windows in a Boeing 747. (Try drawing one around the number!)

125 000 000 000 000

Yet despite the hundreds of books on airplanes, airlines and aviation, no one has apparently recognized a phenomenon (quite unprecedented in the history of humankind) that has occurred in my lifetime: it is now possible for the entire population of not one, but a number of cities (the size, say, of Sheffield or Geneva) to be 30,000 feet in the air at the same time. Since 1960 (and with Boeing 747 jumbo jets carrying 400 passengers from 1970), world passenger traffic has grown on average at 9 per cent a year.

Every day 500 commercial flights move in each direction across the Atlantic. Even if only two-fifths of these are flying simultaneously, each with 300 passengers, this is a veritable superaquatic Atlantis of 60,000 constantly changing inhabitants (maybe we should call them aeronuts), located in a dispersed concentration of some 3,000 miles, give or take a few thousand feet up or down.

Yet for most people, airplanes don't have any separate sense of place, or identity. For their passengers, they're simply suburbs in motion. Laker's Sky*train* and the Euro Air*bus* recognized this as they vapor-trailed their way along country (air) lanes, down well-marked (flight) paths, linking one earth city to another.

Airplane lifestyles are equally suburban. The aim of modern airlines (all 1,200 of them, though only 300 are international) is to *deny* the fantasies, the excitement, of flight – quite unlike the early days when passengers wore heavy overcoats to keep out the cold. My own first trip was in a twin-engined Avro Anson in the mid-1950s. The fuselage was made of corrugated iron (or so it seemed), the wings flapped up and down beating time with the washing-machine motors, and the pilot sported leather helmet and goggles.

In contrast, contemporary jets replace the thrill of speed with the stupor of a couch potato: reclining seats, continuous movies, bland grub, drinks, endless reading, insipid music and any spare moments appropriated for compulsory shopping (Continental's tax-free sales magazine is appropriately named *Sky Mall*).

Socially, airplane life is equally suburban, contact without community. The airplane is just another building, like the church or cinema. The congregation sits in uncommunicative rows, some fingering electronic rosary notebooks, occasionally walking the aisles to take communion, perhaps, with the pilot. Little attention is paid to the service except to join, at start and close, in silent prayer. Or the airplane is a cinema, its patrons denied intercourse by their headsets. As the credits roll, they flock to the twelve separate restrooms as a substitute for inhaling the night air. These familiar environments could be much enhanced. Stained glass windows, a reredos or retable at each bulkhead, the seven stations of the cross at the emergency exits. Or when films are screened, popcorn could be dispensed from the overhead racks.

Little or no imagination has gone into airplane interiors. British Airways could bring in the Laura Ashley organization as consultant, whether for interior, cabin crew uniforms, or optional traveling garb for passengers (William Morris or Voysey). 'Colonial' interiors would suit American Airlines, with red-coated Revolutionary soldiers as motifs in the restrooms and Doric pilaster strips framing the doors.

The language of the skies is equally a denial of flight. Instead of celebrating the difference between planes and boats, everything imitates life on an ocean liner. We are 'Welcomed aboard', shown through 'the cabin', directed to the 'upper deck', provided with a life-jacket (not a parachute) and, as if in the ship's hospital, offered oxygen. Even the titles of the crew (*sic*) are nostalgically drawn from the age of the Queen Mary – Captain, Pilot, Purser, Chief Steward, Stewardess.

This denial of aviation's fantasy will certainly have to change as twenty-first century sky cities become ever more overpopulated. What is needed is not so much a glass ceiling as a perspex floor. Like Amtrack's trains, a see-through roof dome, exposing the constellations, will make everyone their own navigator. Aircraft in flight will be subject to constant passenger surveillance ('Watch out for that one, Captain!'). The natural outcome will be a much greater diversity in how airplanes *look*.

Airplanes are essentially buildings; as such, they're part of the architecture of the city. Like the Empire State and similar structures, commercial airplanes are part of the Modern Movement, children of the 1930s. Nowhere is Louis Sullivan's maxim, that form follows function, more evident than in the sky: stand a Constellation or a DC 10 on its tail, clip off its wings, and it's the closest thing to the Chrysler Building.

Yet unlike buildings, the shape of airplanes has changed very little since the 1930s. Boeing 747s, developed in the 1960s, or the Concorde in 1970, are still pretty much the same as on their first flight. And given their huge cost (about $127 million for a 747) it's not surprising that about 30 per cent of planes in service are approximately twenty years of age.

Yet if the shape can't change, the appearance can. Why not make the planes more part of the city? Abandonment of the drab,

a
a
a

modernistic livery of the 1930s opens up all kind of possibilities: an Airbus could become Coronation Street (complete with chimneys), older planes might be done in High Gothic. Imagine the British Prime Minister taking off – horizontally – in Big Ben.

•AIRPORTS•

by Marc Augé

Today, the airport is the most successful mixture and most typical example of *place* and *non-place*. If we define the place as a space where identities, relationships and a story can be made out, and the non-place as a space where such an attempt is not possible, we have to admit that the airport has an element of both.

Today, the history of aviation is long enough that airports are parts of the collective experiences of the countries in which they are situated. Their names are even considered as quasi-synonymous with the towns they serve. In fact they have a history which is part of their country and with which legendary figures may be associated. From a sociological point of view, an airport is not only defined by its take-off and landing runways, its ability to receive or despatch planes. This function itself requires the regular daily presence of technicians and employees for whom the airport is a place of day-to-day social intercourse. Some companies, such as Air France in Roissy, have located their headquarters in the airport, and, in a manner of speaking, helped to create a sort of town. Moreover shops and various activities happening within the airport (Heathrow is in this respect remarkable) accentuate this aspect: those who live and work daily in this location see it as a *place*. They have relationships, develop friendships, and it may also happen that a regular passenger in the same airport has his own points of reference and habits, and meets familiar faces again.

If we look at it from the average passenger's perspective, the description varies too. International airports welcome passengers whose identities are controlled once they have checked in. Free of their luggage (although some manage by cheating to encumber themselves with heavy and voluminous hand baggage) they are let loose in the area where, free of any further identity control, free of burdens and duty (duty-free) they do a little bit of shopping. At this point, they are merely passengers from the 'n' flight of the 'x' company flying to Hong Kong, Chicago, Rome or Abidjan. Their itineraries pass each other for an instant, in this crossroads of destiny. Soon they will be flying to the four corners of the world. They are waiting for a soft and synthetic voice to invite them to board the plane. Most of the time those who come across each other, those who meet in the same waiting area do not know each other. Screens inform them of any delays. Dialogues and conversations do not exist. The hall or waiting rooms only assemble temporary loneliness.

With a display of shyness or bad manners, a plane passenger could today go around the world, not in 80 days like in Jules Verne – he would really need to take his time for that – but in 80 words, if you answer only by a smile to the air hostesses' attention, and by a nod of the head to the policeman's questions and the politeness of the duty-free cashiers ('Have a nice day'). Regarding the televisions that distract them, the machines that dispense drinks, the voices on the headphones or all kinds of screens, they don't ask for so much.

There is a certain amount of charm to the passenger's loneliness in the waiting area: preoccupied by his imminent departure, he floats in space and time, ready to be guided, settled down, taken away. These suspended minutes between the check-in rush and the boarding call give him a taste of the minutes he will spend in the languid interlude of the flight. The minimum social contact required with his neighbours on this occasion is in itself comforting. But you cannot discount the possibility of someone coming along and making contact; sometimes the romantic nature of travel can stand up to the imposed anonymity of flying. A relationship is struck up and the waiting *non-place* may turn into a meeting place.

Last change on the horizon: the airport of arrival is different from that of departure even if on the way back it is physically the same space. On arrival one must regain one's identity and luggage and take up again the business of the home town, the unfinished project, or the established routine. The arrival is to be amongst the other ones. And even if it is not always Hell, it is a sudden breaking off, a little brutal, with the doubtful but peaceful feeling of innocence that the departure and flying-away *non-space* gives to the airline passenger.

•AMUSEMENT ARCADES•

by Anette Baldauf
and
Katharina Weingartner

> 'You should clean up this city here because this city here is an open sewer, it's full of filth and scum. . . . Sometimes I go out and I smell it, I get headaches, it's so bad. It just never goes away. I think that the President should just clean up this whole mess here, should just flush it right down the fucking toilet.'
>
> 'Well, I think I know what you mean, Travis. But it's not gonna be easy. We are gonna have to make some radical changes.' *Taxi Driver* (Martin Scorcese, 1976)

When Walt Disney's first entertainment park, Disneyland, opened in 1954 in California, it was clearly flirting with urban imagery brought on stage in a safe environment. Primarily for economic reasons, since big profit was made by the nearby hotels and tourist infrastructure outside of the park, Disney's concept for his second mega-simulacrum was geared towards the re-creation of a coherent urban entity. Ideologically embedded in the context of post-World War II, anti-Communism, suburbanization, Sputnik and a strong belief in the technological determination of the future, a constitutive part of this enterprise was the creation of a utopian city: 'The most exciting, by far the most important part of our Florida project, in fact the heart of everything that we will be doing in Walt Disney World, will be our Experimental Prototype City of Tomorrow. We call it Epcot', announced the creator of Mickey Mouse in his TV program in 1966. Under the panoptic surveillance of scientists, people should live and work together in this urban laboratory, demonstrating the implications of new technologies.

When EPCOT opened sixteen years after Walt Disney's death, it added one more theme park to Magic Kingdom and MGM Studios. The dream of a perfect city had to be compensated by the simulation of urban flair. Visually as well as functionally the park presents itself as a coherent urban entity: banks, hotels, hospitals, factories, power-stations and traffic systems mirror a comprehensive urban concept. And in the backyard of the EPCOT center, designers of movie scenarios project bright postcard images of various metropolises on colorful facades. The 'guests', as Disney likes to call its customers, are able to enjoy the phantasmagorias of the world, strolling from Mexico's Rio Del Tiempo to the nearby Norway's Akerhus Castle, the 'Wonders of China' next to Germany's 'Biergarten'. Under the protective eye of surveillance cameras, security guards and guided by an elaborate human traffic control system, cosmopolitan flâneurs safely enjoy city air and imagery: a sanitized version of US-American cities supposedly in crisis and model for inner city restructuring called 'revitalization'.

After an increasing distrust of governmental intervention in urban development gave birth to growth coalitions embedded in the prevalent neo-liberalism, contemporary inner city restructuring is governed by private intervention. The entertainment industry has become a crucial player in this game, using not only urban imagery but also soil as commodities in their project of turning inner cities into tourist locations in hyper reality. The city, more precisely, its sanitized image, cut off from the rest of urban reality, turns into an object of simulation. In a family-oriented urban montage, the city is compartmentalized into bits and pieces, purified, washed free of disturbing elements (e.g. prostitutes, sex stores, the homeless, beggars, etc.) and thrown on to the tourist market, where safety, cleanliness and fun are sustained by an invisible army of minimum wage workers. Confronting the widespread skepticism towards technology and fears about the future, in contrast to the utopian vision of EPCOT center where the future looks rather old today, the master-signifier of this urban restructuring is nostalgia for a cleaned up past.

At Times Square in New York, previously known as the 'crossroad of the world' not only in terms of traffic but also cultural diversity and dangerous love, the so-called revitalization project is a nostalgic retelling of an old story, now turning into its farce. With the help of private police and sanitation workers, some legal changes (e.g. zoning laws forcefully dispersing sex stores) and an annual budget of $6 million, one of its major agents, 'Times Square Business Improvement District', made Broadway shiny again. Rejected were concepts such as an open air information cathedral, noise tower, hip hop hall of fame or video laser shows. Instead, the restructuring focused on the exchange of electric neon-bulbs, the renovation of facades and some foaming cups and camels. Along with Disney, Sony, Virgin Mega Store, All Star Cafe, corporate entertainment has moved in, collectively celebrating neo-fun in franchise-look. The Times Square BID praises itself for reducing street crime by 43 per cent during their four years of operation. Critical theorists see Times Square becoming a mass-produced space simulating the thrill of an urban jungle while in fact police and guards strictly regulate who is in and who is out. Those who stress self-regulation point towards Times Square's history of resisting taming. But both camps agree that global issues are at stake.

In Los Angeles, in the supposedly hygienic and safe enclave of Universal Studios, affluent residents stroll and shop on City Walk, a $100 million simulacrum of urban ambience. The currently discussed concept to 'revitalize' Downtown suggests the implementation of a baseball stadium plus entertainment infrastructure. And Rem Koolhaas, after aggressively attacking New York's 'coalition of moralists, planners, and a nostalgia-driven entertainment giant expelling, as if in some Biblical scene, the unwanted from the city', is working on Universal Studios' new identity. Las Vegas, longing for metropolitan flair, has opened its own 'New York, New York', the hotel and casino theme park which is a miniaturized and safe version of Downtown Manhattan. And in Berlin, Sony is filling the disturbing void on Potsdamer Platz with transnational entertainment items.

History seems to be the time when city tourism was geared towards the consumption of historical monuments such as the Empire State Building, Eiffel Tower or Red Square. Today, the world's biggest record store 'Virgin Megastore' (enough space to accommodate 1,000 New York cabs), the most impressive palace for foot comfort 'Niketown', and Disney's best gift shop all draw on global trotters, which construct the unique sense of place out of a globally distributed vocabulary of spectacles.

Under the floating capital of entertainment giants, inner city imagery is itself turned into 'cheap thrills'. The price has to be paid by those who (used to) use the urban space for living – residents, street artists and homeless people:

> 'When I first lived in Times Square, it was really bad – but there was diversity. Now they took all the pornography out, that's good, but we used to get a quick cup of coffee or go to a bar across the street, where the fellows and girls used to go to have a beer – that's gone. There is no diversity any more. I just don't wonna see Times Square become another Disneyland. Just Disney, you know, where they cater to the trade outside of the community, where you go and have to pay one dollar 68 cents for a cup of coffee. I can't afford one dollar 68 cents for a coffee, I am not a tourist. I am a resident.' (Jeraktine Diver, resident at Times Square Hotel)

•ANIMALS•

by Gargi Bhattacharyya

All the best myths of modern urban living involve animals. Dogs that save their owners from disaster while rearing kittens and barking sausages. Alligators that swim against the tide, back through the u-bend and up into the toilet to snap at unsuspecting bottoms. Boils that burst into spiders, rats that attack babies, pet rabbits that escape to repopulate the urban wastelands.

However much the city forms itself around the fiction of the single species area, placing humanity as not only the central but also the sole inhabitant of these environments, the uncontrollable spectre of biodiversity creeps into the cracks of the story.

In the undergrowth something rustles. Blurry fur darts across the tarmac in front of you. At night you hear small jaws gnashing in the wall, under the stairs, beneath the floorboards.

Outside you step carefully around varieties of excrement – family dogs, police horses, city pigeons – and eat, drink, smoke and watch an urban life chock-full of animals.

As cities expand, the patterns of urbanisation change. Instead of overtaking nature, pushing it under, asserting the powerful culture of humanity, now unruly nature re-emerges as an urban phenomenon. No longer kept at bay by the paranoid structures of modernity, urban vermin brush up against us every day. Who knows the boundary between the domesticated and the feral any more?

We hear of frightening attacks on children – multiple injuries, faces rebuilt – and quake a little at other people's pets. Even in the familiar landscape of the residential urban, we learn that danger is just over the fence. Every next-door neighbour's dog could reassert the boundaries of their species at any moment.

For children, the unexpected return to the bestiality of family pets heralds other horrors. Our wannabe modern culture promises that childhood will be a period of protected innocence, carefully nurtured in the embrace of a domesticated family. This is a domestication that relies on the myth of home, a bounded space within which the unfriendly rivalries of the public sphere do not intrude. Instead, physical proximity breeds affection and its attendant comforts. Household pets are part of this process – perhaps its most exemplary case. The pet enters the family home and relinquishes the troubling otherness of animality. Now the creature is accepted as quirkily human, a holder of likes and dislikes and the particular habits of the eccentric relative. More than human, he or she is now part of the family. For the child, there is no hint of what must be repressed for this to occur. When the animal bites back, the promise of safety through domesticity crumbles away, leaving us greedy and mortal for all to see.

Elsewhere in the city, the urban breeds its own particular vermin. The pigeons of our city centres forget the dangers of the species barrier between us. Now when they see people coming they flock closer, expecting nothing more threatening than another free meal. In a mirroring of the domestic pet, the human/animal boundary is blurred again. And again, this more public version of domestication is inevitably incomplete. Animals may become integrated into our urban landscapes, become part of our families and our everyday expectations, but they bring the risk of a re-entry of undisciplined nature into our barely ordered domestic lives. Instead of becoming safely familialised and familiarised, the urbanised animal may unleash the bestial forces in all our affective arrangements. When man's best friend, the creature we take as another of our offspring, attacks babies in our homes, what is to stop any of us baring our fangs and ripping apart flesh during family disputes? When pigeons mark city streets and squares, colonising some spots as theirs alone, is this an adaptation of their animal ways to the more orderly demands of urban living, or is it another indication that the city is always vulnerable to slow erosion by the feral forces which live in its crevices? As the visions of the idyllic urban of Western ascendancy disintegrate before our eyes, unable to ensure the most basic requirements of human life, even people revert to a feral state which was only ever thinly disguised.

This is the parallel logic of city – those old stories about concrete jungles, unknown dangers, irrepressible but hidden nature(s) lurking in the shadows of urban experience. On the one hand, the city makes us modern – now we are ordered, rationalised, industrialised, able to live in the made-on-the-spot network of relationships with strangers. The city is a sign that now humanity can organise its survival as a large group project; with the city we really start to live with all the benefits and uncertainties of that collective name, society. In the city we begin to know how intimately our welfare is tied to that of others, most of whom we will never know – and we hope that this sense of humility and collectivity will make us more human. But on the other hand, the city increases the animality of modernity – now we are private, anonymous, individualised, freed from the inborn, inbred hierarchies of feudalism. In the city we see the empty swagger of authority and lose our fear in man or god. Instead our destinies become tied to forces beyond the control of any authority, subject only to the vagaries of capital or nature, and in response we become more short-sighted, self-centred, pulled by our bodies beyond the checks of group interest or reason, more animal altogether.

•ART•

by Iain Chambers

We arrived just after nine. John Waters the film director (*Pink Flamingos*, *Hairspray* . . . Divine) was talking on 'A matter of taste'. After queuing to spend a fortune on Dry Martinis we wandered out into the courtyard of this mock French Renaissance hotel – Chateau Marmont – located on Sunset Boulevard.

The occasion was the Gramercy International Contemporary Art Fair. The works on display (and sale) were spread over six floors. Forty-two hotel rooms occupied by galleries and individual artists. Drink in hand, I wandered from room to room observing the paintings and photographs displayed on the walls, the objects strewn across beds, overhearing the sales pitch, the prices: the art market.

In this circulation of bodies and commodities, in this city of dreams and tawdry dealings, the desire for the 'real thing', the instance of 'authenticity', hung on in the passage from one bedroom to another. (Just a few blocks away and across the road is an 'original, down-home juke joint' built from corrugated metal sheets. It has been physically uplifted and transported directly from Clarksdale, Mississippi to become the 'House of the Blues' here on Sunset.) And then this guesting of art in the modern site of transitory experience – the hotel – probably contains little that is different from the salon scene in Paris of a century ago: an equal combination of artistry and artifice, equally fashionable, equally desperate. In both cases, the locale, the city itself, adds value to the works. To exhibit in LA or London, Tokyo or Paris, is already to be 'authenticated' by the international relations of artistic production. Conversely, modern cities also seek their authentication in the art discourse (the bienniale of Venice, Johannesburg, São Paulo, Istanbul). In claiming its role in the public realm (and publicity) of transnational art, the individual city insists on its centrality in a language in which all, even local configurations of poverty, degradation, dystopia and despair, can be translated and transmitted: even the abject can be sustained by the aesthetic. It is rare that art invades the city; more frequent is the obverse. To recognise the role of certain contemporary cities, and to be recognised by them, is to complete a circuit in which business and the beautiful, markets and metaphysics, evolve in an ever tighter embrace.

Yet, in this seemingly abstract dialogue, there emerges an interval within the city itself that obliquely reveals another, a further, city.

Saturday in the financial district of downtown Los Angeles, on Grand, we are going to the Museum of Contemporary Art. The wide avenues, running straight as a die down the man-made canyons between towering skyscrapers, are empty, filled only with the wind coming in off the coast before rising over the San Bernandino Mountains at our backs. After visiting the museum we wander on foot, walking the deserted plazas, accompanied by the gurgle of artificial waterfalls and high-pressure fountains ejaculating their contents skyward every few seconds. A film crew is setting up its equipment, but then LA is a film set – certainly the most filmed city in the world.

We cross over to Flower at 4th and arrive at the Bonaventura – certainly the most theorised hotel in the world. There, after riding the glass elevator up to the thirty-second floor and then, returning with a plunge through the glass foyer roof towards the waters lapping the elevator bases, we sip an expresso at the bar. An hour later we find ourselves on Broadway walking east towards Union Station, Olviera Street and El Pueblo where, according to official accounts, the history of this urban settlement began in the 1780s. Here the sidewalk is crowded; everything – shop signs, newspapers, magazines – is in Spanish. The buildings are nearly all Art Deco and include innumerable cinemas now transformed into giant stores selling cheap clothes. Only the stunning iron, wood and marble interior of the Bradbury Building on 3rd, preserved as a historical monument, and a film set (*Blade Runner*), resists the popular, commercial wave. Between Grand and Broadway lie four blocks of real estate and two separate worlds. The particular day of the week only emphasises the urban proximity of radical division between the abstract space of corporate capital and the persistent place of those who service and sustain it.

So something always escapes both the city plan and the logic that sustain its designs; something always lives on as a disquieting supplement, a potential interval and interrogation. Here in Los Angeles, as the art world prepares for digital galleries – ArtNet's™ proprietary Galleries On-Line system – and electronic art fair catalogues, the commodification of the aesthetic 'aura', the fashionable framing of momentary transcendence, reveals not merely the 'essence' of metropolitan modernity and technology but also, and still listening to Heidegger, the very essence of art as the event that discloses itself while simultaneously withdrawing from view.

•Bars•

Stand-up beer hall

by Walter Benjamin

Sailors seldom come ashore; service on the high seas is a holiday by comparison with the labour in harbours, where loading and unloading must often be done day and night. When a gang is then given a few hours' shore-leave it is already dark. At best the cathedral looms like a dark promontory on the way to the tavern. The ale-house is the key to every town; to know where German beer can be drunk is geography and ethnology enough. The German seamen's bar unrolls the nocturnal plan of the city: to find the way from there to the brothel, to the other bars is not difficult. Their names have criss-crossed the mealtime conversations for days. For when a harbour has been left behind, one sailor after another hoists like little pennants the nicknames of bars and dance-halls, beautiful women and national dishes from the next. But who knows whether he will go ashore this time? For this reason, no sooner is the ship declared and moored than tradesmen come aboard with souvenirs: chains and picture-postcards, oil-paintings, knives and little marble figures. The city sights are not seen but bought. In the sailor's chests the leather belt from Hong Kong is juxtaposed to a panorama of Palermo and a girl's photo from Stettin. And their real habitat is exactly the same. They know nothing of the hazy distances in which, for the bourgeois, foreign lands are enshrouded. What first asserts itself in every city is, first, service on board, and then German beer, English shaving-soap and Dutch tobacco. Imbued to the marrow with the international norm of industry, they are not the dupes of palms and icebergs. The seaman is sated with close-ups, and only the most exact nuances speak to him. He can distinguish countries better by the preparation of their fish than by their building-styles or landscapes. He is so much at home in detail that the ocean routes where he cuts close to other ships (greeting those of his own firm with howls from the siren) become noisy thoroughfares where you have to give way to traffic. He lives on the open sea in a city where, on the Marseilles Cannebière, a Port Said bar stands diagonally opposite a Hamburg brothel, and the Neapolitan Castel del Ovo is to be found on Barcelona's Plaza Cataluña. For officers their native town still holds pride of place. But for the ordinary sailor, or the stoker, the people whose transported labour-power maintains contact with the commodities in the hull of the ship, the interlaced harbours are no longer even a homeland, but a cradle. And listening to them one realizes what mendacity resides in tourism.

•BEACHES•

by John Law
and
Ivan da Costa Marques

Beaches in Rio are like parks in the cities of the northern hemisphere. That is how the people from Rio, the *cariocas*, think of them. Like large public parks. Spread along the edge of their sprawling city. Perhaps everyone has his or her beach. The beach where he or she belongs. A beach which forms an important part of identity. Some beaches are exclusive: not forbidden, but difficult to get to by public transport. Effectively closed to the poor. But Copacabana Beach is different.

At six in the morning they are walking, jogging and cycling. They are out there in the dawn, young people, middle-aged and old. In T-shirts and shorts, in trainers. Cyclists in lycra. Copacabana is one of the densest stretches of habitation in the world. Four kilometres long, three streets deep, people pile themselves into high rise blocks, hemmed in by granite, green hills, water and sun. And half of them are out there first thing in the morning, exercising.

The walkers and the joggers don't go away. But a little later other kinds of people will appear. The aerobics classes start at eight in the morning. Dozens of people, mostly older, largely women, lying out on towels stretching their arms, their legs. And the handball players, or the footballers, almost all male, they don't really start until a bit later either – but will carry on late into the night, playing under the floodlights. And then by mid-morning the hotels are starting to get themselves organised. If you watch you can see the employees of the fancier establishments processing on to the beach carrying folding chairs, towels and parasols, chillers and tables. Areas of the beach are more or less walled off, and the guests lie out, mixing sun, shade and security.

Then there are street people. Early in the morning most are still sleeping. For instance, on stone benches in front of a hotel under construction. There is nobody about, no one to chase them away from this small barren patch on the huge sidewalk. The people who sleep here are mostly young, mostly young adolescent children or even younger – though perhaps they are a family, and possibly there are one or two adults there as well. Here they eat, they drink, they shit, they wash their clothes. Where does the water come from?

Rich, poor. The syntax of dress is not immediately obvious. People say that beaches are 'naturally' democratic places, for if the basic clothes for all are shorts and a T-shirt – or bikini briefs and a top – then the differences between haves and have-nots are more subtle. For instance, trainers as opposed to flip-flops (though some of the wealthy are happy enough with flip-flops). For instance, lighter as opposed to darker skins (though some of the black people are wealthy). For instance, age and colour combined (young and dark and male is a fairly strong indicator of poverty). A complex mix of indicators, none sufficient in itself, distinguishes haves from have-nots.

In the evening, the street people, the street salesmen, are everywhere, the show is continuous. The maps. The gems. The knives. The rolls of peanuts. The circus knives through the head. The balloons. The music. The acrobatics and the contortionists. There is an understanding: the street sellers will eyeball café customers as they sit, three tables in. But they don't approach. Look away, perform civil inattention, and they are rendered invisible and move on. To be replaced, to be sure, by the next performance, the next itinerant salesperson.

And the customers? Many of the people sitting in the cafés are readily comprehensible. Tourists, business people, or perhaps locals, at any rate people who have money, and have come to take the air, to sit in the tropical night, to talk, to eat, to drink. But others? Here is a dark young girl. Almost black. Conventionally attractive. Dressed in a slightly flash way, bright red colours, some makeup, a light jacket draped loosely over her shoulders which covers, but only in part, her midriff, she looks through a little notebook, writing as she drinks her beer. Is she a call girl? Quite possibly. The syntax is unclear. Perhaps she's just waiting to meet a friend.

Copacabana beach is a social construction. A place where urban spatialities and temporalities are rebuilt each day. The hotel parasols, the nets of the volleyball players, the towels of the exercisers, the ropes of the elastic swings, all of these are reassembled every morning. As are the channels through the beach where the little rivulets, otherwise long vanished beneath the tarmac and the high rises, run to the sea. Yes, there are bulldozers remaking those channels each morning, vehicles which clear the rubbish bins, and people sweeping up litter or brushing the sand off the sidewalk.

Some things hold. The great floodlights go off at dawn, but they don't go away. The black and white patterned sidewalk is an invariant, at any rate over weeks. And then there are the kiosks which sell beer, or guaraná, or coconuts. But it takes work. The rubbish has to be cleared up, and the plastic chairs, chained up for the night, laid out again. But trade resumes after only a few hours.

Rebuilding the beach each day. What is less obvious is that the beach is artificial. Like many Rio beaches it is a major piece of civil engineering. Thirty years ago, there was a two-lane road and not the six-lane highway along the sea front. The beach, too, was much narrower. Just a little strip of sand, nothing very special. Children came down with old tyres to play in the water. But now there is a great band of golden sand. And all the high-tech apparatus of beach pleasure. So what happened? The answer is that huge quantities of sand were moved from other places to the existing beach at Copacabana. It's a case of successful civil engineering.

Yes, Copacabana beach may have its problems. With crime. With the quality of the water where untreated sewage competes with industrial effluent (not that this stops many swimming or surfing). But this golden beach is unquestionably one of the great beaches of the world. Except that it is many beaches. Many different beaches to many different people. At different times of day – and night.

•BENCHES•

by Peter Marcuse

b
b
b

Parks and transit stops have benches; buses and subways have seats; plazas have sitting areas; food courts have chairs. One's backside has many choices in the city, it seems.

But hardly free choices. Where and how one sits, with whom, for what, how long, can all be manipulated behind one's back, so to speak. Just look at what Big Brother's design consultants can accomplish.

The textbooks will tell you how to deal with sitting in a plaza. Is the plaza just there because it gives a zoning bonus, so you can add more square feet to your downtown office high-rise? Keep the *hoi polloi* from messing it up with their relaxed bods and their sandwich wrappers, then. Position the enclosures for your plantings at an awkward height to sit on; embed little knobby somethings in them, to dig into any bottom that would set itself down on it; make it narrow, so that balancing on it is a challenge. Above all, make sure all the curves are convex, not concave, and certainly have no right-angles opening out; violating those rules may mean that people sit facing each other or in convenient conversational positions, positions in which they may linger and clutter up the scenery.

If, however, the plaza is intended to be actually used, rather than only admired and passed through, the textbooks remain helpful. Impose order is the watchword. Unruly bottoms can lead to unruly tops; disorderly sitting can lead to other forms of disorderly conduct. So fix the chairs around the tables, mount them on pedestals in the concrete, facing square tables symmetrically in rigid formation. Four properly placed people can sit and eat or talk, but let no fifth join in, and larger groups are certainly unwelcome. And space the tables as far apart as you want; that limits the number, prevents accidental contact, chance encounters, too much and too informal sharing.

Park benches have other purposes, and are supposed to serve other needs. My father thought with pleasure that the park benches he was told about in Hanoi, seating two comfortably but too narrow for three, were symbolic; a presumably socialist society should encourage love and intimacy in a way that capitalist benches for threesomes were unlikely to do. Unabashedly poetic, perhaps; but would not the reality of New York City's current benches have moved his symbolism to reality, if with precisely the inverse meaning? Why do they have dividers separating each seat from the other? Is it to protect privacy? Well, two people sitting next to each other sharing an arm-rest are hardly private. But it does indicate how many can sit on the bench, avoiding the unpleasantness that could result from four behinds squeezing into a space only large enough for three. The contoured seats in the subway cars themselves have that purpose: a gentle and standard bowl-shape, made for the standard behind of the standard citizen. If only standard citizens' behinds were as standard as the seats provided for them, but they are not. Nor, unfortunately, do they come in only two sizes, one fitting a single bowl and the other fitting a double bowl-size seat (with the adjoining edges of the bowl neatly fitting into the crease of the double-behind-size citizen). But then, not all ventures into designed manipulation are successful.

But getting back to benches and their dividers: if the separators don't conform to the unstandardized sitter, are the separators perhaps useful as arm-rests? But if that is their purpose (questionable, indeed; either they are good only for one-armed sitters or they should be two-arm widths across, but then we are back to the romanticism of wartime Hanoi, hardly the capitalist city ethos), why do the arm-rests have cross-pieces underneath them? If they are wooden, as they are in the subways, why are they made of solid heavy timber? These are new designs, making their appearance first in the 1970s, at about the time that homelessness became a troublesome presence in the contemporary city. Yes, the two are related: for if you want to deal with homelessness by arranging it rather than ending it, you would be concerned with moving the homeless from where they are not wanted – parks, subways, bus terminals, train stations – to somewhere, anywhere, else. Only a contortionist could sleep on the new subway benches, on the park benches with their strategically placed wrought-iron 'arm-rests'. Benches in parks are for people enjoying nature, waiting to go home; benches in subway and train stations are for people to sit while waiting to go somewhere. They are not for people with nowhere to go.

But the latest in city seating is a bench that says, loud and clear: 'Don't sit on me!' Made of two tubes two inches in diameter, one bent to form the pseudo-back of a bench, the other parallel but lower, holding a four-inch wide folding metal slab just wide enough to hold a coccyx but none of the flesh around it, these non-benches have been embedded in the non-waiting room of the Port Authority of New York and New Jersey Bus Terminal (they are in the open areas near the gates leading to the bus loading platforms; the terminal has no waiting room near where the buses are). A person can shift some body weight from the feet to the lower middle rear of the body while waiting for the bus, if the bus is coming soon; more than five minutes is hard to take, and sitting it can scarcely be called. In Los Angeles they speak of 'bum-proof benches' – in both senses of the word?

But who said cities were for sitting? Or for everybody?

•BORDERS•

by Mary King

At the border between two countries there's
a fable which they tell,
how all the riches from one side will flow to
the other as well.

Midway separating two cities, on either side of the Mexico/USA border, an evening ritual takes place. As the sun departs to amber in the west, it signals the staccato of engines. Headlights pierce the darkness while footsteps scatter like dry leaves. The Light Brigade is a committee made up of self-elected citizens assembled to halt the flow of illegal immigrants into the USA. As one local summarized, 'It'll be kingdom come when they come over here and take our jobs.' A dead coyote strung up in a tree serves to validate this point.

Here where the children are sifted, the
north from southern dweller,
all the righteous ones shall line up, at the
automatic teller.

Figures for the third quarter signal a potential slow down in the economy based on data culled from factory production, industrial employment and overseas orders. Latin America remains a financial drag on growth as the half-hearted embrace of market reforms by key countries has added to market jitters worldwide. Foreign reserves are shrinking as investors simply do not trust the implementation of meaningful fiscal action. It remains to be seen whether the tough market discipline necessary to avert crisis will be imposed.

b
b
b

The fields shall give forth plenty, a designer
coffee and cigarette
and every exit shall bloom a service station,
a stay-up-all-nite luncheonette.

The current economic upheaval has been particularly felt in Mexico where the peso hit a new low. This has spawned renewed fears that the drop in joblessness and a strong US dollar will increase illegal immigration along the border. As a response, the USA has added to surveillance at critical highway check-points. However, harsh immigration enforcement has resulted in illegal migration under more difficult circumstances. The most dangerous crossings occur over raging water, mountains and desert. Notably, there has been an increase in deaths associated with such attempts, most resulting from exposure.

For even the smallest of faiths produces, a
white bearded saint in red flannel
and what so ever is asked of shall be given
unto two hundred channels.

Due to consumer spending, the economy may just continue to grow. Recent world events have not signaled a serious problem for domestic growth. While consumers expressed some worry about the future of the country and the latest data indicate a cut-back on general spending, increases in luxury consumption, holiday purchases and housing starts have been able to buoy the flagging economy. Whether or not consumers will react to uncertainty by decreasing spending remains to be seen. A crucial question remains: when the going gets tough, will the tough continue to go shopping?

Redemption by consumption, what the
masses are likely to afford,
is to dwell by the golden arches in the
burger palace of the lord.

Increasing polarization between high and low incomes at the global level has led to divisive political repercussions as poor working citizens are pitted against poor immigrants for decreasing economic gains. Immigrants gravitate to where the jobs are, yet people from all places have been thrown together in new ways. On average, a decline in the standard of living has created intense experiential anxiety as workers of all classes come to grips with a more diverse and less equitable social landscape.

Swallowed by what is eaten, the virtuous
understand,
Relax and enjoy the taste, the kingdom is at
hand.

•BOUNDARIES•

by Iain Borden

Boundaries often present themselves to us in cities as if they are natural, physical entities. The river, the shoreline and the escarpment are joined by the wall, façade, window and fence to assemble a city of objects, a city of finite spatial facts that carefully and seemingly unambiguously delineate the spaces either side. In or out, here or there, within or beyond – these are the spaces of the city boundaries, dictating what we can do and when.

In the city, the A to Z map is not a free

movement, not just a free association of emotional, political and cultural desires translated into spatial vectors, but always a series of opportunity constraints. On the one hand there are the walls, barriers, fortifications and edges of all kinds that exclude frontally, which brutally separate the spatial divide. These are boundaries as two-dimensional vertical planes, preventing horizontal movement across the city, boundaries which have no apparent spatial depth but which extend their zone of influence on either side. We all experience these boundaries continuously and everywhere, the spatial which tells us to go no further. On the other hand there are the gates, toll-booths, doors, turnstiles, bridges and thresholds of all kinds which simultaneously deny, control and release spatial movement. These are boundaries as momentary portals, at once exclusionary yet conjunctural, the space-time of connections. These are boundaries which allow us to travel from one space to another (maybe).

And yet, confronted with this spatial facticity, all is not as it might seem. As Georg Simmel noted, the 'boundary is not a spatial fact with sociological consequences, but a sociological fact that forms itself spatially' (Simmel, 1997, p. 143). This suggests that boundaries are not just the cause, but simultaneously the product of social relations and their control. Boundaries are manifestations, not just origins. The physical boundary seen this way emerges as the outcome of a need and desire for social control, as the pervasive extension of power across the city. Every picket fence, every mirror-glass clip-on fenestration system, every ornate door handle, every 'No Entry' sign is disclosed as social not natural, political not innate, prescriptive-proscriptive not neutral.

There is more. Boundaries do more than control frontally; particularly in the postmodern city with its simultaneous increasing paranoia and contradictory desire for a (false) democratic and popular urban space, new kinds of boundaries have begun to emerge which seem open, seem friendly, seem attractive, yet which also seek to control access and bodily movement. Here the tactic is not frontal exclusion but a lateral challenge to the visitor, provoking questions in the mind: am I a welcomed guest, an ambiguous transgressor, or an unwanted trespasser? Do I have the right to be here? Am I the 'right' kind of person to be here? Gates that do not close, public art from Serra to Botero, office receptions with Mies van der Rohe 'Barcelona' and Jacobsen 'Butterfly' chairs, shopping malls entered through Amex, Visa and Bank of America cash dispenser lobbies, heraldic devices with dragons and trumpets, low chain-link fences, changes in 'fashion effect' architectural aesthetics, bollards, posts, parked BMWs and other such threshold signals all combine to create a time-space in which the visitors check themselves, validating their identity against that suggested as normative by the city architecture; as a result, everyone assumes a normative identity and behaviour and becomes a cast-member in the Disney movie of the postmodern city. This is the identity-passport control, the body-customs of the city. Whether it is faced in mirror-glass or historicist pink polished granite, the architecture of the postmodern city acts as a kind of Sartrean 'other's look', a deep mirror for looking into. The boundary emerges as 'thick edge', several metres and seconds in depth, and is ultimately a zone of social negotiation.

There is more still. If capitalism likes to control subjects in space, it also works hard to accumulate capital, extracting money at every available opportunity. As the number of boundary thresholds increases in the city, so does the number of those concerned with charges, taxation and tolls. Along with the old pre-modern city walls, ports and toll-booths we now have not just boundaries of extensivity – airports, transport interchanges – but boundaries of an increasing micro-spatial intensity

– public toilets, entrances to museums, leisure parks, sports facilities and (soon to come) roads with use priced by the mile. These are the boundaries of privatisation, barriers which charge for movement across space; socially, the boundary itself mutates from the modernist conception of crossroads as meeting place to the postmodern interchange where nobody intersects. This is the boundary diffused and dispersed within the city.

References and further reading

Augé, M. (1995) *Non-Places: Introduction to an Anthropology of Supermodernity*, London: Verso.

Borden, I. (2000) 'Thick edge: architectural boundaries in the postmodern metropolis', in I. Borden and J. Rendell (eds), *Inter-Sections: Architectural Histories and Critical Theories*.

Simmel, G. (1997) 'The sociology of space', in D. Frisby and M. Featherstone (eds), *Simmel on Culture*, London: Sage, pp. 137–185.

•BUILDING•

by Iain Chambers

Santa Cruz, California: at the corner of Water and Soquel Avenue, idling at the lights, waiting for the green, your eyes are pulled across the street until they come to rest on a group of semi-naked bodies of both sexes strenuously exercising behind glass. One presumes, given the walls are transparent, that it is permissible to look, to bestow an inquisitive gaze on this corporeal display. After all, the perspiring participants are themselves self-conscious consenting adults, specimens under glass, life-style advertisements for the Nautilus Fitness Center, open twenty-four hours a day.

Yet this public exercise in bodily virtue offers a disturbing spectacle for my European eyes. While certainly more discreet than the public working out at Muscle Beach in Venice (LA) – after all, here there is a screen, an entrance fee to pay, a community to join – there is still something profoundly disquieting. Apart from the intriguing blurring of public and private space, I suspect that it involves the shock of the idea, embodied in the earnest pursuit of an abstract well-being, that there can be no pleasure without exertion, effort, expenditure. In seeking physical definition, moulding the muscles, designing and building the body, a whole series of other definitions are simultaneously being orchestrated. In the political economy of fitness, in its definitions of selfhood, it is as though you can never let yourself go. Here to waste time, that is to spend, sacrifice and consume it without regard for ultimate ends is ritually exorcised. Bataille would not be welcomed. When the body is perpetually caught in a regime – fitness, diet, vitamins – of control, pleasure has to be earned. For all this exercise is not reducible merely to a question of health: you have to work rather than, say, walk the equivalent amount of time. You have to work for, work out, moral redemption. But why does the work-out not work out? Why do you have to return to the task again and again? Is this still an outpost, posed on the edge of the Western psyche, of the far-flung empires of righteous dissidence, those cities on a hill, forced to flee an old world that persecuted them with decadence?

Is it then possible to trace on these bodies, as they fleck their physical and mental frames, and busily mould a libidinal economy, the ubiquitous protocols of production continuing to prescribe the terms of consumption: no pleasure without prior payment, without first earning it? If so, this expression of the body in public life is perhaps inextricably linked to its simultaneous repression elsewhere. It is thus locatable in a specific moral polity that curbs the body whenever it threatens to exit from a productive administration and enter the unchained sea of undirected pleasure. It is perhaps at this border, on this edge, that the sign of perversion and pornography is nervously raised to police and control such traffic.

However, there is also a deviation here that is distinct from the diffuse moral and ascetic figuration of the body in career, of the body in public space and consumption. For body building is also a central occupation among America's incarcerated. Here the punitive violence, the revenge that is inscribed on the frames of those who are invariably poor, underprivileged and often black, sentenced to 'pay the price' for their crimes, is countered in the concentration of power and physical presence, of self-management, of the only thing left to hold on to: your body. There are now noises being made to remove this 'privilege'.

All of this, of course, is also a mirror of my own neurotic assumptions. It no doubt also ignores how many women increasingly occupy and transform this predominantly male space so that the work-out is clearly working something else out. At the same time, coming from a damp island off the edge of Europe that shares some of the cues that are activated in Californian pleasures I am fatally attracted to the seeming exoticism of the programmed pursuit of the self through play and punishment (masochism?). I both recognise myself in it and yet don't quite get the point. It is uncanny, replete with the ambiguous pleasures of losing one's self. And that, in the end, is probably what most disturbs me. The truth of ambiguity has apparently no place here.

There is seemingly no room for the middle-class English (European?) habit, for example, of obfuscating, even hiding, habits of fitness: you must not be seen to be trying too hard, you are expected to play it down, to limit the dis-play to a largely private affair. I could stretch this to a triangular observation, for I live there, and extend the frame to Italy where one is also involved in a continual exhibit of the self. But under the more cynical sun of Mediterranean skies I find myself involved in a deliberate masque, a playing out of a scene, a style, in which each shuffle of the fashion pack can reveal a variant and a knowing subscription to the narrative that allows you to decide what cards to play, which body, which 'you' to expose. As style, such a physical display reveals in the pathos of its language the knowledge of the limits that continually threaten to usurp it. Hence irony.

But then, in this rigorous Pacific heliotopia, in this allegory of the eternal summer where the last wave is postponed . . . forever, to be doubled by doubt, and the threat of dissolution, is perhaps merely to stumble over the stubborn truth of a European male melancholy obsessed with limits, with mortality. Perhaps. It is certainly to confront a diverse configuration of bodies in space.

In their individualised physicality these Californian, predominantly anglo, bodies insistently speak of a city, a locality, a life, that is urgently real. The wager is that reality can be controlled, otherwise it crushes you: this, and only this, is it! The private – as projected body, representative of the individuated control of reality – has here invaded and largely swept away the public. Private . . . property, shopping malls, weaponry and health care have an alarming resonance with tense biceps and beads of sweat. No doubt such scenes proffer the exhilaration of a nihilism that so fascinates European sensibilities jaded with the endless subtlety of the *chiaroscuro*, and the lengthening shadows of a decaying *polis*. The latter's more uneven and diversely informed understanding of place nevertheless persists as an ironic interpellation of the possibilities of modernity. Here physics and metaphysics are continually negotiated and rarely coincide: the individual recites a script she did not necessarily authorise. Such a fracturing, where the city both constructs and constrains the *cogito*, paradoxically suggests a freedom in which the individual is unwittingly conjoined with the repressed in confounding the inevitable. Although seemingly banished, that prospect, that history, surely also accompanies tense bodies in Californian gyms, haunts the lonely freeway driver, as a shadow cast over the beach?

•BUILDING SOCIETIES•

by Deborah Levy

A slice of the cucumber is what friendly societies are all about. Putting a bit aside for a rainy day. However, friendly societies are a thing of the past – perhaps because they were too friendly and susceptible to being held up, not with a slice of cucumber, but the whole damn vegetable.

Certain sections of the English class system are very fond of the cucumber sandwich. Thinly sliced, salted and placed between two slices of buttered white bread, there is no fat to encourage burping in the dining-room, no flesh to arouse lust for the gardeners.

However, unfriendly citizens are prone to abuse the cucumber and represent it in building societies, not as signifier of a delicate and refined palate, but as a sawn-off shotgun. It would seem that other long fruit and vegetables are not collected for the same purposes: for example, the banana, carrot and leek do not have the same suggestibility as the long, dark, green thing, containing as it does, within its crocodile textured skin, plentiful water and seed to panic and torment employees behind the panes of perspex.

No, the criminal imagination has pounced on the cucumber as being 'well hard', often concealing the true nature of their weapon with a cloth.

The cloth and cucumber raids.

It is well known that when babies are teething, their parents encourage them to gnaw on a cool slice of cucumber to relieve their inflamed gums. Perhaps if they were more canny, they would cover the baby in a cloth and hold up a building society with their offspring.

TV Advertisements for Building Societies always make us feel sad.

The advertisements show generations of a family from birth to death and all the things that could go wrong in our lives.

Just as a child is born it is possible our partner might become terminally ill and die; our house might catch fire; we might spend our old age spooning cat food into our toothless mouth – all because we did not think about the infrastructure of our future.

A sound-track, violins and piano, deepens these themes; the tone of the voice-over is kind, concerned, paternal, maternal, wanting us all to be OK: well fed, loved, warm, surrounded by the objects that comfort and make us happy. Deep down, we all want the same things. We are all cousins of the ape.

Life's a bitch. We need someone to look out for us. The man from the Pru. The woman from the Woolwich. The wounds of the daughters of our people wound them too; our fears are their fears; we will weep and they will wipe away our tears. We will howl in phone boxes and enjoy a little singsong in the ambulance afterwards:

> 'the wheels on the bus go round and round, round and round, round and round.'

•Bus-shelters•

by David Sibley

The distance from the bus-shelter to the security camera is about twenty metres. The camera rotates at the top of a tall post, painted dark grey and with a fat base, suitable for posters. The bus-shelter is in the camera's field of vision but the pictures relayed to the police station are unlikely to convey the full social significance of this dismal structure. The shelter is a simple steel frame, slightly bent, with a metal roof and panels, but no windows. The windows gave up long ago and the spaces where the glass once was make ledges for teenagers to sit on. The teenagers are the residents who made the bus-shelter their space, a place to socialize, handy for the chip shop but not such a comfortable spot now that they are unwilling actors in an evening's security filming. For the controllers in front of their TV monitors, such a social use of the street is more of a threat than a cause for celebration. They have not read their Jane Jacobs.

During much of the day the residents are absent, and their ownership is marked only by a few graffiti scratched in the flaking red topcoat. The visitors are the local commuters – students queuing for the bus to the further education college and McDonald's, nurses and cleaners on their way to the city hospitals, a few office workers and the odd academic. This is one of the last stops on the way to the city centre. After the affluent white suburbs where the rush hour bus collects a few environmentally aware office workers and then some more from the less affluent racially mixed suburbs, it is now picking up its heterogeneous passengers from the racially mixed inner city. Like the occasional boarded up shop down the road, the shelter is rather sad. The new brick paving and bollards, the product of the last environmental improvement scheme, contrast starkly with the battered shelter. In its own small way, it contributes to a negative inner city stereotype. The meaning of the shelter as a micro-space of social interaction, a teenage refuge, a structure related to travel and work through the fact that it serves those with no work and nowhere to go, is unlikely to register with most of the suburban bus passengers, whose journeys are briefly interrupted as the black college students pile on at 8.30 a.m.

City centre bus-shelters are marks of civic pride. This might be conveyed in several ways, like saying it with flowers. Baskets of flowers on the roofs of the shelters are nice for people riding on double-deckers. New modern shelters, however, complement new postmodern corporate investments – heavy, stone-clad office blocks with a lot of glass and stainless steel. No place for flowers. Grey-coated, tubular steel structures with curved glass roofs, illuminated advertising panels, and narrow red plastic benches are part of a progressive image. No litter or stale piss.

The public squalor that goes along with private affluence has been removed from sight, at least as far as the shop doorways.

In the city centre shelters, diverse people come together. They are unsorted. The owners of expensive suburban properties and housing association tenants momentarily share the same space. The travellers perched on the benches do converse, but the talk is mostly among elderly women who chat across lines of race and class. The bus to the suburbs then sorts them into various living spaces, moving them from the fragile, heterogeneous and transitory community of the city centre bus-shelter to the places where most passengers probably have more of a sense of belonging. But this is a bus-shelter from the Chicago School, a Burgess bus-shelter, one which harbours occasional congregations of the poor and the affluent who are then assigned to their residential niches *en route*. There are other city centre bus shelters which are nodes in a different urban ecology. They are assembly points for the poor of all ages, young men, women with young children, as well as the elderly with their bus passes, who are dispersed along public housing axes formed by both inner city high-rises and peripheral estates. Their encounters are familiar, untroubled by class difference but marked by a shared experience of high fares and infrequent services.

Although bus-shelter encounters are fleeting, they do bring people together. They have become a residual feature of public life in the western city, however. Close encounters with people of a different racial identity or of a different class in the spaces of a collective transport system can be avoided by all but the poorest, children and the elderly. The protective capsule of the car keeps others at a distance. There is no danger of touch or of verbal abuse or of a brief exchange about the weather. The black hospital cleaner waiting for the bus in the inner city shelter may remain invisible to the white car-borne commuter. The threatening night-time space of the city centre bus-shelter, when the old people have gone home and worries about the drunk and the mad come to the surface, is not threatening for the driver cocooned in car space, insulated from the dangers of the city by Rodriguez' guitar concerto on Classic FM. Reversing the decline in public life in the western city requires more than a programme of bus-shelter renewal. People in eastern Europe, still familiar with a collective existence, who wait at eleven o'clock at night like football crowds for the next number 6 tram on körút Erzsébet in Budapest, for example, manage without tram-shelters (and bus-shelters on bus routes) but they have plenty of trams. The modernist bus-shelters in Leeds or Liverpool city centres possibly signal a revival of public life but the traffic on the ring road tells a different story.

Postscript

'A Hull bus-shelter has been destroyed and the stop moved after it was found that courting couples were making love on its roof' (*Guardian*, 23 July 1997).

•Buses•

by Alastair Bonnett

The heroic scale with which modernist ideologies of liberation have been translated into built form is both compelling and numbing. The drone of traffic, the constant circulation of vehicles, have come to appear as natural as rain. The accompanying promise of atomised free movement, of being able to 'go anywhere', to 'leave immediately', without hindrance, without 'bother', is dripped continuously into our ears, ceaseless proof of the power of technology to emancipate the individual.

First in the USA, and later in other societies, the presence of automobiles became the ultimate symbol both of urban freedom and urban modernity. This chain of association arose in response to, and found confirmation in, images of non-western and pre-modern cities; cities that connoted stasis, the absence of movement, the weight of tradition. Indeed, it is intriguing to observe how the 'state of the roads' and 'the absence of cars' came to obsess western tourists seeking 'remote' destinations (today it is the absence of *new* cars that tends to excite comment). Such features emerged as key markers of the traveller's adventurous encounter with a primitive, ethnic and inaccessible urban landscape. The mini-bus chugging by went unseen, the gaggle of people waiting by the roadside for their bus unnoticed.

The bus is like that. It passes by, and no one looks. Indeed, most histories of the city barely mention it. Classic texts, such as Lewis Mumford's *The City in History* (1961), even Peter Hall's *Cities of Tomorrow* (1988), chart the conquests of highways and cars, but they have little time for the bus. It features only as a soon-to-be-surpassed museum piece. Thus, for example, one of Hall's few references to this form of transit occurs when he is describing the way, in the 1920s and 1930s, New York's Master Builder, Robert Moses, built the bridges on his seaside access routes too low for buses to pass beneath, thus reserving the best of the beach for the middle class.

The bus, or so it once seemed, could not keep up with the twentieth century. It splutters along behind, picking up all those people who will never quite make it into either 'History' or 'Tomorrow'. Perhaps the bus was so regularly missed in such accounts because it mocks modernist illusions of atomised freedom. The bus, after all, *could* go 'anywhere'. It *could* 'leave immediately'. But it doesn't. It is duty-bound to shuttle between predetermined points at predetermined times. A complex weave of routines threads across the city, invisible to newcomers, but a defining feature of every street, every corner, for many urbanites who cannot, or wish not to, use a car. It is a regular and collective phenomenon, as opposed to an arbitrary and individual one. The bus represents technology and technocratic progress just as much as any other vehicle but put to different ends. It is not part of the glorious meta-narrative of the modern city because, in its prosaic, proletarian form, it cannot be absorbed into the propaganda of 'individual choice', of 'free movement'. Not, at least, without a radical redefinition of what these terms might mean.

Of course, over the past three decades or so, the contradictions and unsustainability of atomised transport have re-established the little plastic double-decker, and the little plastic tram, as every trainee architect's favourite models to nudge about on her or his plywood cityscape. This change in emphasis has convinced a lot of people that the dominance of the automobile is coming to an end, and that a new, communitarian and ecological ideology of urban transport has been fashioned. The scale of this transition is, of course, highly geographically differentiated. It is, or so we are told, more 'advanced' in Amsterdam, less so in London, Lima or Lagos.

However, this reassembled version of modernity draws on many of the same ideologies as its discredited predecessor. The new forms of ecological and community transport are, for example, routinely justified in *technological* terms, as part and parcel of western mechanical and scientific superiority. The well-established *colectivos* and other small mini-bus operations common throughout South and Central America, which appear so suitable to the urban scene, play no part in this ideological rearticulation of the bus. For the bus has been reborn as the 'superbus', a bold new departure, a vehicle at the cutting edge of technology, sexier and cleverer than any car. Thus the bus is shedding its dull and pedestrian image by competing on the same terrain as its supposedly despised competitor. This process of technological glamorisation is tied to attempts to tinker with the class symbolism of buses. By emphasising the expense, uniqueness and sleek cleanliness of the contemporary generation of buses it is hoped that *middle-class people will start using them.* Thus 'ordinariness' and 'everydayness' are maintained as the provinces of the working class, and the new transport, like so much of the modern city, is transmuted into a disciplining and educational environment, passing on aspirational values to the lower orders.

It is also instructive that the emphasis on *networking* and *integration* within contemporary transport planning tends to mean networked and integrated with the car. In other words, the bus is being reintroduced as a conduit between, and an enabler of, 'old-fashioned' atomised car use. It is 'park and ride' not 'walk and ride'. Indeed, in city centres already dominated by the legacy of past road building, the parking element of the 'park and ride' equation is often situated in the downtown urban core. This is the case in my own city, Newcastle upon Tyne. The local council argues that the 'park and ride' sites in the middle of Newcastle are part of an 'integrated' transport policy. Indeed, a new road, a multi-lane 'Parisian style boulevard', has been built through the city centre to enable access. Some urban planners, it seems, will stop at nothing to encourage bus use.

Recent reassertions of the bus are more about continuity than change. Appropriating the bus to establish new links between cars or to make proclamations of yet another brave new leap in the west's technological progress, is to remain chained to the myths that got cities into such a congested mess in the first place.

References

Hall, P. (1988) *Cities of Tomorrow: an intellectual history of urban planning and design in the twentieth century*, Second edition, 1996, Oxford: Basil Blackwell.

Mumford, L. (1961) *The City in History*, Harmondsworth: Penguin.

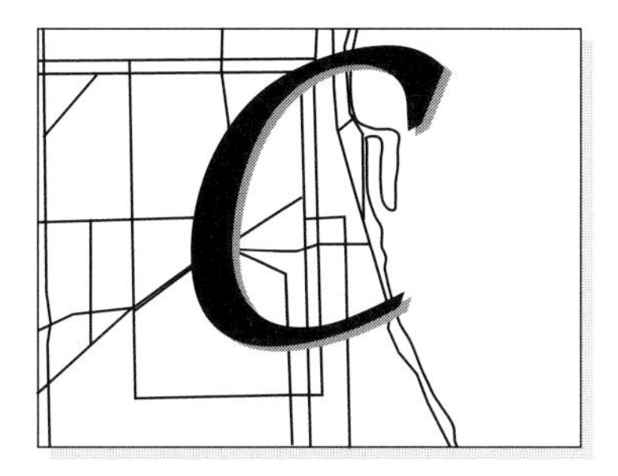

•Calling cards•

by Steve Pile

Cities, it has been observed, are ruthless in their control over time and space. Why? Well, for members of the Chicago School of Sociology, the concentration of so many people in the modern city meant that there needed to be new forms of social co-ordination. No longer was it possible to have people drift through the streets as they pleased. Nor could people expect others to be there when they turned up unannounced, unexpected. Ways would have to be found to organise time and space, to co-ordinate the movements of people. Two technologies became commonplace: the traffic-light and the clock (showing standardised time). Now it would be possible to keep the city moving, to keep a business appointment. Nevertheless, such interventions were not enough and others would follow: the yellow line, the wristwatch, the telephone – and the calling card. The calling card is an intermediary: it contains vital details, such as the name and profession of the caller, with their address, telephone numbers (and now other communicative technologies). The card persists as the afterlife of the visitor – a silent injunction: call me, why wait? Business awaits.

Amidst all these regulations and injunctions, there is still the sneaking suspicion that there are unintended consequences. And, perhaps, the possibility that something quite contradictory is going on. On the one hand, in order to keep the city moving, people would have to wait: for the traffic-light to change, for the time of the meeting to approach, for the calling card to be acted on. On the other hand, as movements became quicker and quicker, as time became more and more precise, people would become more impatient (and less willing to wait): lights are jumped, a second after time and time is up, and a building frustration when people don't return that call. Whether impelled by efficiency gains, by capitalising on investments, or by the need for good order, city life wouldn't quite run in one direction – no one-way street would keep everyone moving in the right direction. Even so, people would have to obey these ruthless principles of city life, or expect to pay the price.

With the advent of newer technologies for arranging appointments (such as mobile phones, faxes, e-mail), calling cards have begun to disappear from the armoury of the caller. Nevertheless, the calling card has adapted to its ever-changing environment. This old-fashioned technology is not yet ready to join the Sinclair C5 in the dustbin of history. They have found a new home: the telephone boxes of central London (and, increasingly, other British cities) have become the advertising hoardings of sex workers.

London's old red telephone boxes are a nostalgic symbol of a vaguely remembered Georgian urban splendour – a time of wealth and civilisation. But, if scratching at London's past uncovers its 'hells', then the 'hells' of today are in-your-face: here, women (and men) offer sexual services. Urban commentators have been fascinated by the figure of the prostitute: for some, a figure that represents the immorality of the city, its sexiness and its sinfulness; for others, the prostitute is the ultimate indictment of a system of wage labour that forces workers – sex workers – to sell their bodies to live; for yet others, the prostitute is simply an immoral woman or, maybe, she's just fallen. But the sex calling card has a slightly different story to tell, for it suggests that the oldest of professions has professionalised. The calling cards, like those for any business, announce a name and give numbers to call.

7724 is an exchange in the Paddington area of London. Paddington, of course, has a railway station, and the areas around them are traditionally inhabited – in popular imagination, if nothing else – by the low of life. It is also one of the exchanges that figures most prominently on London's calling cards. They tell potential customers what tastes are catered for. Yet men's desires are not uniform (except when they are) or static (except when told to be). At the same number different services can be offered, while different numbers offer different services. Like?

Well, a super sensual versatile mistress offers 'Electrics, Baby Service, Humiliation, Domination, Bondage, Correction, TV Wardrobe, Training, Rubber, Leather & PVC, Torture, Caning, Troilism, Water/Hard Sports, O and A levels'.

Or you could 'Pick Your Pleasure' from these: Fantasies, Toys, Feminisation, TV's welcome, Enforced Cross-dressing, Foot Worship, Kinky Rubber, Hot Wax, Videos, Nanny Service, Infantilism, Nurse . . . and much, much more.

But you might not want such an easy time: 'Madam Gladiator / Is my name submission / holds are my game, Exciting Female Wrestler'; or perhaps something more direct would get you calling: 'Madam Gladiator [Genuine Photo] Wants to Overpower You and Sit on Your Face, YOU WIMP.'

You might be in a 'Schoolroom, Dungeon, Rubber or Baby Room' – with or without a TV (note: a television, in this case). There are bubble baths and VIP visits, sizzling lessons, perhaps with spanking, it could be 'every man's dream'. And not just every man's. 'Couples are welcome.' As are women.

Like any expanding business sector, quality is a good sign. So, the calling cards have become ever more sophisticated. Initially, the cards were barely worthy of the name – poor photocopies on cheap paper cut out by scissors and stuck up with the cheapest Blutack substitute. From one telephone number, you'll have the choice of technologies for arranging those precisely timed (time is money) appointments, from mobiles to web-sites. Even the Blutack is better now, as the cards are increasingly colourful and well produced.

Such a production costs money, not just for the cards themselves, but for their constant renewal and distribution (the authorities have tried various ways to stamp out the calling cards, but have yet to find a legal or effective way to do so – and commodification wins its usual battle over morality). More than this, these personal services require investment: rubber, uniforms, S/M gear, chains, dungeons that are not cheap to install, nor maintain. Increasingly, too, it is women who are taking control, making the decisions. Men pay for this, of course. Mostly, it's men's money – and there's a lot of it. It is estimated, for example, that the worldwide telephone sex trade is worth £2 billion a year and that London's card-carrying sex workers make £180 million a year.

But the professionalisation serves only to hide the grim realities of the profession. The colour, the money, the calling cards present a beautiful face, your every desire, your fantasies fulfilled, when in fact what happens is that women (and men) have sex for money with men (and women) they despise. Like the city, prostitution is ruthless – especially if you pay extra.

•CAPITAL•

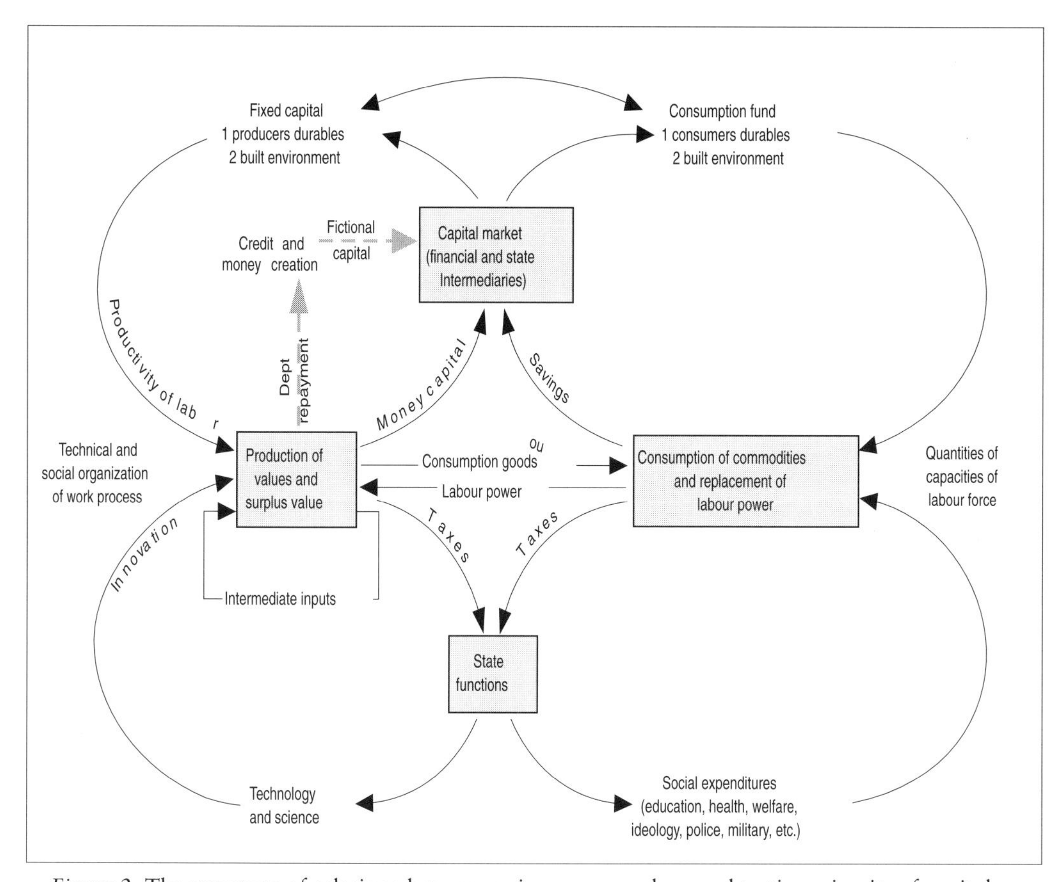

Figure 2 The structure of relations between primary, secondary and tertiary circuits of capital

Source: D. Harvey (1978) 'The urban process under capitalism: a framework for analysis', in M. Dear and A. J. Scott (eds) (1981) *Urbanization and Urban Planning in Capitalist Society*, London: Methuen, p. 99.

•CAR CRASHES•

by David Bell

Perhaps the most exceptional scene from David Cronenberg's movie of J. G. Ballard's novel *Crash* is set at the scene of a motorway pile-up. The central, crash-obsessed characters – James, Catherine, and Vaughan – wander through the wreckage, gawping at the disfigured, the damaged and the dead. Vaughan takes photographs of wounded cars and wounded bodies, coaxing Catherine to pose among the carnage. All around them, workers from the emergency services cut people from their wrecked cars. The scene is perverse, bewildering, carnivalesque.

In many ways this scene is a distillation of every urban dweller's fascination-revulsion for the spectacle of car crashes – a condition which is itself an extension of our relationship with the car, that uneasy symbol of modernity. Contemporary urban culture is saturated with the car, which has become a prosthetic extension of the body, either as a shield (car ads emphasising safety, 'Baby On Board' bumper stickers) or as a weapon (ram-raiding, road rage). Car crashes are thus especially emblematic of technocultural corporeality – car crash victims, as depicted resonantly in passages of Ballard's novels, become mutant cyborgs, fusing meat and metal.

In another of Ballard's novels, *The Atrocity Exhibition*, the central character, Travers, organises an exhibition of crashed cars, while Vaughan in *Crash* dreams of dying in a car smash with movie star Elizabeth Taylor. Similar images resonate through Barbet Schroeder's remake of the 1940s film noir *Kiss of Death*, which he recasts around the fetish of the car. From its opening scene of a breaker's yard onwards, the car dominates this movie:

> It is a fine, bright day and the camera holds everything in view. The slow pan along concrete supports of the 'raised' freeways might be the opening shots of any film, but, instead of moving onto new terrain, the close-ups linger. Gradually, as the camera holds the same ground, huge rusting heaps of scrapped automobiles come into view, mile after mile, just below the surface of the freeway. Modernity is being placed alongside its 'normal atrocities'. . . . So this is where the American enlightenment leads. It is the end of the yellow brick road. It is hope abandoned. (Munro, 1998, p. 185)

The dead car – wrecked, abandoned, burnt out – is itself a powerful and evocative symbol, then, of the 'normal atrocities' of urban modernity. I can't help but think of Philip Zimbardo's 'A field experiment in auto shaping' (1973) in this context. Published the same year as *Crash*, Zimbardo's essay tells his tale of 'observing [and then re-creating] the ritual destruction of the automobile – the symbol of America's affluence, technology and mobility, as well as the symbol of its owner's independence, status, and . . . sexual fantasies' (p. 85). His record of the scavengers who descend on and strip abandoned cars, and the vandals who engage in acts of 'random destruction' (breaking windows, slashing tyres) on the remaining hulk, is incredible enough (a 'test car' placed as part of this experiment on a New York street by Zimbardo endured twenty-three separate 'attacks' over three days, by which time it was no more than a skeleton); but that pales alongside his discussion of abandoning a second 'test car' on the campus of Stanford University in California to repeat the 'field experiment'. This abandoned car lay forlornly untouched by scavengers and vandals for over a week, so Zimbardo decided it needed more visual cues – to state more plainly its status as abandoned (he'd left the hood open to signify abandonment, but some kind soul closed it during a rain storm). He dispatched some (male) students with a sledge-hammer to rough up the car a bit. At first they were unable to do it: the moral code of the human–car relationship clearly forbade such violence, even though they had been sanctioned to carry it out. After much awkward standing around, one student tentatively dealt a blow. And then another, harder one. Once that threshold had been crossed, the students' violence erupted:

> they all attacked simultaneously. One student jumped on the roof and began stomping it in, two were pulling the door from its hinges, another hammered away at the hood and motor, while the last one broke all the glass he could find. They later reported that feeling the metal or glass give way under the force of their blows was stimulating and pleasurable. Observers of this action, who were shouting out to hit it harder and to smash it, finally joined in and turned the car completely over on its back, whacking at the underside. (Zimbardo, 1973, p. 89)

As Rolland Munro writes of *Kiss of Death*: 'in performances of masculinity, there is a magnification of the self that goes on to disparage the very technology that serves [as] its prosthetic extension' (1998, p. 186). The car's the star, but it also has to be wrecked, disparaged, precisely because of our dependency upon it, because of its prosthetic function. The fascination for car crashes, then, is about this love and loathing; a cool fascination born out of modernity and urbanity.

(*With thanks to David Fox*)

References

Munro, R. (1998) 'Masculinity and madness', in J. Hassard and R. Holliday (eds), *Organization-Representation*, London: Sage.

Zimbardo, P. (1973) 'A field experiment in auto shaping', in C. Ward (ed.), *Vandalism*, London: The Architectural Press.

• CATS •

by David Sibley

Cats, like window-boxes, complement buildings. They are furry accessories which reposition themselves occasionally on ledges and walls. In the city, there is a place for cats, like the feral cats of Fitzroy Square in London which were celebrated by T. S. Eliot in *Old Possum's Book of Practical Cats*, admired by Virginia Woolf and popularised in Andrew Lloyd-Webber's musical *Cats*. These were black ones with white faces and white waistcoats. When not supine on flat roofs in the sun, cats impress with their physical skills, their ability to jump, climb and balance. They bring nature to the city but the domestic variety will also sit on your knee for a stroke – the pleasures of touch and the pleasures of animal nature. Like Topcat of Officer Dibble fame and Corky the Cat, they are also streetwise and smart. They are well adapted to the unpredictable features of city life. They control territory. They have highly developed survival skills, an uncanny ability to seek out fish heads. They have agency.

The celebration of cats in verse, musicals and comic books suggests that cats have an easy relationship with cities but, in fact, urban cats are viewed with ambivalence.

A British Cat Action Trust newspaper advertisement, appealing for funds, featured wretched little Darcy, a feral kitten, stepping from a pitied existence in the wild to a 'normal' domestic life as a pet, rescued, like Victorian missionaries rescued prostitutes. Living in an abandoned building or on wasteland, a feral animal may be associated with dereliction and disease, a polluting presence that is in need of sanitising and domestication. Thus little Darcy will be made suitable for apartment living, as unobtrusive as a cushion, by spaying, frequent shampooing and, maybe, de-clawing. Wild nature has to be tamed and rendered suitable for the settee in front of the TV. Yet nature cannot be controlled. When the domestic cat goes through the flap, it returns to a city increasingly populated by transgressive animals – cougars in southern California, wild boar in Barcelona, foxes in most British cities. These scavenging species bring the wild to the city and frustrate attempts to draw a line between urban civilisation and wild nature. The city cannot eliminate the wild. Particularly at night, it moves back in, leaving a trace in the form of ravaged bin bags and chicken bones. But the wildness of cats is not invariably viewed with distaste. They hunt rats and so confine them to their subterranean spaces. One source of repulsion in the city helps to contain a more potent one.

In a world where many species and habitats are threatened, television nature programmes provide gratification and some reassurance. Gambolling wart-hogs can be enjoyed by city dwellers who have little or no direct experience of this kind of wildness. Similarly, the wild qualities of feral cats are a source of pleasure for some people. Contrary to the image projected by the Cat Action Trust, feral cats are seen as beautiful as well as resourceful. These cats may be admired at a distance – on wasteland or allotments – but cities also provide feeders, people who have daily contact with feral animals and who establish a close (but not usually tactile) relationship with them. In Fitzroy Square in the 1970s, the cats congregated inside the railings of the gardens at about 9 o'clock each evening and waited for the feeder to arrive. Her name was Mary. When she arrived, these cats, which had no other human contact, approached her with tails upraised and some even rubbed themselves against her legs. She did this for many years. Such people have an intimate knowledge of cat spaces – buildings, derelict gardens, cemeteries, wasteland, a hidden geography of the city. In Europe, it is Mediterranean cities which are best known for their feral cat spaces, notably the Colosseum in Rome, but there are similar public places colonised by cats in Naples and Barcelona. Northern European cities have largely relegated their feral colonies to marginal spaces, away from the public gaze, but these are spaces familiar to their supporters and detractors. In the 1970s, the Royal Navy Dockyards in Portsmouth, a walled, closed space, had a colony of about three hundred cats which were descendants of ship's cats from the early eighteenth century. The animals in this enclosure were known to dock workers and fishermen who fed them, but cat welfare people and 'catbusters', environmental health officers, have cleaned up the more public cat territories.

Southern European cities seem more accommodating, in some respects more tolerant places than northern cities, with a relaxed attitude towards cats (and children). The animals share space and time with people rather than being rescued from the wild or tidied away to marginal spaces. Cosmetic, purified nature is generally accepted, in city parks, for example, but encounters with wild nature in public spaces, encounters with weeds, feral cats and foxes, reconnect people with other species and remind them of the fragility of the urban boundary as a marker of civilisation. This mixing of the wild and the civil makes people anxious. In the medieval city, the walls helped to keep out wolves as well as other undesirables, but the boundary between the city and nature has never been that distinct. Thus cats in the city are inevitably viewed with ambivalence. Their odours, residues and caterwauling are polluting but other aspects of the species, their movement, their coats and their cleaning behaviour enhance the urban scene and make them acceptable pets, the more so when purified through medical intervention. Thus the presence of the cat contributes to the permanent tension between wildness and civilisation, order and disorder in the modern, western city.

•CCTV•

by Iain Borden

Moving back along the historical wire, we discover the logic of Closed Circuit Television (CCTV) the first time anyone looked at anyone else and thought to maintain a watchful eye. Surveillance thus has its origins in any system of social control. From the spatially spread complex governor-province system of ancient Rome, the confidant-networks of Chinese Emperors, the dispersal of churches and monasteries across medieval Europe and the roving maritime navies of the Dutch, Spanish and English, to the more local spy networks of such operators as feudal lords, guild masters and revolutionary enforcers, the need to keep an eye on what everyone else is doing has been paramount. Looking as a form of expectancy, looking as a noticing of the unusual, of the different – such are the origins of surveillance, the roots of CCTV.

But the rise of the capitalist city, particularly at the end of the eighteenth century, brought with it new problematics – notably that there were far more people to keep in sight, that most of them were of unknown name and origin, and that conventional systems based on a degree of social and spatial intimacy were now far too expensive, not to say impossible.

It is in this context that the technical systemacity of surveillance first emerged in Jeremy Bentham's panopticon prison, an architectural construction that simultaneously served as model for the controlling mechanism for urban social relations. Here, a central warden could view many prisoners silhouetted against the glass walls of their cells, subjecting them to ever-present supervision in a supremely efficient manner. Asymmetry was all-important – hidden behind narrow slits, the warden was invisible to the prisoners, who consequently were never sure if the warden was indeed looking at them, and so were forced to conclude that he always was – the warden as omnipresent God, whom they could never see. One warden, hundreds of inmates. Power effected through the threat of a glance.

This then is the surveillant logic of Haussmann, of open spaces, of the flattening of medieval courts, dark alleys, labyrinthine rookeries and narrow streets of the nineteenth-century city. Fresh air and the city of hygiene brought with it the city of surveillance, the birth of the police force and the ever-present eye. Urban streets became schools, hospitals, prisons, factory floors and parliamentary debating chambers – places where people were assessed by their visibility. For the newly dominant middle classes, the system of the eye offered control of the Other, a defence against the threat of the urban proletariat and the mysterious customs of immigrants.

So the arrival of CCTV in the latter half of the twentieth century has seemingly been little more than an ever-increasing and ever-heightening extension of the surveillance principle into the everyday lives of metropolitan dwellers. Road junctions, shopping malls, underground stations, gas stations, city pavements, shops and office lobbies are all surveyed, all part of the proliferation of star systems of guardian cameras. They are sniper zones of vision, waiting for the 'unusual' as much as for the criminal, immediately shooting its evidence down on to tape. Every space is a panopticon prison, for we are never sure who is watching us. And they maintain asymmetry, for while such places survey all on to tape at all times, photography by customers and visitors is commonly banned.

CCTV extends outward too, through computer videocams, video-mail, video-conferencing, Shockwave and ISDN – CCTV goes global, creating ever wider closed circuits of vision. We can be closed to the public but open to the corporate world. Ultimately, we are heading for the disembodiment of the eye, the eye of power, the state as cyborg, the privilege of vision as a technical device.

Yet, in contrast to the panopticon prison, CCTV also seems to enjoy the support of its inmates. In February 1993 a two-year-old boy, James Bulger, was led away by two ten-year-old boys from a Liverpool shopping centre, escorted two-and-a-half miles along city streets to a railway line, where, as dusk fell, he was battered to death with household bricks and an iron bar. Much of his terrible last journey was captured on CCTV, first in the shopping centre and then peripherally by various company security cameras along the route. Their evidence helped identity and convict the culprits. Since then, CCTV has been welcome in the UK, viewed almost universally by the public as a panacea to urban crime and violence. Municipal authorities like Knowsley spend their last £5 million on a CCTV system to attract respectable family stores like Marks & Spencer and Sainsbury to their high street, while students in Oxford, threatened by muggings and assaults, demand the extension of CCTV from the main streets into the back alleys of its medieval fabric. Special television programmes use police and company CCTV footage as a form of entertainment.

Furthermore, we are increasingly welcoming CCTV technology into our own lives. Beyond home security cameras, the use of camcorders, palmcorders and VHS playback surveys not just baby's first faltering steps but the behaviour of potentially errant child-minders and even partners and spouses. In the Neighbourhood Watch schemes of the inner and outer suburbs, the camera is eradicated altogether; the twitch of the curtain substitutes for the firing of the shutter, the ever-watchful network of 'community' participants replaces the wires and tapes of the CCTV system. Where once the camera replaced the eye, the fear of the unknown now replaces the camera.

Further reading

Bogard, W. (1996) *The Simulation of Surveillance: Hypercontrol in Telematic Societies*, Cambridge: Cambridge University Press.

Cummings, D. (1997) *Surveillance and the City*, Glasgow: Urban Research Group.

Dandeker, C. (1990) *Surveillance, Power and Modernity: Bureaucracy and Discipline from 1700 to the Present Day*, New York: St Martins Press.

Foucault, M. (1979) *Discipline and Punish: The Birth of the Prison*, Harmondsworth: Penguin.

Lyon, D. (1994) *The Electronic Eye: The Rise of Surveillance Society*, Minneapolis: University of Minnesota Press.

Staples, W. G. (1997) *The Culture of Surveillance: Discipline and Social Control in the United States*, New York: St Martins Press.

Tabor, P. (2000) 'I am a videocam', in I. Borden, J. Kerr, J. Rendell with A. Pivaro (eds), *The Unknown City: Contesting Architecture and Social Space*, Cambridge, MA: MIT Press.

•CELEBRATION•

by Anette Baldauf

'It's a small world after all'

Figure 3 Celebration montage by Alice Arnold

'In a true sense we are cast members because we are also introductory people to Disney because this is a Disney town . . . just without Mickey Mouse running around.' (Jerry, resident)

'It's like living in Downtown, Main Street USA at the Magic Kingdom . . . the security, the cleanliness, the orderliness of it all. It's an opportunity to live in Fantasyland.' (John, resident)

Expanding successively from the production of movies to fun brought on stage in fenced-off areas of entertainment parks to finally the simulation of a perfect life – US-America's leading dream factory designed, constructed and now administers the realization of a utopian town: 'Celebration' is built on 4,900 acres of swamp land in Orlando, only one highway exit away from Walt Disney World (Florida), where 8,000 residential families are going to praise the return to 'good, old times'.

Family Hancock was living in a gated community when they decided to participate in the building of the American future community, constructed upon the five cornerstones of education, health, technology, community and a sense of place. Their new home, built in neo-colonial style, is surrounded by blooming flowers, the front gate is wide open, the spacious balcony framed by white columns. Inside, antique nostalgia, enriched by an ambitious terrace house aesthetic, matches new technology like the fibernet 'Intranet', connecting Celebration's parts to a virtual whole.

'Basically it's an outdoor mall but it's different. It's an old town and its got every technological benefit of the twentieth century.' (David Hancock, 12 years old)

'When we are outside at night on the street walking down the street or riding the bike, you almost feel that you are in a movie set. . . . It's just very unusual.' (Mrs Hancock)

'I love it. . . . Everything is so close, you don't even have to go out of Celebration to do anything. It's all like a small, little world.' (Candice Hancock, 8 years old)

Celebration is indeed 'a small world after all'. Its downtown 'Market Street' is modeled after a traditional retail business district of an imagined American 'small town'. The bank (designed by Robert Venturi), town hall (Philip Johnson), post office (Michael Graves), movie theater (Cesar Pelli), and retail offices (Robert Stern) construct a visually coherent center, romantically posted in front of an artificial lake. Downtown, owned by the Disney Corporation, leisure zones such as the pool and golf place (Robert Trent Jones sen. and jun.) as well as residential areas (currently 350 homes designed by Robert Stern) are all within walking distance.

'The philosophy is that we all have the same philosophy. . . . What Disney and everybody is trying to get at is to re-create the early communities . . . where we had less headache and problems, stress, and hate and people knew everybody. You came out of your community and walked to the grocery store, the post office and did not have to drive 12 mile to get to one of them. That's really what the whole idea is all about.' (John, resident)

Architectonically and morally looking towards the past, technologically into the future, Celebration is supposed to represent the ideal model for urban planning in the context of 'New Urbanism'. Every weekend, thousands of tourists march to the precursor of 'contemporary life-style' to look at the composition projected on the surface of a 'small town' of the pre-Second World War period, promising a fast forward return to the past. Fabricated by the leading 'imagineers' of the country, 'community' is the magic wand of this urban transformation.

After the US-American white middle-class nuclear families had turned their backs on the inner city in the 1950s and, guided by fear and desperate for protection, took the newly paved road to suburbia, their driving dreams only partly came true: the desired community of fraternity occasionally turned into fratricide (Sennett) and the intended safety zone itself became a major war location. The successive spread of 'gated communities', the suspicion of 'Neighborhood Watch' and signs like 'Armed Response' in the front yards increasingly mark suburban space eaten up by highways and angst.

Prototypical *lebensraum* in contemporary US-America is the so-called 'mallopolis' (Sorkin). Nearly half of the US-American population is living in this gap between shopping mall, highway and parking lot. Small towns between New York and Los Angeles are increasingly becoming incorporated into an expanding megalopolis and successively vanish from the US-American geographical map. While in 1970 about one in ten Americans lived in a small town, by 1990 under sixteen million, or about one in sixteen, still live there. Statistically, small towns are marked by a high proportion of welfare dependency, single mothers, violent crimes, anti-immigrant and pro-death penalty attitudes as well as a tremendous mistrust towards the Government (General Social Survey, National Opinion Research Center, 1994). Nonetheless, since the rural configuration 'small town' is slowly dying out, it increasingly offers a perfect matrix for romanticization.

For the residents of Celebration, 'small town' generally indicates the protective social net of a community, i.e., neighborly care, intimacy and safety in the context of a Disney-controlled and administrated environment. The foundation of this 'community' is constructed upon two antagonisms: the image of Gotham-City and its threatening triad of 'crime–prostitution–drugs', and the Lynchian-suburbia of cut-off ears in the flower beds. Perfidiously, Donald Duck's logic of profit is appropriating these dystopian urban imaginaries, partly co-produced in its own house, using the mythographically produced fear for purposes of real estate business and image politics. Individually, the residents of Celebration dream of a well-protected childhood ('It reminds me of my parents growing up . . .') which some connect with the locations of Disney's fairy-tales; collectively, they not only trust the Disney Corporation more than the Government (e.g. Disney associations more than a democratically elected major) but they also share a desire to be 'among people of my kind'.

This foundation might explain the striking social homogeneity of the 'open' (i.e., not gated) town which, despite its rhetoric (e.g. a media-conscious drawing that supposedly elected the future property owners) and the availability of cheaper rental apartments ($800 a month upwards), marks the future community. Besides the economic selection based on real estate prices which are comparatively expensive by Florida's standards, the social set-up of the town, its subtle dynamics of exclusion and segregation, is organized around the symbolic boundaries of life-style and race.

> 'Well, I wouldn't wonna live here. I'm a hard-working man, I'm a laborer. When I come home from work I am tired, I wonna sit down in front of the TV, have a beer. I am not interested in community or shopping and this kind of stuff.' (Bricklayer working in Celebration)

Once a week, the Disney Corporation introduces new residents into Celebration's superstructure. Duplicating the obligatory introductory courses for Disney employees in the parks, the intention of the seminar called 'Foundation' is to construct a common reference for residents of this town built from scratch. The residents learn about the history of Walt Disney World, the land Celebration was built upon and to finally travel collectively into the imagined common future in town.

Celebration makes sense to its residents; it fits their desires and contains their anxieties. Like most of the residents Jody, who works in Epcot Center (Walt Disney World), spent every holiday and all her vacations in Disney World. With the realization of Celebration, a dream has come true. Work and leisure, childhood memories and present merge, organized around Disney themes: Her living room is the empire of the seven doves, her sleeping room a farm, and the Lion King is taking care of her kitchen.

'My family spent a lot of time at Disney. We went so often I couldn't even count. And it's just where we made our memories. There are twelve grandchildren in our family now and we meet several times a month in Disney World. We just go and enjoy our time together . . . I have been collecting Disney things for a long time because for me it was a reminder of home and a piece of my childhood. And for me living and working here now is just perfectly right.' (Jody, resident)

•Cinema•

by James Donald

First, cinemas in the city. Initially, at least in the USA, they were mostly to be found in working-class and immigrant ghettos. Between the mid-1910s and the 1930s, the explosion in the number of cinemas, and their glorious transformation into picture palaces, took place not only in central entertainment districts but, especially, in the new suburbs springing up around major cities as mass transportation made travel easier. This was a key factor in making cinema a genuinely mass medium. It still offered different pleasures to different audiences, though. In the film journal *Close Up* (1927–1933), the novelist Dorothy Richardson charted cinema's functions in various parts of London – entertainment in the West End, a 'civilising agent' in the slums, a sanctuary for weary mothers on a Monday afternoon in a North London suburb. In all these cases, Richardson finds in cinema a public of urban spectators being educated for modernity: 'everyman at home in a new world', thanks to the movies. 'They are there in their millions, the front rowers, a vast audience born and made in the last few years, initiated, disciplined, and waiting. What will it do with us?'

One thing cinema did with 'us' was to teach us ways of seeing the modern city. Second, then, the city on film. Street lamps reflected in the rain-slick streets of film noir. The obstreperous utopianism of New York in *On the Town*. History saturating Berlin in Wenders' *Wings of Desire*. The deliriously violent Hong Kong of John Woo. London and Paris uneasily multicultural in *My Beautiful Laundrette* and *La Haine*. Random screen images, but in none is the city just a backdrop; nor is representation really the issue. To borrow a distinction from Lefebvre, film presents urban space as itself representational, as simultaneously sensory and symbolic. Hence Walter Benjamin: 'The camera introduces us to unconscious optics as does psychoanalysis to unconscious impulses.' The fragmentation of modern urban experience created the need for cinema. 'Perception in the form of shocks was established as a formal principle'; montage offered a 'new law' for assembling the 'multiple fragments' of both city and film. Hence, then, cinema's analytic and epistemological power:

> Our taverns and our metropolitan streets, our offices and furnished rooms, our railroad stations and our factories appeared to have us locked up hopelessly. Then came the film and burst this prison-world asunder by the dynamite of the tenth of a second, so that now, in the midst of its far-flung ruins and debris, we calmly and adventurously go travelling.

Benjamin's optimism captures the spirit of the 'city symphonies' in the 1920s: among them *Manhatta* (1921), a Whitmanesque celebration of New York by the photographer Paul Strand and the painter Charles Sheeler; Alberto Cavalcanti's *Rien que les Heures* (1926), a portrait of Paris structured around two 'ordinary' people; and, best known, Walter Ruttmann's *Berlin: Symphony of a Great City* (1927) and Dziga Vertov's *The Man with the Movie Camera* (1928). Both use the 'day in the life of a great city' motif to capture a dynamic of traffic, machines, work and leisure. Ruttmann links rhythmic editing and a modernist eye for abstract shapes and formal juxtapositions with a sequence of private dramas played out in public: children going to school, people chatting in cafés, pick-ups, street performers in silly costumes, a woman's suicide. Vertov's constructivist city, in contrast, offers neither portrait nor record. No one place, it combines footage of Moscow and locations in the Ukraine to demonstrate the structure of vision embodied in the 'eye' of cinema. *The Man with the Movie Camera* makes strange our normal perceptions of the city by laying bare the device of cinematic perception.

Rather like Vertov, Benjamin saw in modernity a store of contradictory and so potentially dialectical images which needed to be unfrozen from their appearance as 'dream images'. Turn from Vertov's Factory of Facts to the Dream Factory of mass cinema, and how then does the labyrinth of the modern city appear?

Here popular fears materialise in the form of celluloid cities unbuildable without the artifice of the studio lot or the ingenuity of special effects. 'The city' observes Peter Wollen, 'is perceived as a kind of dream space, a delirious world of psychic projection rather than sociological projection' (1992, p. 25). Although it is easy to be mesmerised by their architectural imagery, the narratives of these films also suggest an archaeology of urban experience. The look of *Metropolis* (1926) was inspired by Fritz Lang's first visit to New York in 1924, but its critique of the cult of the machine is articulated in a story about forces which, once unleashed, have an unmanageable capacity for destruction: the proletarian mob and the technology that can enslave it, but above all an untrammelled female sexuality. The image of King Kong astride the Empire State Building may have become an icon of New York, but it also condenses a morality tale about nature commodified to make a Broadway holiday. The discussed-to-death architectural hotchpotch of *Blade Runner*'s imagined Los Angeles provides a cyberpunk agora for a philosophical debate about the limits of the human and what we can hope for.

Perhaps the key to these fantastic cinematic cities can be found in the comic-book grotesquerie of the Batman films. Gotham City is less a dystopia than the primitive city enigmatically discussed by Michel de Certeau: an infantile spatial practice always threatening to undo the rational semiotics of the planned city. De Certeau quotes Kandinsky: 'a great city built according to all the rules of architecture and then suddenly shaken by a force that defies all calculation' (1984, p. 110). So cities in cinema turn out to be built on the fault line in modernity. Here too rationalised representational space is haunted by a specifically modern uncanny; in Baudelaire's image, 'the savagery that lurks in the midst of civilisation'.

References

de Certeau, M. (1984) *The Practice of Everyday Life*, Berkeley: University of California Press.

Wollen, P. (1992) 'Delirious projections', *Sight and Sound*, August.

c
c
c

•CITIZENS•

by James Donald

Aren't citizens simply people who live in the city? Not the way republicans, communitarians and most political theorists tell it, they're not. At least as far back as the Greeks, being a citizen was not just about where you live. When Aristotle said that man is by nature an animal intended to live in the polis, he meant that, even if the space of the city has to be shared with others, citizenship as an aspect of the good life is an ethical status achievable only through active and responsible participation in its public deliberations.

Two thousand years later, Rousseau was still using citizenship as a criterion by which to judge cities and their inhabitants. 'In a well-conducted city, everyone rushes to the assemblies', he comments in his *Social Contract* (1754). 'Under a bad government, no one cares to take even a step to attend them.' Citizenship means rational engagement in public life, as opposed to the affective virtues and consolations of domesticity. And being rational, this is a public life dedicated to political deliberation, not surfing the city's pleasures and perils, the anarchic sociability and subtle rules of street life. These are what make the city a dangerous place. The life of Paris streets and salons is a masquerade. 'In a big city, full of scheming, idle people without religion or principle, whose imagination, depraved by sloth, inactivity, the love of pleasure, and great needs, engenders only monsters and inspires only crimes.' In a small republican city like Rousseau's idealised Geneva, by contrast, every citizen is forever subject to the surveillance of government, to the gossip of women's circles, and to the inner voice of conscience. Stripped of the seductive masks of fashion, politeness and anonymity, people are forced back on their natural inner resources, and this provides a more fertile soil for the cultivation of human nature and virtue.

Rousseau's dream of a transparent city purged of all zones of obscurity or disorder is thus linked to a belief in citizenship as the ethical formation of the self, a technique for making the self transparent. Another two centuries on, this compelling but rather odd idea continues to haunt champions of both deliberative democracy and communitarianism. Here, as a random contemporary example, is David Marquand:

> the central message of civic republicanism is that the Self can develop its full potential and learn how to discharge its obligations to other selves only through action in the public realm of a free city – that politics is both a civilised and a civilising activity; that it is, indeed, the most civilised and civilising activity in which human beings can take part. (1991, p. 343)

Western conceptions of the good life are permeated by the language of the city (political from polis, civilisation from civis) and by urban imagery: the republican city, the city as public sphere, the radiant city of rational planning. Equally, of course, the city also mythologises the hubris and pathos of such aspirations to sociability: the fate of Babel and Babylon, the grimy reality of an alien and opaque environment associated with disorder, irrationality and corruption.

This conceptual geography of citizenship is not much help when it comes to finding practical answers to the question posed by the city: how can we stroppy strangers live in a shared space without doing each other too much violence? Citizenship isn't about gents in frock coats talking politely to each other; nor is it just about debating different political opinions. The city presents us with more diverse and challenging enigmas of sociability than that, and it demands more diverse and ingenious arts: reading the signs in the street; adapting to different ways of life right on your doorstep; learning tolerance – or at least, as Simmel taught us, indifference – towards others and otherness; showing responsibility, or self-preservation, in not intruding on other people's space; exploiting the etiquette of the street; picking up new rules when you migrate to a foreign city. It is through social, cultural and semiotic negotiations like these, rather than the disciplines of republican participation, that the modern urban self is routinely formed.

Start from this profane geography of the city, and where does that leave politics? In *Justice and the Politics of Difference*, Iris Marion Young defends city life as a normative political ideal. Her city is based not on transparency but on 'an openness to unassimilated otherness', and that entails a view of politics as 'a relationship of strangers who do not understand one another in a subjective and immediate sense, relating across time and distance' (Young, 1990, pp. 227 and 234). Cosy old Gemeinschaft is not an urban option. City life understood as the being together of strangers offers different pleasures and possibilities: the permeability of symbolic boundaries; the open-mindedness and eroticism of spaces; a public life which is culturally as well as politically diverse, agonistic rather than consensual, seriously playful rather than earnestly civic.

Young acknowledges that her portrait may lack shadows. The city is also, inescapably and scarily, a place of aggression, violence, and paranoia: 'on city streets today the depth of social injustice is apparent: homeless people lying in doorways, rape in parks, and cold-blooded racist murders are the realities of city life.' (p. 241) But hold to both that reality and the promise of urbanity, and you see glimmers of a pragmatic new political imagination. The good(ish) life of today's citizenship recalls the worldly, slightly melancholy cosmopolitanism which Salman Rushdie invokes when he defends his novel *The Satanic Verses*. In place of community, identity and change through deliberation and consensus, the sociability he celebrates involves 'hybridity, impurity, intermingling, the transformation that comes of new and unexpected combinations of human beings, cultures, ideas, politics, movies, songs'. Rejecting the absolutism of the Pure in favour of mongrelisation, he foreshadows a citizenship for global times: 'Mélange, hotchpotch, a bit of this and a bit of that is how newness enters the world. It is the great possibility that mass migration gives the world, and I have tried to embrace it.'

References

Marquand, D. (1991) 'Civic republicans and liberal individualists: the case of Britain', *Archives of European Sociology*, XXXII.

Young, I. M. (1990) *Justice and the Politics of Difference*, Princeton, NJ: Princeton University Press.

• CITY •

by Adrian Passmore

Eyes flash as he drinks the city.
Subterranea. Magma boils: he bites his tongue and reaches for names. He craves. For gutters that course with rained worlds, for crawlable kerbstones, for vehicles, motors and drives. For bricks and the cracks between them, for crazed paving, trick shadows and unpaintable ceilings; for the pretending, fighting and the bitten lips. For the undertow, the statues, the speed-lost present. For the coalition building, and the common. For the explosive mortar. For all of it, there's nothing else. Not that he can see.

He bites again.
Songs untie and manacles bind special-offered wrists. Lush rhythms seep through neighbouring walls; nearby, the silencing culture of living close to others. These are the raw material, the object, and the white noise. No, that's the work, the drudge, the throwaway line, paper and life. Its hushed sweating. The smell and age of others, the stench of silence in a deafening world. Rich, delicate, cultivated hunger. And a loathing so strong you will it to stay.

Slap.
He turns a neat phrase for the force of incorporation, for the violence that makes some bleed and others laugh. For ransacked land, and the crack of a green belt. The hill, the kop, the vestige, the cheap cul-de-sac. For the intricate conversions of nature, and for blank-bus faces. For all of it.

He reaches for himself and looks away ruined.
City seek's words in which everything is found. The numbers, bleeps, envelopes and folds; the digits and the opposing thumb. The calls, the barcode, the postcode, the telegraph wire, the radio, the programme, the maker, the pull to the centre, depot and warehouse. The lavish care of the manicured finger and the point of broken spectacles.

Ruses told.
Heart-felt plea, charity, tendered old, wizened baby, and the interminable youth of the sun. The flaps about the edge, the fold and the tuck. The shapes that make the outline, the colours that buildings contract, and one-way streets crossing like enemies. The traffic-lights shot, the corners turned and the vegetables too. You think again of all the baskets, the cases, the markets, all of the futures. All of it.

Hurricane breath.
Latitude thinking. The growing trope and the old stories of penetration. Blurring, systems, networks and threads. Muses. The arches, the bent back, and the contorted glaze of a deadened fish eye. The surplus heat of the cold latitudes, the spring wherever, the leap in the dark, the front doorstep and the back passage, the need and ecstasy. The terrible weight, the sudden, all of a sudden. The old world in the new, the dust of white trucks, the looming cruel punchline of the pin-making machine. The finished article.

Time dries to a language for his city.
Loose change. No, forced labour, tight schedules, rigid formulas, agendas, great expectations, promises, the biggest, the furthest. Town-dripping and village-damp. Smoke curl and tyre rolling. Sugar, slave architecture and master plans. Robbed style, borrowed lyrics, baroque-unoriginal and the excellent view. The pace of it, the slowness and speed, the incorruptible ethic, the absolute principle and the flex of learnt muscle in the sweat-sweet gym. The

subordinate universe, the order, the command, control and world, the scale, and again the speed tarried. The telly still and the inky rush of the future. The loose effect.

He's a memory ram.
Remember numbers. The bliss of cut stone and caned eyes. Wicker chairs and hot verandas. Base ten, base habits and electric base two. Recollect the furniture, the architecture and the mattress beatings. The counter pain. The trauma lack; he's the logistics, the model, and again the plan. The scene of the possible, the birth and the arrival, the station announcement, the train. The decadent steel and insistent rubber, the broken skin of a foreign body. The same body thrown to the wind in the same way again and again. The relentlessness, the ongoing, the churning, the love of it and earning. The belief, contract and pact. The surety and bolster. The pillow talk, the underwriter and the undertaker: the insured and the ensured. The shamelessness of a crowd. The swallowing gulp on a vicious pissed note. The intricate tears, bent knees and the pyjama punch. Ruinous names and seeping speakers. And the last of it.

Unheedable vice and the gaudy ticket.
Stolen kisses of plonk prose and bankers' decanters. The swap, promise, and again the same promise; the detail screen and the big sweep. The bought baby, white camera, north, west and wasted. Powder, nose, pillbox and machine-gun. The irrepressible, unaccountable, bent, crooked and pious. The hooked, the story, the line and the fish. The model plane that flies in your face. The surety, certainty and absolution; the planted kick. The grip and the loss. The slide and the phoenix, the battery, the bird, the pigeon hole. The chicken, the fear, the rupture. The flowing banked semen. The certitude of an untraceable leak: the night. Again the night, dream and promise. Mirror-stage corporate home. Surface stillness, bitten tongue. The pinpoint timing of the second coming. The tranquillisers and tranquillised. The adjective stalk. For all of the city.

He has a nerve.
Trigger reaction. Underground stream. Demonstration. Example. The same bankers and decanters: the same streets without names. The pledge. Paradise circus and adventure capital. White goods. The green, the brown, the autumn, the unnameable foil, the fag passed, the pinned eyes, the machinery for living, and the reference. Reflex habits.

Care lacquers a careering tongue.
Hanging's back. Kicks and straps. The honed definition of the refugee; statistical nightmares. Tightened belts. The read screen and gold. The frantic awake and sleep-staring eye. The prison guards who have not seen a prison, the train drivers who didn't drive trains, the world travellers that never saw no one. Here, the raw duty of expression. The rasp encounter. Bliss again of caned eye, caned pupil and caned teacher. The testament and its enforcement. The statute, no! the slide-rule and the economy of swift calculation. The numb and the uncountable. The outliers and the normal. The tube of life. Mute home, loud bar. The stain and the salted name. The face, taut bladder and hollow urethra.

Drawn air features.
Old habits writ large on his kisser. The repeat action mechanism, the colt at its birth, the frontier. The ship, crew and voyage. The counterfeit sentence and the panic cleaning. The spunked cash, the paved street, the tarmacked road, the pebble dash, the suburban flight, the subordinate, the chief thing, the main principle, the ethic, no, the principle, no, the flex in the veined wrist of the eye-patched captain. The bishop. The copper in your pocket, the note on the door and the villainous party. The knock of the atom polis, my friend.

Death launch.
Recall the joys of coercion, in fact the unadulterated joy. The gatecrasher, the lost shirt and the need for an intimate verb. The possibility of the unprotected: exposure, hypothermia, villainous cold soup. Dead warm fur. The response of the profligate. The ends of an agreement. Communicable dis-ease and spitting words for the polis orificer.

Bench marks.
Spilling to the ground. Crying red tears: rueful at being discovered. 'Sticks and stones will break my bones' he sings in an underground echo.

•CONCERT HALLS•

by Susan J. Smith

Every city has one . . . or more.

Concert hall: ancient or modern, glassy or domed, grand, powerful, steeped in the past, reaching for the future, resonant with sounds of the city.

Concert hall: symbol of civilisation, source of civic pride, stamp of saleability; a microcosm of life and times on the town.

Concert hall: container, laboratory, marker, and boundary.

Boundary? Yes boundary.

The arbiter of what is art and what is not, the line between music and noise, culture and nature, control and disorder, civilisation and wilderness.

> Arrive a fashionable 10 minutes before 7.30.
>
> Check hair, adjust clothes, look round expectantly.
>
> Lower your voice, and prepare to be quiet.
>
> Try not to cough or fidget, and *please* have your sweets unwrapped in advance (if you must have them, and can bear to make all those sucking sounds at the least appropriate moments).
>
> Don't be the one to moan along with the tune, a quarter tone flat – you will attract sidelong glances, throat clearings, arm crossings and indignant humphings. (No one will actually say anything – not for some time, anyway.)
>
> Keep quiet throughout.
>
> Pay attention to the appropriate moments for clapping – listen for the knowledgeable person in the front circle to start you off (but don't be led on by the ignorant enthusiast at the back of the stalls, or everyone will stare at your noise through their own silence, and you will feel very small).
>
> It *might* be appropriate to whistle, cheer or stamp your feet at the end. Hissing and booing is out, though, unless you really know what you are doing. Unless you have secured a place among the arbiters of taste and behaviour.

In the concert hall, we are a caricature of ourselves.

The spaces of the hall are segregated, by price, like the spaces of the city. Listening takes place; the structure of social relations is on show and undisturbed. As Theodor Adorno observes, tracing out the natural history of the theatre:

> In the gallery we find the irrepressible enthusiast, naively worshipping the tenor . . . sitting next to the starving expert who unremittingly follows the inner parts in the score. (1994, p. 68)

> The stalls are the home of the bourgeoisie. . . . All are seated on the same raked floor, carefully separated from one another by the arm of the chair. Their freedom is that of free competition . . . Limits are set to their equality by the hierarchy of seats and prices. (1994, p. 69)

> The boxes are inhabited by ghosts. They have not bought any tickets, but are the owners of prehistoric subscriptions, yellowing patents of nobility. Sometimes the ghosts provide champagne suppers. (1994, p. 71)

In the concert hall, we substitute creature comforts for outdoor life. Concert hall is the epitome of urbanism. This listening space is a place where nature is tamed, wilderness domesticated, unknown territory transported to the heart of the city. Why venture alone into the uncertainties of the world? Through composers, conductors, and performers we can experience them vicariously, at leisure, after supper, filtered through that small glass of chardonnay, and just a stone's throw from the car.

The concert hall adds some fine tuning to the world; it is an arena for the civilisation of noise; a space which makes sense of the sounds of the city. Lewis Mumford puts it like this:

> the division of labor within the orchestra corresponded to that of the factory . . . the collective harmony, the functional division of labor, the loyal co-operative interplay between the leaders and the led . . . the timing of the successive operations was perfected in the symphony orchestra long before anything like the same efficient routine came about in the factory. . . . Tempo, rhythm, tone, harmony, melody, polyphony, counterpoint, even dissonance and atonality, were all utilized freely to create a new ideal world. . . . Cramped by its new pragmatic routines, driven from the marketplace and the factory, the human spirit rose to a new supremacy in the concert hall. Its greatest structures were built of sound. (Mumford, 1934, pp. 202–3)

The concert hall makes sense of silence too, all 4'33" of winds stirring, raindrops pattering, and people chattering or walking out.

And if the blurring of environmental noise with musical style is the hallmark of the twentieth century, it is the concert hall which polices the boundary.

> Are the sounds just sounds, or are they Beethoven? If sounds are noises but not words, are they meaningful? Is a truck passing by music? (John Cage, 1973)

In the concert hall, we know that sound is a cultural product, saturated with meaning. Any sound is music, if that's what the composer wants. Any noise is civilised providing it is meant to be heard within the confines of this special space.

It is in the Concert Hall, too, that we learn the art, the practice, the politics of performance. The concert hall is a world within worlds; a city in microcosm.

In the old days, it was more important to be seen than to listen, much less to hear, as James Johnson shows:

> In the Old Regime, attending the opera was more social event than aesthetic encounter. (1995, p. 10)

> Opera glasses were *de rigueur*, necessary not so much for seeing the stage as for observing other spectators. (1995, p. 16)

But today, the stage is the centre of attention:

> The inconceivable in 1750 – that spectators might not be a necessary part of the spectacle, that chatter and appointments need not be accepted as normal – was now conceivable. (1995, p. 60)

Why, Johnston asks, did audiences stop talking and start listening?

> Spectators listened to the wealth of harmony that flowed from the stage and gradually made sense of it. . . . This purely musical element so entranced and captivated them that breaking the spell of the music with visits or discussions about other spectators was unthinkable. To listen with the attention these listeners claimed the music required was all-consuming. (1995, p. 256)

But there is more to it than this: musical experience is never just musical.

> The pace of change in listening is dependent upon the pace of change in political ideologies, social structures and musical innovations. And because there are always possibilities for musical innovation and changes in patterns of sociability, listening continues to evolve. (1995, p. 281)

Today, the concert hall is where we listen to music, the music that Edward Said has shown is so vital to the production and reproduction of social life.

> The concert hall is the place where performance matters, as music becomes 'a mode for thinking through or thinking with the integral variety of human cultural practices' (1992, page 105).

> The composition and performance of music is a crucial activity in civil society which 'is in overlapping, interdependent relationship with other activities'. (1992, p. 97)

The concert hall is a space where the contribution of the musicians is not merely to advance their art, but to secure the elaboration, or maintenance, or rolling along, of society, 'giving it rhetorical, social and inflectional identity through composition, performance, interpretation, scholarship' (1992, p. 70).

In the isolated, segregated, boundary-policing, city-epitomising world of the concert hall, we experience a discontinuity with everyday life, which takes us away from ourselves, yet reinserts us into society. It is a world within a world which connects us to the world. It is a space for sounds:

> whose pleasures and discoveries are premised upon letting go, upon not asserting a central authorising identity, upon enlarging the community of hearers and players beyond the time taken, beyond the extremely concentrated duration, provided by the performance occasion. (1992, p. 105)

The concert hall *is* the city!

> Those who sit . . . at the furthest remove from the stage know that the roof is not firmly fixed above them and wait to see whether it won't burst open one day and bring about that reunification of stage and reality which is reflected for us in an image composed equally of memory and hope. (Adorno, 1994, p. 67)

References

Adorno, T. (1994) *Quasi Una Fantasia*, trans. R. Livingstone (1992), London: Verso.

Johnson, J. (1995) *Listening in Paris*, Berkeley: University of California Press.

Mumford, L. (1934) *Technics and Civilisation*, New York: Routledge & Sons.

Said, E. (1992) *Musical Elaborations*, London: Vintage.

• CRIME •

by Eugene McLaughlin

Although I am a criminologist it is not necessary for me to carry out research, read through mountains of official crime statistics or the findings from victimisation surveys to tell you that the city is the dominant 'crimescape'. How do I know? I could direct you to the sheer number of films, set in a bewildering variety of different geographical locations, that open with heavily underscored images of rain-slicked, neon-lit 'mean streets' which turn out to be the setting for an elemental struggle between streetwise criminal gangs and 'hard-boiled' law enforcement officers. And of course the hard, menacing soundtracks accompanying these visually exciting representations deepen and authenticate the dystopian mapping of the crime-ridden, disconnected city. However, no matter how sophisticated or novel the story-line of these films are, they (probably because they have to create a readily accessible narrative spine, their own consistent worlds) never quite manage to capture the sheer complexity and fragmented nature of crime in the city.

This is why I have decided to narrate a much more local story about the impact and meaning of crime in my part of London. To do so I will start with my local weekly newspapers which I guess, like almost every other, generate a graphic and often alarming mapping of what has been happening in this seemingly inexhaustible space of crime and disorder. In recent weeks I have read the stories of pain, loss, anger and confusion and periodic references to cherished memories of safer times associated with headline stories such as:

- fifteen burglaries per day in the borough;
- nurse weeps in court as she recounts her terror of being sexually assaulted by a man whom she used to pass on the street on her way to work in the mornings;
- shop traders and residents are having to cope with drugs, petty crime and racial harassment;
- the killer of a well-liked local pub landlady has been traced through the internet;
- drug dealers, junkies, drunks, pimps and prostitutes have driven children out of local parks;
- awards for the local police officers for acts of bravery and dedication 'beyond the call of duty'.

You will not be surprised to learn that different parts of the borough are being radically reshaped in response to residents' fear of escalating crime. A variety of strategies have been devised during the past five years by the local authorities to reclaim this particular crimescape for law-abiding citizens. Police officers in local crime 'hot spots' have, for example, been told to target the kind of offences normally disregarded in the more exciting war against crime. Begging, public drinking, spitting, littering, street trading, rowdy behaviour, busking, graffiti are also to be cleared off the streets. Police officers have also been working with borough planners and businesses to 'design crime off the streets' by 'target-hardening' urban surfaces. High prestige office blocks, as a result of the City of London imposing a 'ring of steel' around its perimeters, have attempted to cordon themselves off spatially in a frantic attempt to neutralise the possibility of terrorist 'spectaculars'. The main shopping streets are being architecturally refashioned so that they do not provide obvious hideaways and escape routes for criminals, and street furniture has been removed or redesigned to make it more difficult for the homeless and drunks to loiter. Streets, markets, car-parks and public transport stations have been flooded with surveillance cameras in an attempt to deter and/or catch criminals, provide hard evidence for courts to convict offenders and assure the law-abiding public that the vigilant electronic eye of the authorities is watching over them. Periodically, police officers wielding video cameras roam the local parks in an attempt to crack down on petty crime, 'cruising', 'cottaging' and indecent exposure. Virtually every bar and club on the main streets now employs a new breed of private security guards to keep out 'undesirables', and the local telephone companies employ teams to clear out telephone boxes which are full of 'illustrated' advertising cards for prostitutes. Then there is the haphazard reclaiming of unoccupied or derelict buildings in traditionally rough areas which is being achieved through constructing bunker-like 'walled and gated' residential enclaves which can architecturally co-exist cheek-by-jowl with run-down estates.

The other side of urban renewal is ghettoisation, represented by the borough's notorious sink estates where to all intents and purposes the writ of the local authorities has ceased to have meaning. These are effectively 'no-go areas' with a pervasive aura of decay where many public services have either been withdrawn or are pared down to the minimum. Left to their own devices they have turned in on themselves. Certain residents have felt it necessary to transform their homes into DIY fortresses with reinforced doors, grills and guard dogs. Local police commanders know that responding effectively to residents' complaints about a lack of police presence could spark serious outbreaks of disorder and violence. In many respects, power, authority and respect within these clearly demarcated estates passes back and forth between the local authorities and the young and the strong. It is they who have the physical capacity and visible presence to enforce some form of order, act as a conduit between the legitimate and criminal economies and provide role models and deliver career opportunities for local youths. Nearby respectable estates use what are in effect vigilante tactics to seal off their neighbourhoods in a desperate attempt to prevent them from becoming red light areas and/or home to those marginal populations who have been 'moved on' as a result of police clamp-downs.

However, there are limits to this local war against crime. Guidebooks to London tell readers that the heart of this borough is the 'coolest place on the planet' and that this is primarily due to the mesmerising variety of disorderly street rhythms and diversity of life-styles on display and the unsegregated 'rough around the edges' atmosphere. The local authorities are only too well aware that these images have delivered tourists from virtually every part of the world as well as considerable economic growth, and that heavy crime-busting initiatives and the excessive spatial engineering required to produce risk-free zones could kill the goose that has laid the golden egg. The uncomfortable reality is that like all other seemingly contradictory and infuriating aspects of life in the borough, residents and the local authorities have little choice but to internalise the fact that the buzz and the aggro of street life are inextricably linked. This is why we have no choice but to continue to develop our own rough-and-ready mental maps of dangerous places, times and people and deploy a range of scanning skills and tactics, which allow us to negotiate the risks associated with street life in (what has been called) 'the coolest place on the planet'.

Further reading

Jencks, C. (1993) *Heteropolis*, London: Academy Editions.

Jacobs, J. (1961) *The Death and Life of Great American Cities*, Harmondsworth: Penguin.

Sennett, R. (1996) *The Uses of Disorder: Personal Identity and City Life*, London: Faber and Faber.

•CYCLING•

by David Sibley

Fantasy cycling. Every July, people in about one hundred and sixty countries watch the *peloton* of the Tour de France flowing along minor roads and heaving up legendary mountains like L'Alpe d'Huez and the Col du Tourmalet. Beyond France, it excites viewers in Colombia and Ethiopia and the Tour becomes a daily diet for some in the core of 'cycling Europe' – in Holland, Belgium, Italy, France and Spain. But this extremely physical confrontation with wild nature, assisted by a caravan of team cars, dieticians, masseurs, mechanics and commercial sponsors, is also an urban spectacle. Rouen, Bordeaux, St Etienne, and other French cities which made successful bids for the 1997 Tour, interrupt normal life for its passage. Paris gives it a day. Eight or ten circuits in the city centre with a finish in the Champs-Elysées – a flash of road machines, shaved, oiled legs and company logos. Like marathons in London, Boston, New York and Tokyo, the Tour has secured a space and a time in a large metropolitan centre and contributed spectacle to city life. La Grande Boucle is a celebration of the male body (women's tours don't get the TV coverage), of the French landscape (and bits of adjoining countries) and capitalism. The teams are sponsored by Banesto, Gan, TVM, ZGMobili, Casino – banks, insurance companies, furniture manufacturers, supermarket chains and so on. The tour is easily accommodated in the capitalist city. It is not about the humble bike as a benign mode of urban transport but the promotion of cities and products, the pleasures of consumption combined with the voyeuristic pleasure of masculine athleticism. City centre cycling events in other European cities are similar, if less dramatic, spectacles to the Tour's circuits of Paris. Closing the city centre for a few hours is a sound investment. City businesses benefit and it's good for Coca-Cola.

Subversive cycling. Information and money circulate electronically but packages and letters still need to move across town. Dispatch riders in the more sclerotic city centres, like lower and mid-town Manhattan and central London, speed up the flow but they cut their own subversive paths through the city. Time is money, so weaving in and out of the traffic, crossing pavements and riding against the tide of cars is OK by the employer, and as it's casual labour you don't have to worry about insurance. However, dispatch riders have style. High-specification, all-terrain bikes and racing gear go with anarchic riding. ATBs go anywhere, they are not confined to normal channels of circulation, they escape the discipline of regulated streets. The riders are working for big business but they are marked as different, a small community, with a shared experience of injury and death, that gets a buzz from risky riding. Dispatch riders adjust to other traffic and even use it for quicker journeys by hanging on to trucks (in New York City). Other adults and children, less glamorously attired, also ride on pavements where they unnerve the elderly. They make their own space but do not effectively challenge dominant modes of travel.

In the city there is a contradiction. Cycling is hazardous but the cyclist may also be seen as a hazard. The bicycle is acknowledged by local authorities to be a desirable mode of transport – healthy, consistent with the idea of the sustainable city – but it has to be disciplined – no riding on the pavements. So, British cities, unlike Dutch cities, make space for the bike in places, with an occasional green or orange strip marking cycle space, a narrow refuge along the side of busy streets but going from nowhere to nowhere. Dutch cities have for a long time created networks for bicycles although without much effect on the use of the car. The indifference of British local authorities invites direct action, like Critical Mass, making the streets safe for cycling on the first Saturday of every month. The bicycle can be a potent weapon of resistance, a machine for reclaiming the streets, but effective collective action by a small minority of radical cyclists is difficult to achieve. It is easier to cycle round the problem.

Green cycling, cycling with the greens. On allotment gardens in British cities, well-oiled black Raleighs with chain guards and Sturmey Archer gears still survive. They are used by older allotment holders, who wear bicycle clips if they are men, to take home the cabbages and onions and carrots. The bikes have baskets and boxes for the vegetables. They belong to a period before style became central to mass consumption, when local bike shops lent you the tools for replacing your cotter pins. This is the bicycle as an unglamorous form of freight transport, like barges on canals, or the bikes used in India for carrying cans of water or other heavy goods. In India and elsewhere in Asia, however, style is intruding. The bicycles of Delhi and Ho Chi Minh City, which provide a high level of mobility, are being superseded by motor scooters and mopeds and in Kuala Lumpur motor scooters are rapidly giving way to Mercedes. In the city, personal transport is a fashion statement and an item of conspicuous consumption, so the prospects for the basic bicycle are not good. Perhaps the survival of the bike as an efficient mode of urban transport will depend on fantasies and fetishes, on the kinds of images projected by the Tour de France or the images of freedom and physicality associated with mountain biking in the wilderness. Carbon fibre frames, eighteen gears and Lycra shorts are hardly necessary for getting you down the street, but they are flash – like the customised motor scooters favoured by the Mods in the 1960s. Commodity fetishism might in this case be harnessed in the cause of sustainable development and a civilised city.

d
d
d

•DANCE HALLS•

by Helen Thomas

Where older people come into visibility in the city

'The first thing you have to know about ballroom dancing,' says Juliet the teacher, 'is that it is a contact sport.' Muffled sounds of embarrassed laughter ripple all the way round the hall. 'The other thing you have to know,' Juliet continues, 'is that on the dance floor the man is always right.' The majority of the male and female partners in the class have never met before. Now they are standing awkwardly face to face in close proximity, holding each other in a semi-closed ballroom dance position. Juliet pushes the partners closer together gently but firmly, so that their upper bodies are in direct contact all the way down, making sure that there is no visible gap between them. This is a most disconcerting experience for the uninitiated 30- to 40-something participants of this south-east London inner city ballroom dance class. But this is not surprising. After all, they do inhabit a predominately heterosexist culture where there is an unspoken rule that if you have to come into close bodily contact with strangers in public places, as is somewhat inevitable in an urban culture, for example, a crowded tube or train, you should turn your body slightly away from them and tense your muscles so as not to give the impression that this might be a pleasurable experience. This is particularly so if the contact is with a member of the opposite sex. Moreover, the cultural codes entail that you should avert your gaze, so as not to invite the possibility of unwanted face-to-face communication at this close proxemic, interpersonal level where touch, breath, sight and olfaction are vividly present. Close face-to-face interaction is usually reserved for communication between intimates, preferably in private. To flaunt the codes is to invite communication, and some individuals and/or groups actively subvert the codes, transforming this rule breaking into rule-guided behaviour. Close proxemic behaviour among strangers in public places, then, as writers like Erving Goffman (1971) have demonstrated, is a managed experience.

The 'closed' ballroom dance hold is also highly managed, gendered and rule bound. The inability of the 30- to 40-something participants to deal with it partly relates to the fact that the majority of people from the post-Second World War 'baby boomer' generation onwards are simply unfamiliar with the codes of social ballroom. For the past thirty years or so bodily contact on the social dance floor has been the exception rather than the rule. In the discos and clubs you might dance in a group or with a partner, but the relationship is one of non-contact. Moreover, as a consequence of the women's movement, younger women have come to question the traditional notions of male dominance, which makes the idea of men leading in dance appear somewhat problematic and outdated.

Since the emergence of solo social dancing and the rise of club culture in the UK in the 1960s, representations of social dancing have been split between images of the old and the young. The more positive images have come to be associated with young people expressing themselves, although these have also been sites of 'moral panic' over the years. The negative versions are largely allied to images of older people dancing in a more uniformed, gendered, stereotypical ballroom dance style, at tea dances in the few ballroom dance-halls that remain, or in the numerous church halls and civic centres up and down the country where a considerable number of dedicated older social ballroomers can be seen tripping the light fantastic every week. A stranger entering into any of these more formal dance settings might be forgiven for thinking that s/he had stepped momentarily into a BBC costume drama set or that the Chance sisters, from Angela Carter's *Wise Children* (1991), had suddenly come to life and were travelling across the dance floor in their best frocks. But it would be a mistake to stay with that image.

Despite the fact that positive images of older people are dominant within contemporary consumer culture, numerous negative stereotypes, handed down via popular culture, continue to be pervasive. Elderly people are almost invisible in the streets of the city. It is not that they are physically absent. Rather, in a consumer-oriented culture dominated by the visual, they tend to go unnoticed as they go about their daily business. The outer body appearances of the elderly do not always conform to or confirm the demands of consumer culture to express the self and radiate physical attractiveness. This is largely because the dominant representations of physical attractiveness and individuality are suffused with the ideology of youth. Although ageist stereotyping affects men and women, women tend to be doubly stigmatised in later life. The combination of ageism and sexism means that women are characterised as being past their 'sell-by date' at an earlier age than men.

On the dance floor, however, the elderly participants of the tea dances in our south-east London dance-hall move out of the shadows and into the light. Their bodies are fluid and expressive as they glide around the floor, displaying the skills they learned in their youth in the magnificent dance-halls and palais that were so popular before dance crazes like the twist finally put the nail in the coffin of couple or 'touch' dancing. Furthermore, the intrigues of romance are not restricted to the young and may also be witnessed on the dance floor on a fairly routine basis. It appears that some of those men and women who danced regularly together at the tea dances did not tell their 'real life' partners in the 'outside world' that they frequented the dance-hall.

The visibility and vitality of the older people on the dance floor stands in stark contrast to their invisibility in the outside world. On the dance floor they are not 'has beens' or 'nobodies', rather they are skilled practitioners of a phenomenon that ironically has witnessed a resurgence of interest in countries like the UK, USA, Japan and South Africa in recent years. Our 30- to 40-something novices could only gasp in admiration at their seniors on the dance floor and wonder if they would ever be able to reach such heights of proficiency, and thus transcend their own present sorry state of 'dancing by numbers' and appearing to be made out of cardboard.

References

Carter, A. (1991) *Wise Children*, London: Vintage.

Goffman, E. (1971) *Relations in Public: Microstudies in Social Order*, Harmondsworth: Penguin.

•DANCING•

by Nicola Miller

On a patch of grass in a municipal park a small child raises the palms of her hands and with out-stretched arms begins to turn slowly, distractedly at first, face lifted to the sky, fingers rippling in response to rushes of air. As she gains momentum her circles travel, spiralling outwards in a less meditative, more exhilarating motion. Her arms cut through currents of warm breeze, her tracings lose their regularity becoming more chaotic and angular and just as the momentum is taking over . . . she drops to her knees. There she remains for a while swaying, intoxicated, watching the virtual swirling of the grass. Gradually, the ground stills. She jumps up and runs off.

'Community dance' has come to refer to a formalised approach to the provision of dance which is usually either state-subsidised or at least state-managed. While it appears as an inclusive category lacking precise boundaries, there are discernible themes in the instances of its practice. Dance is here offered as a social experience and so classes tend to rely on non-competitive, interactive styles of dance and teaching, such as improvisation and social dance forms. Community dance is often presented as the route through which dance may be offered to those who are generally excluded from it. The emphasis is on safe dance and links are frequently made between dance and alternative approaches to health care. Through its practice, community dance seeks to build self-confidence and release tension. Although dance-in-education is often dealt with separately in terms of funding, there are certainly strong links between it and community dance. For example, community dance workers are often responsible for initiating workshops and residencies in schools and colleges. In the same way, one of the priorities of community dance is the provision and encouragement of experiences outside of the western theatre tradition. In the context of the community, dance is considered as a means of active self-expression and it often aims to create works relevant to specific audiences. Finally, and perhaps most idealistically, there is an emphasis on self-help in many of the projects, a hope that the animateur/dance worker will eventually be replaced by local teachers or organisers or the groups themselves.

The practice of community dance resonates with the ambiguities which the concept of community evokes. What is inferred by community? The 'community' which is referred to in 'community dance' may, partially, be reduced to its geographical definition. Since most of the community dance workers will be funded by organisations or bodies with a geographical brief, this would tend to impose physical boundaries on the scope and size of dance projects. However, community dance also seems to present a nostalgic version of a 'community lost' in the 'anomie of contemporary living' (Thomson, 1989) tempered by the promise of its being 'regained' through this work. It is a related dualism which splits the experience of metropolitan life from that of rural living. The former emphasises individuality and typifies relations as being anxiety-ridden, faceless contracts. The associated goals are those of material goods and immediate gratification. In contrast, rural life is presented as a slower paced environment where people have names and belong to families and villages. In this context there is time to enjoy process and less pressure to move rapidly between activities. In this context community dance becomes a 'treatment', a means of redressing the balance, at once offering group activity as an antidote to the anonymity of the city, and advancing a measure of metropolitan sophistication to those remote from the city.

No criticism is intended of the individual dance workers, who are often charismatic people who work tirelessly to spread the experience of dance within the boundaries enforced upon them, and often with little remuneration. The concern is rather with the limits which are inevitably imposed once a movement, which emerged historically as the efforts of a few individuals, is transposed into a state-managed category of 'provision'. Too often the state, in its desire for equity of provision, reduces an activity to a quantifiable set of rules. This can be seen in its physical manifestation as the straight lines, the immalleable concrete and the colourless grey of the contemporary city. However, this is only a façade, since this is an environment populated by real people resisting the reductive, categorising tendencies of the modern state.

Dance, as a creative, bodily centred activity, is as perennial as the grass which grows between the paving slabs of all but the most alienated streets. It occurs as a consistent feature of community gatherings, weddings, parties, festivals. Tea dances are a quietly popular and regular event for many older people; younger generations are more likely to follow the crazes of Ceroc and salsa, but people also dance spontaneously in their everyday movement. The projection of dance as something to be taught in a bleak studio, removed from everyday living, may not encourage a sense of community so much as alienate individuals from their own desire and capacity to dance.

Dance is an expression of vitality. If there is not enough participatory dance occurring in contemporary society, this speaks of a broader malaise, one which is not easily rectified by importing or 'providing' community dance in 'target' areas. Perhaps it is best approached by a re-emphasis of the local, once the antithesis of the metropolis. A new sense of co-existence seems to be emerging whereby people create a new set of ideas, meanings and activities around their immediate environment. Insofar as these are positive, perhaps they will include dancing.

Reference

Thomson, C. (1989) 'Community dance: what community, what dance?', *Dance and the Child International*, 3 (4).

•DETECTIVES•

by James Donald

Forget those country houses and those daft English amateurs. The detective is a creature of the city. More than that, it is the detective who makes the modern city thinkable.

It is not just a question of detective stories as topography, re-creating cities in writing or, more likely, writing imagined cities into being. True, variations on the urban landscape are one of the attractions of the genre. Think of the Los Angeles variously conjured up by Raymond Chandler, Walter Mosely and James Ellroy; or Simenon's Paris, or Charles Willeford's Miami, or Sarah Paretsky's Chicago, or Elmore Leonard's Detroit. Yet however precise the mapping, however dirty the realism, verisimilitude remains in the end a sop to the reader's nostalgia for naturalism, a loss leader. The real space of the detective story remains insistently and stubbornly symbolic. Why so?

By the nineteenth century, the city had become a new problem: a problem of visibility. Too big to encompass in a single glance, populated by illegible strangers rather than familiar neighbours, the city threatened to exceed the capacity to envisage, represent or – key point – govern it. To get around or behind this opacity required a new way of relating to the city: investigation. Hence all those administrators, reformers and journalists braving the nightmare depths of the city to shed light on its terrible secrets. Edwin Chadwick, Friedrich Engels, Henry Mayhew, Charles Booth and Jacob Riis all in their own ways embodied a new breed of anthropologist-explorer-detective deploying new techniques for getting the measure of the city: statistics, environmental health, physical and moral taxonomies, photography.

The cities of nineteenth-century fictional narrative share this will to visibility. Pedagogic as much as representational, they teach ways of seeing. Walter Benjamin noted the historical link between new techniques for fixing identity, and so creating individual records as a feature of urban surveillance, and the enigmatic city of the detective story: 'The detective story came into being when this most decisive of all conquests of a person's incognito had been accomplished. Since then the end of efforts to capture a man in his speech and actions has not been in sight.' What was essential in such stories was less the crime than the urban staging: 'the pursuer, the crowd, and an unknown man who arranges his walk through London in such a way that he always remains in the middle of the crowd.'

This is the London not only of Poe's 'Man of the Crowd' but of Dickens. His great image is of fog everywhere, mixing with smoke to blot out the sun, allowing only fitful illumination by gas lamps, peopled by foot passengers anonymous and distracted beneath their umbrellas. Dickens' London is not just huge, but monstrous and alien. The pedagogic reaction is then to wrestle it into a narrative. Read the clues, however trifling, the novels teach us; make the links between them, however obscure. There are still subterranean networks and secret social connections at work which make the city ultimately systematic – and so narratable, and so thinkable. This will to a knowledge and a rationality capable of penetrating the fog of social relations is figured in the Victorian detective. Sherlock Holmes, though, is not purely a creature of reason. His ability to read clues is often little more than a party piece. Solving cases, when not a matter of chance, as often involves his mastery of disguise, his ability to move anonymously or illegibly through the crowd, to observe unseen, and so to exploit the opacity of London. Holmes thus prefigures a popular modernism in which T.S. Eliot's 'unreal city' is recycled as the blank urban space of twentieth-century detection, even if that space is filled out with distinctly pre-modern forms of subjectivity. Here, in all its boozily sentimental machismo, is Raymond Chandler on the private eye as hero:

> down these mean streets a man must go who is not himself mean, who is neither tarnished nor afraid. The detective in this kind of story must be such a man. He is the hero, he is everything. He must be a complete man and a common man and yet an unusual man. He must be, to use a rather weathered phrase, a man of honour, by instinct, by inevitability, without the thought of it, and certainly without saying. (Chandler, 1962, p. 14)

And so he rhapsodises on, having little more interesting to say about the city than that its meanness provides both occasion and backdrop for this display of disenchanted, down-at-heel integrity. Although talking about the gangster as tragic hero, not the detective, Robert Warshow's similarly overheated image does at least get at something significant about the city as the imagined environment of modernity.

> The gangster is the man of the city, with the city's language and knowledge, with its queer and dishonest skills and its terrible daring, carrying his life in his hands like a placard, like a club . . . for the gangster there is only the city; he must inhabit it in order to personify it: not the real city, but that dangerous and sad city of the imagination which is so much more important, which is the modern world. (Quoted in McArthur, 1972, p. 28)

Detective and gangster share a style of being in the city which has less to do with deciphering its signs or solving its mysteries than with using its space for (in their case) nostalgic self-creation.

Lefebvre offers a clue to the regulated and disciplinary nature of that space: 'Space lays down the law because it implies a certain order – and hence also a certain disorder (just as what may be seen defines what is obscene)' (Lefebvre, 1991, p. 143). The flâneur domesticated spatial disorder by turning the city into an interior. Although crime stories may equally disavow the city's aggression and violence by restaging them as a comedy of manners, however rough, the detective reveals the city as a space of desire and prohibition. What the detective makes thinkable is not cities as places, but the narrated city as an experiential, often paranoid, and yet curiously ethical space.

References

Chandler, R. (1962) 'The simple art of murder', in *The Second Chandler Omnibus*, London: Hamish Hamilton.

Lefebvre, H. (1991) *The Production of Space*, Oxford, Blackwell.

McArthur, C. (1972) *Underworld USA*, London: Secker & Warburg.

•Dogs•

by Deborah Levy

Nietzsche had a name for his cancer. He called it Dog.

Every time he was in pain, he screamed at the Dog. It is possible that all the dogs in Britain are pain dogs. Migraine Dog, Depression Dog, Backache Dog, Loss Dog. Dog is called back, scolded, cajoled, kept a watchful eye on. Dog is owned, groomed, manicured, whipped, fed, stroked. Pain is man's best friend.

Pain Dog.

England is a Nation of Dogs. A bulldog represents the nation during national elections – this is democracy with balls. Small but perfectly formed, muscular, fierce, loyal. Bulldog's balls. When the monarchy is finally abolished it will become a Republic of Dogs. The Dog Coast. The United Church of Dog. Dog Mansions. Dog Café. Dog University.

A man gets on a bus in the city, his dog tucked under his arm. Four stops later, he takes the dog into the supermarket and buys two frozen lamb chops from New Zealand. Cavemen and their dogs hunting for meat in the city.

•DREAMS•

by Steve Pile

> There were a hundred thousand slopes and substances of incompleteness, wildly mingled out of their places, upside down, burrowing in the earth, aspiring in the air, mouldering in the water, and unintelligible as any dream. (Charles Dickens describing a suburb of London in 1848)

In his novel *Invisible Cities* (1972), Italo Calvino imagines a meeting between Marco Polo and Kublai Khan. In the course of their conversations, Marco Polo conjures up images of many fabulous and incredible cities. At one point, however, the Great Khan challenges Marco Polo. He has begun to notice that these cities have begun to resemble one another. The Khan's mind now sets out on its own journey. Interrupting Marco Polo, the Khan begins to describe a wondrous city. And he wonders whether it exists. But it seems the Khan had not been paying attention, for it seems that Marco Polo had been telling the Khan about precisely that city. Intrigued, or perhaps in disbelief, the Khan asks Marco Polo the name of the city:

> 'It has neither name nor place. I shall repeat the reason why I was describing it to you: from the number of imaginable cities we must exclude those whose elements are assembled without a connecting thread, an inner rule, a perspective, a discourse. With cities, it is as with dreams: everything imaginable can be dreamed, but even the most unexpected dream is a rebus that conceals a desire or, its reverse, a fear. Cities, like dreams, are made of desires and fears, even if the thread of their discourse is secret, their rules are absurd, their perspectives deceitful, and everything conceals something else.'
>
> 'I have neither desires nor fears,' the Khan declared, 'and my dreams are composed either by my mind or by chance.'
>
> 'Cities also believe they are the work of the mind or of chance, but neither the one nor the other suffices to hold up their walls. You take delight not in a city's seven or seventy wonders, but in the answer it gives to a question of yours.' (Calvino, 1972, pp. 37–38)

As with cities, so it is with dreams. Marco Polo's analysis is clear: the randomness of cities – their absurd or deceitful realities – has an inner meaning, an inner rule, a perspective, a discourse, in the same ways as dreams. Underlying the production of cities are the hidden workings of desire and fear. In other words, cities are desire and fear made concrete, but in deceitful, disguised, displaced ways. It is the same with dreams. Kublai Khan cannot accept this interpretation, either of dreams or of the city. And, surely, dreams and cities have nothing to do with one another. Dreams are illusions, unreal. Cities are very real, the work of the conscious mind, not the random, absurd juxtaposition of astonishing images. Out of these elements, it might be possible to discover the deceitful discourses, to uncover the hidden desires and fears, to dream again of/about the city.

Cities are like dreams, for both conceal secret desires and fears, both are produced according to hidden rules which are only vaguely discernible in the disguised and deceitful forms (of dreams; of cities). There is of course a difference between the world of dreams and the waking world: to begin with, the world of dreams pays no attention to other people – a rare luxury in waking life! The mind, far from operating in completely incompatible and unrelated ways, in sleep and awake, works in parallel ways. Simmel was the first to suggest that mental life in cities is characterised by indifference, reserve and a blasé attitude, but this only reinforces the idea that mental life in cities is characterised by displacement, condensation and the use of images to represent *and effectively disguise* desires and fears – as in dreams.

But to what purpose is all this musing? Perhaps, as Walter Benjamin suggests, we should strive to shock modernity into waking up. A revolutionary idea and one that might just meet with the Khan's approval. No such option would appear to exist in Freudian thought, for we could never fully shake off the dream dust. Marco Polo would appear to agree. On the other hand, it is possible to draw other lessons from Benjamin, Freud and Co. In this, we can think again of the paradox of dreaming: that it occupies both our sleeping and waking worlds. Through dreaming, it might be possible to imagine different transformative possibilities. Thus, instead of waking (to realise those secret wishes), or, instead of returning to the dream (to find those hidden messages), the significant move may be to pursue with greater enthusiasm the unconscious logics of the city. These will not be singular, nor universal, nor capable of being circumscribed by a master narrative of urban development. Instead, we would be forced to recognise that cities will have contradictory, and perhaps incommensurable, logics. And perhaps this is why cities are like dreams, both because they are never simply works of the mind or of chance, and also because they embody paradoxical and ambivalent elements.

It still feels like musing, all this talk of dreams. The alarm bells are ringing loud and clear: cities are wrecked by earthquakes, riots, (not so) smart bombs, pervasive disease, abject poverty (and that was only this week). The problems confronting cities are so vast that they seem absurd: western-dominated neo-liberal economic strictures force people off the land and into the shanty towns of the poorest countries of the world, so cities of twenty million plus are created where there isn't enough food. But it is important to remember that neo-liberal dreamings are not the only ones, nor the inevitable ones. Perhaps the scale of the problem explains why it is so easy to forget what the dreams of the city are all about – what it means to live in cities, their freedoms and opportunities, their new communities and cosmopolitanism. This suggests new urban practices that pay as much attention to imagining and mobilising better stories as to shocks to the system. Collapsing neither into the waking world of rationalisations and instrumental logic, nor into the dream world of barbaric desires and satisfying fears, the transformation of urban space would instead necessitate an understanding of vicissitudes of the dream city.

References

Calvino, I. (1972) *Invisible Cities*, London: Faber and Faber.

Dickens, C. (1848) *Dombey and Son*, Harmondsworth: Penguin.

•DRIVING•

by Harvey Molotch

Although driving in most cities is often thought of as a zone of conflict and aggression, it is really the proving ground of civility. Nowhere else in society are people on such an equal footing. Toughs can bully other kids on the street corner; the rich can use money to go in front of everyone else in restaurants. But if you push too hard in the car, you risk your own well-being. In fact, the more expensive or valuable your vehicle (perhaps because of your own labor customizing it, if not the initial cost), the more you have to lose.

The tension of driving, both in degree and type, varies between places. Boston and New York are famously the US cities of verbal driving with maneuvers syncopated to expletives, usually to oneself, but sometimes to other drivers or pedestrian bystanders. Los Angeles' driving society is, in my experience, also oral but with food. As US food in general has moved beyond burgers toward more architectural and chopstick-related scenarios, so has behind-the-wheel consumption. Some keep a towel or sheet in the car to cover the lap to avoid arriving with one's smart white linen looking like Pollack canvas. San Franciscans make fun of Los Angeles drivers for always being in the car, but at least they go from one place to another rather than circling the same blocks looking for someone to die so they can inherit their parking space.

In Britain, of course, there is nothing oral about anything, driving included, although muttering occurs. What is left of the British road to socialism is reflected in drivers' willingness to sometimes share a lane with one another. In the USA, one 'owns the lane' such that if two cars are heading toward each other, perhaps because one is passing a slower vehicle, the game of chicken sometimes has to be decided by crash (democracy in action). But in the UK, I have seen drivers make way for those proceeding down the middle, moving a wee to the side. This is a nice life-and-death pleasantry, a vestige of crowded tea-room etiquette perhaps.

Sometimes the work requires less rather than greater attention to the needs of the other. In my hometown of Santa Barbara, California, courtesy can become so contagious that a four-way stop can turn into a near-permanent condition. It would just be rude to go first. Because decorous people have no way to communicate audibly car-to-car, there is no way to say 'after you.' This means gridlock. Just as thinking about collision when on foot generates awkward choreographies of approach-avoidance, road urbanity also depends on a mindful thoughtlessness which makes things possible that conscious consideration precludes.

The same lesson is taught at the opposite end of the intensity scale. In Taipei, a driver pursues every possible opening and this more or less works, because every driver knows every other driver will do the same. Pedestrians are also in the know, which keeps them alive as drivers pay no heed to boundaries like curbs, sidewalks, or other hardware ephemera. The ballet of near-miss enables far more cars to move at faster speeds than on more demure streets of the world, increasing holding capacity. But it demands total commitment with no time out: no talking at all and finish your food *before* leaving the restaurant.

In places where much of this has not yet been worked out, as in the newly automobiled cities of the People's Republic of China, fatality rates are astounding. But where the veterans have had time to establish the local conventions, starting with the arbitrary decision of which side of the road to drive on, the world's streets display how people take care of themselves and achieve coherent order. As long as folks know deep in their social psyche what one another do, there is life on the road.

•DUST (1)•

by Alastair Bonnett

> When you are surrounded with shadows and dark corners you are at home only as far as the hazy edges of the darkness your eyes cannot penetrate. You are not master of your own house. (Le Corbusier, 1987, p. 188)

For Le Corbusier the modern city was to be a city without dust, a city where 'everything is shown as it is'. The Law of Ripolin (i.e., a coat of whitewash) would be strictly enforced, each citizen being required 'to replace his hangings, his damasks, his wall papers, his stencils, with a plain coat of white ripolin'. In order to 'protect them from dust' (Le Corbusier, 1986, p. 117) such items should be stored away and brought out only when required.

Modernity's war against dust has been a long and exhausting struggle. Within the modern city suspicion of decay is unremitting. A state of constant vigilance has to be taught and learnt if the *complete* visibility, the total visual and intellectual availability of the urban environment is not to be threatened.

If only everything could finally be seen, once and for all. And the filth of the hidden mopped away. For once 'the malevolent phantoms that disgust cleanliness and logic', the 'dismal covers of dust [that] ceaselessly invade our territorial abodes' have won, then 'there will be nothing more to save us from nightmares' (Bataille, 1976, p. 197).

This busy neurosis has been sponging and smoothing, simplifying and rationalising cities across the globe for many years now. It is a never-ending task. 'Dust-traps' are pinpointed and expunged. Old buildings and streets are demolished or remodelled. 'Brighter', 'cleaner' spaces are, invariably, promised in their place. The contamination of the hidden is identified for each new generation, each new decade, and the city is made healthier, more hygienic, over and over again.

The modern imagination sees a clear and obvious link between dust and disease. It is a largely irrational but zealously guarded association. Hospital buildings provided one of the first arenas where such concerns were put into architectural practice. From the late nineteenth century onwards reformers succeeded in establishing the notion of a 'dust-free' environment as central to hospital design (see e.g. Burdett, 1893). The Law of Ripolin was applied and all possibilities of 'hiddenness' ruthlessly identified and expunged. The new hospitals, with their endless 'spotless' corridors, their regular white rooms and expanses of window, proved to be a prototype for the modern city.

Perhaps the attempt to connect dust and ill health is related to the unnerving discovery that most house dust is simply dead human skin. The first layer of the human epidermis is composed of the same material. What we see of people is, by and large, dead matter. Dust. To dust. The fallen evidence of our hidden living state, when mixed with the detritus of construction sites and industry is transmogrified into grit, city dust. Of course, whether inside or outside, all those frantic, rushing attempts to clean up and scrap away merely give rise to more clouds of particulate matter, more grit to irritate the eye. The verb 'to dust' implies as much, denoting as it does a simultaneous process of removal and distribution.

The scale of the onslaught on dust is becoming ever more impressive. Amidst the miles of glossy marble effect flooring, where giggling toddlers can roll around in safety; amidst the immaculate heritage sites, stinking of wood preservatives, 'wipeable' cities are emerging. In these new environments dust must be kept permanently airborne, in eternal circulation. Dust cannot be destroyed, but it can be prevented from settling. The effort involved in achieving this is both vast and fetishistic. Indeed, the pleasures of cleanliness, of full disclosure, of transparency, are growing in complexity, becoming strangely labyrinthine. In the dust-free mall are dust-free checkouts put together with dust-free components made in dust-free factories by workers wearing hairnets and face masks. The surfaces and areas that must remain uncontaminated have multiplied; they have grown crowded and busy. Thus the desire to 'see everything' is becoming imperilled by its own rampancy: a new clutter is being born, the millions of pathways and elements of the dustless city swarm beyond sight.

Le Corbusier looked forward to giving the walls of his buildings a rub down with whitewash. It was a gesture against chaos and for simplicity, against ignorance and for information. Today it is the agents of cleanliness who seek to confuse and confound us. A baroque blandness has been created, difficult to resist, impossible to fathom. And, as a consequence, the cultural meaning of dust is shifting, albeit subtly. Dust is becoming an object of nostalgia and a symbol of resistance. To live with dust, to admit that it will continue to fall, continue to settle, suggests an ability, a desire, to live with the irrational, with history, 'the darkness your eyes cannot penetrate'. Gleaming surfaces may once have promised liberation, a utopian adventure. But no longer.

Acknowledgement

The translation of Bataille, and citations of Le Corbusier, are derived from Eric C. Puryear's (1996) 'Dust in the machine', *Inventory*, 1(2), pp. 10–29.

References

Bataille, G. (1976) *Ouevre completes de Georges Bataille, vol. I*, Paris: Gallimard.

Burdett, H. (1983) *Hospitals and Asylums of the World: Their Origin, History, Construction, Administration, Management, and Legislation*, London: Churchill.

Le Corbusier (1986) *Towards a New Architecture*, New York: Dover Publications.

Le Corbusier (1987) *The Decorative Art of Today*, Cambridge, MIT Press.

•DUST (2)•

by Ross King

> The dust of the bourgeois era obscures the traces of its own origins, even of revolution. (Benjamin, 1982, p. 158)

The story has it that when a wing of the Louvre, abandoned after the somewhat inconsequential Bourgeois Revolution of 1830, was re-entered some dozen years later to prepare it for a royal wedding (of the son of Louis Philippe, '*le roi bourgeois*'), the re-enterers were confronted by the chaos and turmoil deposited, all those years before, by a fleeting illusion of the past – a shadow *ancien régime* – all blanketed with the dust of the succeeding age. The story appealed to Walter Benjamin, with its image of the dust of an age obscuring its origins (Benjamin, 1982, p. 158; see also Buck-Morss, 1991, p. 402, n. 81).

Dust (and language?) is the great deluder, thereby also the seeming decoloniser. In Jakarta one can visit the site of Sunda Kelapa, the great port of the ancient Pajajaran dynasty, the last Hindu kingdom of West Java – and still a great port. In 1527 Sunda Kelapa was submerged by Jayakarta, the 'victorious city' of the colonising Muslims, in turn obliterated by Dutch Batavia – a recurring theme in that long succession called history. Batavia was never truly obliterated, just lost in the dust (red mud and black mould) of Jakarta, its name changed to Kota (Malevich, 1920, p. 297: 'We bring new cities. We bring the world new things. We will give them other names.' Leningrad – St Petersburg, Batavia – Kota). Kota is now Chinatown. After 'the troubles with the Chinese' in the 1950s came the next decolonisation: Chinese language and culture were suppressed, Chinese signs banned. Kota today presents the extraordinary spectacle of a Chinatown without language and characters; gambling – a Chinese fetish – is banned; the race-track (Koningsplein) long ago became Merdeka Square. Dust and language.

Indonesia's Great Bourgeois Revolution was in 1965, with the bloody overthrow of the Communists – colonisers of the spirit overthrown by the world of the commodity. There is still an old Jakarta, jumbled, small scale, a one- and two-storey high labyrinth of diverse paths, houses, kampungs (the original 'urban villages'), and canals – though occasionally cross-cut with the relict avenues of the Dutch – and everywhere bougainvillaea, the stench of the wet tropics, sticky mud, and dust (or is it spattered mud that adheres seemingly to every built surface?). Yet even in old Jakarta the neocolonialism of the global economy is ubiquitous – there may not be Chinese characters in Kota, but there is McDonald's, Pizza Hut, KFC, and everyone seems to drive Toyotas, albeit locally badged. However, the new appropriation is even more insistent: threading its way through the old is the new Jakarta, of wide boulevards, freeway-linked, beads of lushly landscaped gardens from which rise the glass and polished granite towers of today's hotels and offices (with their staggering debts written, as always, in American dollars), and the well-mannered logos of transnational corporations. To them no dust adheres.

The old colonisers (Batavia) are greeted today almost with nostalgia – the monuments of past dominations accrete romance (castles, cathedrals, palaces, the prisons of the past). Masters of the present, whether manifest or suspected, are greeted with fear (Kota), and so there are still 'troubles' with the economically dominant Chinese. The next masters – America, Japan – are perceived but not comprehended (Jakarta) – the signs are allowed (seized as signifying progress), but lost in the rainbow mists of the future rather than the grey dust of the past. But all these *spaces* – Batavia, Kota, multiple Jakartas – occupy the same *place*. Layered dust parallels layered worlds and meanings.

All colonialisms destabilise language (signs). Benjamin bequeathed us a little joke:

> I saw in a dream barren terrain. It was the market-place at Weimar. Excavations were in progress. I too scraped about in the sand. Then the tip of a church steeple came to light. Delighted, I thought to myself: a Mexican shrine from the time of pre-animism . . . I awoke laughing. (Benjamin, 1979, p. 60)

A joke on a joke: something primitive buried under the sand (dust) of the bourgeois world; but is the (anti-)monument Christian, or animist, or indeed even pre-animist? Did not the Christianising of New Spain involve the spiritual subversion of Christianity itself by the very system that it set out to destroy? So what is it that lies beneath the dust? For Benjamin, history only 'begins with awakening': in the dust of the bourgeois market lies unrecognised that other origin, in colonialism and imperialism; and all exist in the absurd delusion that they have indeed conquered, and that resistance has long been overcome.

Just as colonialisms insert new language (and both architecture and the constellations of names and signs of the city are part of language systems), so decolonisation also works through such languages. Languages are also the medium of resistance – like Deleuze's metaphoric rhizomes, endlessly bifurcating, horizontally spreading to occupy the crevices and fault lines, the weeds that infiltrate and undermine. The two Jakartas are mutually rhizomatic: the glass and granite spires along the new ten- and twelve-lane bifurcations eat into the old city as it collapses into its mud and dust, ending a richer, more complex world; but the alleys and mud and the informal economy – the great eroding threat to the high capitalism that is represented in the boulevards and freeways – is only half a block behind the façades, and the titanic struggle between opposed economies and their opposed spatial worlds goes on. The dust may disguise not only origins, but also the future: the desolation of old Jakarta may well mask the greater stability of its informal economy and the possibility of its resurgence. For there is another rhizome – the eroding rot of corruption will assuredly threaten the corporate towers and grand boulevards far more than the alleys and street vendors. Currencies collapse, regimes end, nepotisms liquidate. But none of that for now.

In the old Staatshuis, the town hall of the Dutch Batavia, is a museum of maps, where the old names are left to disappear with the rotting paper. Not only are the old names of that earlier world falling away to dust and mildew, but also those of the 'emancipation' by Japan, and the revolution and decolonisation.

References

Benjamin, W. (1979) 'One-way street', in W. Benjamin, *One-Way Street and Other Writings*, London: New Left Books.

Benjamin, W. (1982) *Gessamelte Schriften: Das Passagen-Werk, Vol. 5*, ed. R. Tiedemann and H. Schweppenhuser, with T. W. Adorno and G. Scholem, Frankfurt: Suhrkamp Verlag.

Buck-Morss, S. (1991) *The Dialectics of Seeing: Walter Benjamin and the Arcades Project*, Cambridge, MA: MIT Press.

Malevich, K. (1920) 'UNOVIS – the champions of the new art', in L. A. Zhadova (ed.) (1982), *Malevich: Suprematism and Revolution in Russian Art, 1910–1930*, London: Thames and Hudson.

•EARTHQUAKES•

On the Fault Line

by Dolores Hayden

He loads fresh batteries, Type D, three pairs,
into dead flashlights we've ignored for years.
I pack dried food, canteen, and heavy shoes,
work gloves, thick socks, a radio for news.
Prepare: earth's plates change gears, groan, scrape, and grind,
dissolve (in thirty seconds) peace of mind.
We've crowbar, hammer, pet food, leashes, brandy,
plus blankets, camp stove, bandaids, chocolate candy,
some traveller's checks (the cash machines might close),
phone lists (long-distance only, we suppose),
and fifty cans of non-gourmet cuisine
(odd veg and meats our child has never seen).

So L. A. couples ponder shaking sills:
our safe deposit boxes hold both wills,
our 'Quake Awake' alarms (the cost be damned)
give '*up* to thirty seconds warning,' and
our engineers have shown us where to stand,
holding our toddler daughter by the hand.

'On the Fault Line' was originally published in *Playing House* (1998, R. L. Barth)

•ECONOMIC ASSETS•

by Ash Amin

There seems little doubt that cities, certainly those of the developed world, were the workshops of the industrial age. They housed the large factories, mass workforces, infrastructure, know-how, services and capital that facilitated industrial expansion and modernisation after the mid- to late nineteenth century. The city was a factory, and the industrial smog that enveloped its hustle and bustle provided the visible proof. Nowadays however we are less sure of the economic role and distinctiveness of cities. For well over two decades in the West, manufacturing on a large scale has been disappearing from the metropolis owing to a number of factors, including the rise of the post-industrial service-based economy, the decentralisation of production to the outer city, peripheral regions and the developing world, and the ascendancy of smaller factories owing to the availability of labour- and space-saving new technologies. As a result of this process of deindustrialisation – marked by the many derelict or converted factories and industrial zones to be seen in cities – we no longer think of factory life as the economic motor of the city.

Instead, the urban economy has become a mixed bag of activities. Perhaps the most decisive shift has been the emergence of services as the core economic sector. Today's urban hustle and bustle is associated with activities linked with consumer services such as shopping, leisure and tourism, producer services such as banking, insurance and accountancy, welfare services such as education, health care, sanitation and transport, and services related to local and national public administration. The urban economy increasingly generates employment, income and profits by producing and exchanging services.

But the city also continues to undertake three other important economic activities. One is trade, that is, its millennial role as a marketplace and commercial centre for the distribution of goods and services around the world and to the rest of the regional and national territory. Another is continuing manufacturing through the enduring sweatshops and small firms working long hours to produce traditional consumer goods such as clothing, design and innovation-intensive productive activities in revamped inner city locations, and the headquarters and research divisions of international corporations in central business districts which provide access to specialised expertise and services. The third activity – perhaps the mainstay of the urban army of poor, underpaid, unemployed and marginalised people – is the informal economy, composed of a variety of precarious and perhaps often illegal ventures (from fly-by-night contractors to petty criminals).

What are the central assets of today's city of economic variety? What gives the city its economic vitality? What makes the economic base of the city different from elsewhere, especially in terms of its competitive advantage? These are not insignificant questions, since in recent years, against the backcloth of urban deindustrialisation, and escalating urban poverty and social conflict, many have come to see the city as a symbol of economic decline and an impediment to economic dynamism. The critics stress the high cost of urban premises as well as the problems of urban congestion (from traffic jams to housing shortages and class sizes in schools). They see them as an economic burden on firms, employees and urban dwellers. They remind us that the threat of crime and other urban disorders such as begging and street peddling is a powerful deterrent on investment as firms seek to protect their assets, employees and clients. Parts of the city which house the permanently unemployed and other social groups confined to economic inactivity (e.g. the disabled) are described as sinks in which welfare and other expenditure is poured down, without any productive returns. The litany is endless.

Without denying any of these impediments, it is still possible to conceptualise the city as a source of economic dynamism. Underlying the contemporary economic variegation to which I have alluded, there are three sets of assets which contribute to the continued economic salience of the city. These are assets which, as we shall see, are quite different from the urban factors that explained the age of the city as factory. Then, urban density and urban agglomeration played a vital role in providing the labour, capital, infrastructure and markets to satisfy the voracious appetite of the mass production factory and make location in the city profitable. Today, the advantages of urban proximity are quite different, and less narrowly economistic.

The first asset is the clustering of strategic capacity that makes the city a centre of corporate control. The large metropolis in particular is a key command and control centre within global industrial and corporate networks, financial markets, producer services industries, and other associated service industries (telecommunications, business conferences, media, design and cultural industries, property development, etc.). The shining towers of its central business districts house the headquarters and nerve centres of the corporate world. It seems that the dispersal of the productive capacity of transnational corporations over increasingly global distances has produced a parallel territorial concentration of high-level functions in knowledge- and power-rich cities. In such contexts, the face-to-face work environment, the personal relations between corporate leaders, and the struggle to improve access to changing know-how, innovations and industry standards, are crucial for maintaining competitive advantage.

The second asset consists of two aspects of the city as an economic motor: as a knowledge-base, and as a source of vital agglomeration economies. The economic competition based on consciously anticipating and shaping markets (reflexivity) finds its sources of innovation, information and knowledge in the elements of a city's knowledge fabric linking headquarter functions, media, cultural and arts industries, education and information services organisations, and research, development, science and technology institutions. In addition, metropolitan life offers rich transactional opportunities, benefits of agglomeration of specialised firms, and advantages of interpersonal proximity, to facilitate innovation-based adjustment for the emerging economy of volatile and design-conscious demand.

The third asset of cities is their centrality in cultural production and creativity. The city remains the centre of cultural activity, through its attraction for the media, entertainment, sport and leisure industries, the education and research organisations it spawns, the cultural demand it meets through its theatres, restaurants and cinemas, and the pull that urban life seems to have upon the creative professions (from artists to musicians). The 'creative city', encouraged by policies of urban revitalisation based on reclaiming public spaces and animating them with diverse cultural activities, is increasingly seen by civic leaders around the world as an important source of urban economic renewal and comparative advantage. The economic growth is anticipated not only from the expenditure of an ever-increasing global demand for cultural consumption, but also the potential for economic innovation offered by the creative people housed by the culturally active city.

Clearly, this is a rosy reading of cities based upon the experience of a small number of usually global cities. It does not speak to the city of poverty, hardship, informal economic activity, and hand-to-mouth survival. And yet, there can be no question that the city also remains the economic motor of post-industrial society. Indeed, and this is the paradox of urban life, some of the dynamism actually draws upon the 'less dazzling' features, as we can see, for example, from the desire of high-powered global professionals and business people to 'experience' the delights of ethnic restaurants in districts of extreme poverty and social degradation.

•ENTRANCES/EXITS•

by Allan Pred

The high-rise apartment building

AT A CENTRAL CITY LOCATION,
at a hillside site on Cherry Road (*Körsbärsvägen*), but fifteen minutes' walk from the very core of Stockholm,
at a landing paved with foot-wide cement squares, flanked on two sides by a 19-storied apartment building – one of the city's tallest structures – and on a third by a sprinkling of cherry trees that slope some few yards up to their namesake street,
at a small bicycle-strewn plateau which serves as the sole entrance- and exit-route for those residing in the adjacent 'skyscraper', for those university students and their families who are predominantly of Middle Eastern, Latin American or other non-European origins, predominantly immigrants and refugees who have entered Sweden to stay rather than visiting enrollees who have come so as to soon again exit,
at a limited space repeatedly traversed by those who – via their daily entrances into city life – have quickly learned their social and cultural geography lessons, have quickly discovered the spaces to which they are limited, have quickly become aware of the cartography of their Otherization, have quickly memorized the map of meaning conveyed by the difference-confirming gaze, the unfriendly stare, the leer of desire/disgust,

'Many shopkeepers think that you come into their shop just to steal. Its noticeable how they watch you and check wherever you go.' (Unidentified migrant (Erland Bergman and Bo Swedin, *Vittnesmål-Invandrares syn på diskriminering i Sverige* (Stockholm: Liber Förlag, 1982), p. 131))

'Foremost among Africans, but also among Arabs [sic] *and Latin Americans . . . in large measure every other respondent reports that s/he had been subjected*

the look of suspicion,
the uneasy glance of fear,
AMIDST THE ROUTINE PRACTICES AND OCCURRENCES OF EVERYDAY AND EVERYNIGHT LIFE,
amidst the sporadic foot traffic,
amidst the every now and then passage of elsewhere dwelling people on their way to nearby research institutes or the Royal Institute of Technology,
amidst the occasion flit-by of the sweatsuit-clad bound for the jogging paths of *Lill-Jans skogen* (Little Jan's Woods) or an indoor tennis-court,
amidst the intermittent comings and goings of building residents, of young women and men returning from classroom or library visits, of parents or grandparents ushering small children, of the bike dismounters and plastic-bag carriers having completed a shopping trip or some other heart-of-the-city excursion,
amidst the temporally scattered building entrances and exits of young women and men who more or less frequently fret about what does or does not lie ahead, who more or less frequently translate past experiences of discrimination and current word of mouth into dread of the future, who more or less frequently are gripped with insecurity about the world of (non)opportunities they will enter upon exiting from their current student status,

publicly to threats, insults and other forms of harassment at least twice during the past year owing to their foreign background.' (Anders Lange (*Invandrare om diskriminering II* (Stockholm: Centrum för invandringsforskning, 1996), p. 1))

'The Swedes were nice and smiled friendly the first years I lived here. But behind all that one could sense uncertainty and fear. The last year [1990–91] *attitudes have become different. Suspicion, hostility and aggressiveness have become increasingly common. Migrants get the blame for unemployment, narcotics crimes, the housing shortage – even that it's been warm and snowless Christmas Eve.'* ('Bettina', high-rise resident, medical student, migrated from Iran in 1984 (*Dagens Nyheter*, November 11, 1991))

'Violence against migrants has been refined and become more effective. The attacks have become more numerous and this creates greater stress and anguish among the migrants.
I know Swedish, I know how to talk and debate and even avoid clashes. I am secure and feel myself a Swede. But even I have begun to wonder if it was right of me to get children. Previously one was able to live in economic, humane, cultural harmony. I see a danger of society becoming polarized'.
(Mahmoud el Hissi, psychologist and social worker, migrated from Palestine in 1961 (*Dagens Nyheter*, November 10, 1991. (The reference to avoidance of clashes is an attempt on el Hissi's part to indicate that he is fully 'Swedified', as conflict avoidance is widely regarded by Swedes as the most common of their personality traits))

VIOLENCE MAY ERUPT
WITHOUT PRIOR
WARNING,
a particular conjuncture of
seemingly ordinary
circumstances may, in the
blink of an eye and the
bending of a finger,
become extraordinary in
the deadliest of ways,
the translation of economic
discontent into the
scapegoating of racialized
migrants and refugees,
may result in physical
action,
the festering boil of imagined
threat may burst open,
releasing 'normal'
aggression or paranoid
outburst,
the current moment of danger
may become a site-specific
corporal reality rather
than a general state of the
hypermodern present,
as around 6.30 p.m. on November
9, 1991,
when late afternoon was
completing its exit and
early evening was
commencing its entrance,
when Jimmy Ranjbar, a
34-year-old Iranian refugee
student and father of two,
was about to enter the
high-rise,
when a man stepped out from
behind one of the trees,
when a red laser beam issued
from a rifle sight, briefly
flashing on its target – as it
had on four separate
previous occasions in which
a student of Ethiopian
origins, a student of Iranian
background, an indigent
Greek migrant, and a
musician of Brazilian birth

'The victim is a migrant, he has no criminal past.' (Opening words of a police statement to the press (*Dagens Nyheter*, November 9, 1991))

'Hostility towards foreigners must be opposed wherever it turns up.' (Yet another joint proclamation issued by establishment organizations, including the Social Democratic Party and the Lutheran State Church (Politicians, as well as the majority of Swedes, use 'hostility towards foreigners' and a variety of other terms to deny the widespread presence of cultural racism. That presence runs counter to central elements of national identity, including a view of the country as a bastion of equality and a moral superpower. Only skinheads, neo-Nazis and other extreme right-wingers may be labeled as racist, even though they do not – even by the wildest stretch of the imagination – have anything to do with the extreme labor-market discrimination, residential segregation and social apartheid to which non-Europeans are subject. See Allan Pred, *Even in* Sweden, 2000))

'I have lived in Sweden twelve years. I work together with Swedes, have Swedish friends and feel a sense of loyalty towards my new homeland. But societal developments have made me feel

were each non-fatally
wounded,
when the 'Laserman' struck
again,
when a bullet made its
entrance into the back of
Ranjbar's head,
when he crashed to the ground
in a fast widening pool of
blood,
when life began to exit him,
when
AT ANOTHER TIME AND PLACE IN THE CITY

scared. Hostility towards foreigners has become obvious and partly accepted by society and the politicians, and lately I almost never go out alone.' ('Saddam', a close friend of Jimmy Ranjbar (*Dagens Nyheter*, November 11, 1991. Ranjbar died a few hours later. The 'Laserman' was apprehended in January, 1992, after wounding another six non-European migrants. He turned out to be a German-born resident alien))

•EXCHANGES•

Select stock exchanges: market size, 1990

	Market value (US$ millions)		*Listed companies (N)*		
	Stocks	*Bonds*	*Domestic*	*Foreign*	*Member firms (N)*
New York	2,692,123	1,610,175	1,678	96	516
Tokyo	2,821,660	978,895	1,627	125	124
United Kingdom (mostly London)	858,165	576,291	1,946	613	410
Frankfurt	341,030	645,382	389	354	214
Paris	304,388	481,073	443	226	44
Zurich	163,416	158,487	182	240	27
Toronto	241,925	–	1,127	66	71
Amsterdam	148,553	166,308	260	238	152
Milan	148,766	588,757	220	–	113
Australia	108,628	46,433	1,085	37	90
Hong Kong	83,279	656	284	15	686
Singapore	34,268	98,698	150	22	26
Taiwan	98,854	6,551	199	–	373
Korea	110,301	71,353	699	–	25

Source: *Tokyo Stock Exchange 1992 Fact Book*, April 1992. Tokyo: International Affairs Department, Tokyo Stock Exchange. Cited in S. Sassen (1994) *Cities in a World Economy* (Thousand Oaks, Pine Forge Press), p. 24.

•Fire alarm•

by Walter Benjamin

f
f
f

The notion of the class war can be misleading. It does not refer to a trial of strength to decide the question 'Who shall win, who be defeated?', or to a struggle the outcome of which is good for the victor and bad for the vanquished. To think in this way is to romanticize and obscure the facts. For whether the bourgeoisie wins or loses the fight, it remains doomed by the inner contradictions that in the course of development will become deadly. The only question is whether its downfall will come through itself or through the proletariat. The continuance or the end of three thousand years of cultural development will be decided by the answer. History knows nothing of the evil infinity contained in the image of the two wrestlers locked in eternal combat. The true politician reckons only in dates. And if the abolition of the bourgeoisie is not completed by an almost calculable moment in economic and technical development (a moment signalled by inflation and poison-gas warfare), all is lost. Before the spark reaches the dynamite, the lighted fuse must be cut. The interventions, dangers and tempi of politicians are technical – not chivalrous.

•First aid•

by Walter Benjamin

A highly embroiled quarter, a network of streets that I had avoided for years, was disentangled at a single stroke when one day a person dear to me moved there. It was as if a searchlight set up at this person's window dissected the area with pencils of light.

•FLOODLIGHTS•

by David Gilbert

In *Koyaanisqatsi*, Godfrey Reggio's 1984 cinematic essay, nightfall reduces the routine movements of an anonymous modern city (Los Angeles? Osaka? São Paulo?) to a hypnotic abstract of patterned and flowing lights. As captured by time-lapse photography, the lights of the city become a visual accompaniment to the minimalism of Philip Glass' soundtrack: the headlights and tail-lights of traffic become rhythmic sequences choreographed by traffic signals, each sequence similar in form yet modulated in detail. The city at night must be one of the iconic images of modernity. For the Hungarian modern architect Zoltán Kósa (1968), 'we have a feeling that the city in daylight is more a pool of traditions, a touch of nostalgia, while its night appearance urges us more towards the goals and wishes of tomorrow.'

The lights of the night-time city show the double-sidedness of the modern city. The patterns and flows of light are the result of thousands and thousands of immediate individual decisions, desires and urges; the patterns and flows of light are also the result of the purposive decisions of architects, politicians and planners. The real city of people, with its dirt and clutter, its compromises and improvisations, frustrates the grand designs of modernist planning, but architects and planners are able to express their fantasies in the virtual city of lights. From the earliest experiments in electrical floodlighting in the mid-nineteenth century, lighting has been used to make the city appear more rational, more ordered, more perfect. It was no accident that the transformation of the night-time cityscape of Paris in the late 1920s was directed by a floodlighting firm called 'L'Eclairage Rationnel'. Textbooks for architects, planners and 'illuminated engineers' have long made the connection between public lighting and illumination of the structure and order of the city. William Lam, in his standard reference work *Perception and Lighting as Formgivers for Architecture* (1977), argued that the main purpose of public lighting was to 'give the greatest possible information about the structure and hierarchy of the urban environment'.

But floodlighting has always been about more than just the rational expression of urban order. The earliest experiments with electrical carbon arc-lighting took place inside the Paris Opera as early as 1846, some twenty years before their widespread adoption for external lighting. Throughout their history floodlights have been used to turn the cities into stages or film sets. Perhaps the greatest of these transformations was for the Barcelona Exposición Internacional of 1929, where Carlos Buigas created a magnificent luminous fountain as the centrepiece of a display which anticipated the cinematic spectacles of Busby Berkley. The *New York Times* said simply that 'to see the lights of Barcelona is to see heaven' (quoted in Horváth, 1989). In cities prized for their historical architecture, the rules of the heritage game change at dusk as historical verisimilitude is exchanged for unashamed display. At night the skylines of cities like Prague or Florence are transformed into archipelagos of shining sights in a sea of shadows. No one claims that the historic cities ever looked like this; rather, visual leitmotivs are pulled from their daytime contexts to serve as symbols of their city.

Of course such symbols are often contested, and the decision to illuminate or to leave in shadow can be an exercise of power as well as civic promotion. In Warsaw before 1989, the giant central Palace of Culture (a 'gift from the Soviet people') was visible enough during daylight, and a constant, grating reminder of Poland's political subjugation. At night the brilliantly floodlit building became even more intrusive and threatening, as much of the rest of the city was left in relative darkness. In general the more powerful the illumination of parts of the city, the darker become those left in shadow. In Warsaw these shadows were expressions of a geography of political power, but in other cities relative darkness is a sign of economic inequality. In many cities the brightly illuminated centres, markers of 'progress' or 'development' dazzle, while slums and shanties remain hidden.

The term 'floodlighting' covers two rather different forms of illumination. The first is perhaps better described as façade illumination or public spotlighting, where individual buildings or their details are picked out in light. Other kinds of floodlighting have less to do with the deliberate composition of the visual cityscape than with a simple need to be able to see after dark. But these utilitarian floodlights reveal other dimensions of the symbolic order of the city. Wolfgang Schivelbusch in *Disenchanted Night*, his classic history of the industrialisation of light (1988), tells the story of the Colonne-Soleil, an electrical sun tower 360 metres high which was proposed in 1885 as a way of providing Paris with permanent daylight. Sébillot, the architect and engineer of the ill-fated project, spoke of light penetrating into houses and flats, and it was clear that the motivations for the tower included surveillance and public control of a revolutionary city as well as simple illumination. The tower itself would have stood as a shining symbol of the power of the triumphant bourgeois state. The lighting tower may have been a dead-end in architectural history, but the late nineteenth and early twentieth centuries saw the construction of giant lighting pylons to push industrial activity into the night. While public spotlighting was being used to create a symbolic order in city centres, or to transform views of historical landmarks, the brightest place in the industrial city was usually to be found at the docks or the railway marshalling yard. These bright lights displayed the energy and dynamism of the twentieth-century city, and became another of the industrial virility symbols in the competition between capitalism and state socialism. Deindustrialisation has seen the closure of many marshalling yards, and many port activities have been moved away from the cities. The greatest symbols of civic pride are no longer places of worship or government offices; the brightest lights at the beginning of the twenty-first century flood the sports stadia of the world's cities, from the Melbourne Cricket Ground to Chicago's Soldier Field, from the Maracana in Rio to the Stade de France, and, of course, from the Stadium of Light in Sunderland to Benfica's original Estadio da Luz.

References

Horváth, J. (1989) *The Floodlighting of Budapest*, Budapest: Hungexpo.

Kósa, Z. (1968) *Az Esti Városkép*, Budapest: Folyóirat.

Lam, W. (1977) *Perception and Lighting as Formgivers for Architecture*, New York: McGraw-Hill.

Schivelbusch, W. (1988) *Disenchanted Night*, Oxford: Berg.

•FLOODS•

by Mary King

Stories are like wild grass that spreads rapidly. Some become more tangible with growth, gaining legitimacy through repetition. It often happens that over time a tale garners a credibility substantial enough to supplant the initial event that inspired it. Such was the case for the city of Boston as the First World War drew to a close and people found themselves in a time of rapidly rising costs, long work days and low pay. War produces both riches and misfortunes. This one funneled unprecedented wealth into industry, utilities and banks. But, for workers, more than any time in memory it was a winter of strikes, poverty and civil unrest. At the onset of frost an influenza epidemic struck. Reaching across the entire city like a gray hand, it caused such morbidity that makeshift circus tents were erected to house the caskets awaiting burial.

f
f
f

As the year turned a corner, the weather grew strangely warm. By the second week of 1919 the worst of the disease had abated. January unfolded calmly, a month of unanticipated warmth. Snow melted in trickles flowing towards the gutter and wooden crates expanded, stretching with the sun. January 15th rang in sympathetic and mild, placid as the moment before an advancing gale. People on the streets participated in daily activities, enjoying the unseasonable delight of fair weather.

Across from Cobbs Hill, on the ocean side of Commercial Street, the Purity Distilling Company had erected a giant, modern storage tank. Towering over the surrounding neighborhood like a newly cut tooth, it stood out as a harbinger of progress and a symbol of the firmly established wealth of the city. This day found it filled beyond capacity, heavy with thick molasses. As the temperature rose in the unnatural heat of the day, the tower creaked and groaned.

Promptly at midday it exploded. Plumes of cloying smoke burst into the air and thousands of candied sparks showered to the ground below. In one calamitous instant the ruptured tank spouted more than two million gallons of syrup.

Molasses which is neither liquid nor solid but something in between cascaded in slow motion down the cavity of Commercial Street. The swell was fifteen feet high and as wide as three city blocks. Under the weight of the wave, structures were torn from their foundations and warehouses and businesses instantly collapsed. People stood watching in disbelief, as their homes, drenched in juice, floated towards the bay.

Horses transporting merchandise to and from the market were swallowed whole, head to tail in a brown sugar horror. Women, their skirts swirled in syrup, were swept away. Workers carrying bundles toward the dock found themselves flailing in the coagulated liquid. Several soldiers home early from the war became unexpected casualties. All told, 150 people were injured that day. Twenty-one suffocated to death, the majority quite literally drowning in commercial excess.

When the wave finally receded to a depth of three feet, the city deployed regiments of firemen to hose down the streets with saline water in the hope that the salt would neutralize the sweet. Syrup dripped from cornices and banisters as workers sprinkled sand on to the molasses, frosting the road with a granular concoction. Their efforts, however, did little to remove the sticky fluid which seeped into the crevices of walls and the cracks of sidewalks and clogged the overfilled sewers.

The coming summer months would witness the worst infestation of flies and vermin the city had ever known. Warm days for many years wafted the surfeit odor of molasses across the channel toward the neighborhoods of the South End, the aroma perceptible even to this day.

During the night the weather abruptly turned cold, and toward dawn a light snow began to fall. Silently it dusted the ground, forming shallow deposits over the molasses, turning it to slush then hardening it with freezing. When new morning broke, it was as if the streets had turned to taffy. Molasses coated everything. Brown and clear as aged church glass, it clung to buildings, glazed the road, and hung from the tops of utility poles.

Gingerly, the residents of Cobbs Hill made their way on to the street. Using ice picks they cracked loose wide chinks of the candied landscape. The clank of shovels and picks resonated throughout the neighborhood as people hoisted slabs and bricks of molasses into sacks, baskets, and carts. Displaying the resourcefulness of the poor, the people of Boston labored all day long, prodigiously carrying the crystallized syrup indoors to melt into sugar for their cupboards.

•FUN•

by Gargi Bhattacharyya

Everyone knows that cities are fun. Bright lights, big city . . . when I've got worries I can always go – downtown. City-slick café society urban landscape. Before the city, fun seems silly, naive, childlike – all folk festivals and not-quite-pagan dressing up. To us, mega-urban modernities, non-city fun looks like a sad making do, the desperate leisure of those who have to get up early in the morning and have no knowledge of consumerism, cosmopolitanism, public culture as a perpetual parade.

For us, on the other hand, the possibility of fun is another much appreciated mark of our (waning?) modernity. We know that fun is the name of a certain section of the day, of a certain travelled-to location. After work, fun begins. Away from home, fun is possible.

More than landscape, fun is possible because of people. In the city, who knows whom you might meet? Every time we step outside our doors we enter the realm of adventure. Perhaps our workplace encounters are too familiar for excitement, maybe we can predict who will occupy our domestic space day by day, even our neighbours might be dulled by proximity – but beyond that, the city offers us streets full of strangers. Every new face which turns to us unexpectedly might signal a new start, a life-changing revelation, the one we have been waiting for, the beginning of the rest of our lives.

As if it is all about sex?

Lots of talk about what is fun in the city assumes that you are young and funky with time to spare. All that strolling, sitting, watching, snacking, flirting, chatting, all that pleasure-leisure hurly-burly of the urban funscape, it all assumes a certain kind of person. Tired and wrinkly with too much to do hardly fits the role.

Isn't that the constant complaint – that cities are only fun for the young, free and single? That for the rest of us cities just mean crowds and dirt and bad public transport? And, of course, life will always be sweeter and easier for those who are younger, richer, more beautiful, in whatever location they find themselves.

But, deep down and whatever we may

say, we all know that the city is different. Whoever you are, and despite the ongoing discomforts of social inequality, cities are fun, at least sometimes.

Here are some of the reasons why.

1 *People.* As mentioned above, this is, of course, the main source of pleasure and excitement in the city. All those people, from all over and just around the corner and who knows where. Every one of them an adventure waiting to be had. None of which assumes anything sexual – but we can dream.
2 *Games.* Children play games, step out of their homes into the endless playground of the world. The city makes this experience available to all once again, far beyond childhood. In the city, everyone is wrapped up in their own make-believe – talking in voices, assuming characters, mimicking passing strangers. We all hop, skip and jump in and out of entrances and conveyances, run our fingers along walls and railings, all the while careful not to step on the cracks in the pavement. The possibility of danger and the danger of possibility combine to spread the aura of the playground throughout the city and the riskiness is part of what is fun, like walking on walls.
3 *Buildings.* Even for the most antisocial, cities offer so much to see. Years of human endeavour are layered up in the city landscape. From the most visionary achievements to the barbarism they rest upon, everything can be seen in the mishmash of buildings of any city centre. Whatever the programmes of erasure brought by planning or conquest, cities invariably reveal their histories in their physicality. All this can be seen for free by anyone who cares to look.
4 *Shops.* Too much has been said already about the ambiguous pleasures of modern consumption and, of course, buying requires cash which many, if not most, do not have. But looking still costs nothing and cities have become perpetual great exhibitions, displaying the produce and innovation of all corners of the globe to anyone who catches the bus into town.
5 *Sensation.* All environments are packed with sensual experience, but the city crams these experiences together more densely. In the city we see a dozen points of interest on every street, hear the assorted noises of the city's assorted life, smell the smells of a global assembly and taste the produce of this mixed gathering. We reach out to touch textures which span the range of creation. What can be more intensely fun than this?

There are other things which could be mentioned – the assortment of leisure activity, the bustle of public life – but these are split into more particular tastes, with each person tied into their own habits of leisure and the public. Points one to five belong to us all, if we care to take them – so this is the kind of fun I want to promote.

•FUNCTIONS•

Functional types of cities

City type	*Characteristics*	*Examples*
Global cities	Accumulation of financial, economic, political and cultural headquarters of global importance	London, Paris
Growing high-tech/ services cities	Modern industrial base, national centre of research and development, production-oriented services of international importance	Bristol, Reading, Munich
Declining industrial cities	Traditional (monostructured) industrial base, obsolete physical infrastructure, structural unemployment	Metz, Oberhausen, Mons, Sheffield
Port cities	Declining shipbuilding and ship repair industries, environmental legacies, in the south burdened by additional gateway functions	Liverpool, Genoa, Marseille
Growing cities without modern industrialization	Large informed economy and marginalized underclass, uncontrolled development and deteriorating environment	Palermo, Thessaloniki, Napoli
Company towns	Local economy depending to a large degree on a single corporation	Leverkusen, Eindhoven
New towns	New self-contained cities with overspill population in the hinterland of large urban agglomerations	Milton Keynes, Evry
Monofunctional satellites	New urban schemes within large agglomerations with focus on one function only (e.g. technopole, airport)	Sophia-Antipolis, Roissy
Tourism and culture cities	Local economic base depending on international tourism and cultural events of European importance	Saltzburg, Venice
Border and gateway cities	Hinterland divided by national border; gateways for economic migrants and political refugees	Aachen, Basel

Source: K. R. Kunzmann and M. Wegener (1991) *The Pattern of Urbanization in Western Europe, 1960–1990*, Report for the Directorate General XVI of the Commission of the European Communities, Institut für Raumplanung, Universitat Dortmund, Dortmund. Cited in United Nations Centre for Human Settlements (HABITAT) (1996) *An Urbanizing World: Global Report on Human Settlements 1996* (Oxford: Oxford University Press), p. 62.

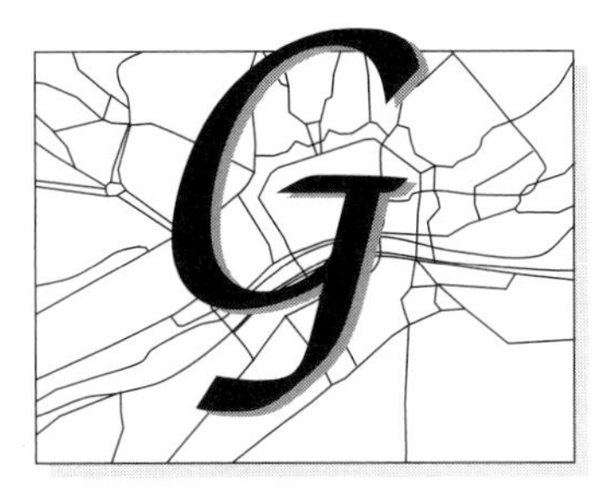

•GATES•

by Ash Amin

When we think of gates, nothing particularly extraordinary comes to mind. We see the gate as a barrier that separates two spaces and controls passage from one to the other: the gate of a house, which secures the privacy of private space; the park gate that protects public property at vulnerable times, and so on. The gate is an ordinary thing with an everyday purpose. Or is it?

There is another older, medieval understanding of the gate, symbolizing the enclosure behind high walls of entire communities from strangers and 'barbarians' on the outside. Many urban commentators claim that this kind of gated community is returning to the contemporary city. Those who visualize cities of crime and fear such as Los Angeles as a model of the future argue that the spaces of the city are increasingly being re-engineered by urban leaders, planners and architects as 'enclave' spaces. Within these spaces surrounded by impenetrable gated walls and guarded by security forces and various technologies of surveillance, ordinary folk and the well-to-do can feel protected from the threat posed by various urban undesirables such as criminals, beggars, hawkers, drunks, drug addicts, night revellers, and so on. The impetus behind gating of this sort is the public perception that urban life is becoming more and more nightmarish because of the consequences of rising unemployment, family and social collapse, intrusion by migrants and outsiders, and breakdown of law and order. The gate becomes the symbol of social withdrawal from public space and public encounter.

Consider this description by David Dillon, as American architectural journalist, who succinctly captures the phenomenon:

> One of the most familiar sounds in the US these days is the clanging gate. Not the garden gate, or the alley gate, but the gate that closes off the street, the block, and increasingly the entire neighborhood. . . . Terrified of crime and worried about property values, Americans are flocking to gated enclaves in what experts call a fundamental reorganization of community life. . . . Walls are only the beginning. Inside may be surveillance cameras, infrared sensors, motion detectors, and sometimes armed guards . . . Hidden Valley, a private

> community north of Los Angeles, installed anti-terrorist bollards two years ago to keep non-residents at bay . . . the bollards rise up to impale vehicles that try to defy them. The tally so far: 25 cars and four trucks. . . . Ironically, the rush to gated communities coincides with widely reported decreases in violent crime statistics. . . . Developers of gated communities exploit . . . anxiety by marketing their projects as safer, friendlier . . . Their ads and brochures are sprinkled with words like 'village', 'community', and 'cozy' to suggest a friendliness and manageable scale that is supposedly missing outside. (1994, pp. 8–12)

Urban commentators write about similar trends throughout the world, developing and developed. It seems that gating is becoming an increasingly common feature of our cities, regardless of whether, as Dillon notes, the fear of crime and urban breakdown is supported by real developments. As long as the perception of threat remains strong and is sustained by the offer of enclave solutions, the trend will escalate.

In the context of the walled community, the gate has lost all its ordinariness, now symbolizing a much more important process of social fortressing. This sense is again aptly captured by David Dillon:

> This desire for control could have dangerous consequences for American cities – and for the world, says urban critic Jane Jacobs . . . Jacobs sees gated communities as a new brand of urban tribalism that will pit races and ethnic groups against each other. 'It's a gang way of looking at life, the institutionalization of turf. And if it goes on indefinitely, and gets intensified, it practically means the end of civilization.' (1994, p. 12)

This is truly a bleak portrayal, as people retreat from public life into the comfort and safety of homogeneous communities, living in fear and mistrust of other 'tribes' beyond the walls. But is this the only possible interpretation of the gated community? Is its separation from the outside that total and uncontaminated?

Consider two ambiguities. First, behind the stark visual image of clinical separation portrayed by gates, walls, security guards, bollards, sirens, and so on lies the 'ordinary' family that sees living in the gated community as something not too different from the much older and more familiar urban trend of living in a 'problem-free' suburb. Like the suburban family, it is keen to ensure its own safety, and its seclusion does undeniably sustain a culture of fear of public spaces. But it too is able to detach itself from the hype of total enclosure, if only because its links with the rest of the city and beyond remain strong through such activities as shopping, commuting, eating, and leisure. Of course this is not to suggest that contact with the outside world involves exposure to, and interaction with, the undesirable tribes. It does imply, however, that the psyche of people in gated communities cannot be reduced to simply the consequences of seclusion. We cannot, simply on the evidence of gating, condemn either the insider or outsider to indifferent or hostile barbarism.

Second, even the most apparently 'homogeneous' of gated communities is full of internal contradictions and impulses which challenge its seclusion and security from external contamination. Think of the many sons and daughters who would much rather be among the 'barbarians' or the spaces occupied by them. Their boredom or dissatisfaction with the homogeneity of their gated community might actually serve to encourage cross-tribal transgression through mingling with some of the so-called urban undesirables, or it might take them to sites in the city which they consider to be places of social heterogeneity and mixture. Either way, it is clear that neither the homogeneity nor the social closure of the gated community is that easily preserved.

Whether gating will lead to the end of

civilization, its significance as a development that breaks decisively with the idea of urban public space as a shared space among diverse social groups has to be acknowledged. Gating is consciously intended to privatize and homogenize public space, with powerful interests willing to sustain this vision of urban life.

Reference

Dillon, D. (1994) 'Fortress America', *Planning*, June, pp. 8–12.

•GENTRIFICATION•

Grandmother Evicted in Echo Park

by Dolores Hayden

At first the shapes recall her wedding day.
In festive stripes of pastel pink and white,
in salmon, silver, aqua, gold, pale gray,
wide canvas tents stretch up her street's green heights.

It's May, all homes in escrow now – or later –
require pest wrapping. First she sees the hearse,
advertisement for an exterminator.
Two Anglo boys in black get out, rehearse

mass kills, shake toxic vials at termite skulls.
Latino boys dressed all in white advance
the rites one street away. They doctor: kills
cure structural rot, they drive an ambulance.

Oh spring, with houses sold and tents in bloom,
season of rituals, lives rearranged.
In real estate's mad roller coaster boom
her bungalow's been sold for carnival change.

The rickety ride swoops down too fast by half.
She chokes one predictable sob in the back of her throat –
those boys with doctor coats! She has to laugh.
This is the end, the end, the end, all out.

•GRAFFITI•

by Susan J. Smith

Whose city is this?

Corporate identity shapes the skyline; commercial products line the streets.
Faceless thousands surge through nameless spaces.

Whose place is this, and how do we know?

Look to the 'twilight zone of communication'.
The signs in the streets, the measures, the markings, the meanings, the movement. . . .

Graffito: A drawing or writing scratched on a wall or other surface.

What's wrong with graffiti?

Graffito: . . . scribblings or drawings, often indecent, found on public buildings, in lavatories, etc.

g
g
g

What's wrong with graffiti?

Tricia Rose knows, she writes in *Black Noise*:
By the mid-1970s, graffiti emerged as a central example of the extent of urban decay and heightened already existing fears over a loss of control of the urban landscape. (1994, p. 44)

And that's not all, as David Ley and Roman Cybriwsky observe in 'Urban graffiti as territorial markers':
A zone of tension appeared, which is located exactly by the evidence of the walls. . . . Diagnostic indicators of an invisible environment of attitudes and social processes . . . far more than fears, threats and prejudices, they are a prelude and a directive to open behaviour. . . . The walls are more than an attitudinal tabloid; they are a behavioural manifesto. (1974, p. 503)

Graffiti is what they call an urban incivility. A signal that the social order is breaking down; a sign that some still scorn the forces for social control. Graffiti is the tip of the iceberg. If the writing is on the wall, it is only a matter of time before something worse will clamber from the hidden depths of the underworld, to threaten our well-being, steal our property, harm our loved ones, challenge our lifestyle. Graffiti is in our space, in our face. A blot on the landscape. Full stop.

Graffito; It. Graphein: to write, write: to record: to decree or foretell: to communicate.

A loss of control?

I started writing . . . to prove to people where I was. You go somewhere and get your name up there and people know you were there, that you weren't afraid. (Cool Earl)

Words out of place?

Spray painting . . . is another solution to the incessant search for recognition, identity and status in the inner city. (Ley, 1974, p. 127)

Graffiti: words of despair . . .

. . . or signs of hope?

Words function . . . like chants, spells, incantations, curses, cheers, raps, expletives. . . . In some instances the most profound thing that can be uttered is the most obscene thought one can think of, or the most violent. (On Jean-Michel Basquiat, cited in the book of an exhibition mounted at the Witney Museum of American Art, 1993, New York: Harvey N. Abrams, Inc.)

Is graffiti vandalism . . .

. . . or could it be art?

Art: application of skill to production of beauty.

Writing names, symbols, images; working top to bottom; spreading colour along the length of the subway.

What is more interesting? Bare, twisted, rusting metal; bland advertisements, dusty floors and peeling walls? Or something surprising, colourful, words to hold your attention, changing each day, breaking the monotony of your life and its environment?

'Relax', says Tim Cresswell, as he explores the crucial 'where' of graffiti, and enjoy the visual bonus that comes these days with the purchase of a subway token (1996, p. 35). But city power brokers loathe these signs of life. Graffiti undermines their authority, usurps their space, encroaches on their territory. And so 'the graffiti problem was reconstructed as a central reason for the decline in quality of life in a fiscally fragile and rusting New York' (Rose, 1994, p. 44).

But theirs is not the only view; and soon:

The level of municipal hostility exhibited towards graffiti art was matched only by the SoHo art scene's embrace of it. (Rose, 1994, p. 46)

Art matters . . .

Aesthetics is the geography of every day life. As bell hooks puts it, 'more than a philosophy or theory of beauty, it is a way of inhabiting space' (1991, p. 104)

Art provides glimpses of otherness and elsewhere, frees desire from fantasy.

bell hooks again: 'in a democratic society art should be the location where everyone can witness the joy, pleasure and power that emerges when there is freedom of expression' (1995, p. 138)

Perhaps we should invest our radical aspirations in art, look to aesthetics to nurture the spirit, create the will to change the world?

That's what the street artists have done.

Experiencing art can enhance our understanding of what it means to live as free subjects in an unfree world. (hooks, 1995, p. 9)

So *Art* can tell us whose city this is, and why, and what it means!

Take, for example, the life and work of Jean-Michel Basquiat; graffiti artist and star; projecting signs from the street onto walls in the gallery. The more hostile New Yorkers became to urban graffiti, to words out of place; the more the art world gathered up these words, framed them, tamed them and accommodated them privately.

Not that Basquiat himself pandered to this distinction:

> In keeping with the codes of the street culture he loved so much, Basquiat's work is in-your-face. (hooks, 1995, p. 36)

> Basquiat often placed his work next to Soho art galleries on the night before an opening. Graffiti challenged the rigid *de facto* segregation of American cities by placing the work of outsiders where it could be seen by everyone. (Mirzoff, 1995, p. 164)

> 'Basquiat's work is in your face . . . [it] holds no warm welcome for those who approach it with a narrow Eurocentric gaze. (hooks, 1995, p. 36)

Jean Michel Basquiat, American son of Haitian and Puerto Rican parents, dared to be 'black', dared to tap into the deep anxieties of his time, into the fear that 'the city was being "lost" by "us"', that is to say by white European-Americans, and was being taken over by "them", the African-Americans, Latinos, Chinese and others who were supposed to be neither seen nor heard, except in the appropriate venues' (Mirzoff, 1995, pp. 163–164).

> To make sense of Basquiat's language you have to first respect African people as language manipulators of the highest order, to respect the complexity of African cultures as series of overlapping texts, tongues and dialects, ranging from the in-joke to the alienated, from the colloquial to the schizophrenic. (On Jean-Michel Basquiat, cited in the book of an exhibition mounted at the Witney Museum of American Art, 1993, New York: Harvey N. Abrams, Inc.)

Basquait dared to write, to paint, to project his words and pictures into the heart of whiteness: 'Basquiat's painting challenges folks who think that by merely looking they can "see"' (hooks, 1995, p. 36).

You want to clean this graffiti up?

Get rid of this obscenity, this mess, this thing that is talked about as if it were a disease spread by madmen, these words that are nonsense, violent, barbaric, nonsense?

No civilised metropolis would endure a rash of graffiti; no decent city would let itself be so contaminated; what thinking person would want to read it, or reflect on it, or add to it?

Clean it up, then!

> Basquiat journeyed into the heart of whiteness . . . a savage and brutal place. (hooks, 1995, p. 43)

> He tried to imagine a body that was not marked by race, while being constantly reminded of the racial mark inscribed on his own body by others. (Mirzoff, 1995, p. 189)

Like a graffiti tag sprayed on to the side of a subway carriage, the name 'Basquiat' circulated too publicly, too fast and across too many boundaries.

Vandal: a wilful or ignorant destroyer of anything beautiful, venerable or worthy of preservation.

Vandalism: ruthless destruction or spoiling of anything beautiful.

Whose city is this?
This place where we know the cost of so much and the value of so little?

Whose city is this?
We must learn to see.

References

Cresswell, T. (1996) *In Place / Out of Place: Geography, Ideology, and Transgression*, Minneapolis: University of Minnesota Press.
hooks, b. (1991) *Yearning*, London: Turnaround.
hooks, b. (1995) *Art on my Mind*, New York: The New Press.
Ley, D. (1974) *The Black Inner City as Frontier Outpost*, Washington, DC: Association of American Geographers.
Ley, D. and Cybriwsky, R. (1974) 'Urban graffiti as territorial markers', *Annals of the Association of American Geographers*, 64, 491–505.
Mirzoff, N. (1995) *Bodyscapes*, London: Routledge.
Rose, T. (1994) *Black Noise*, New England: Hanover University Press.

•GRAPHS•

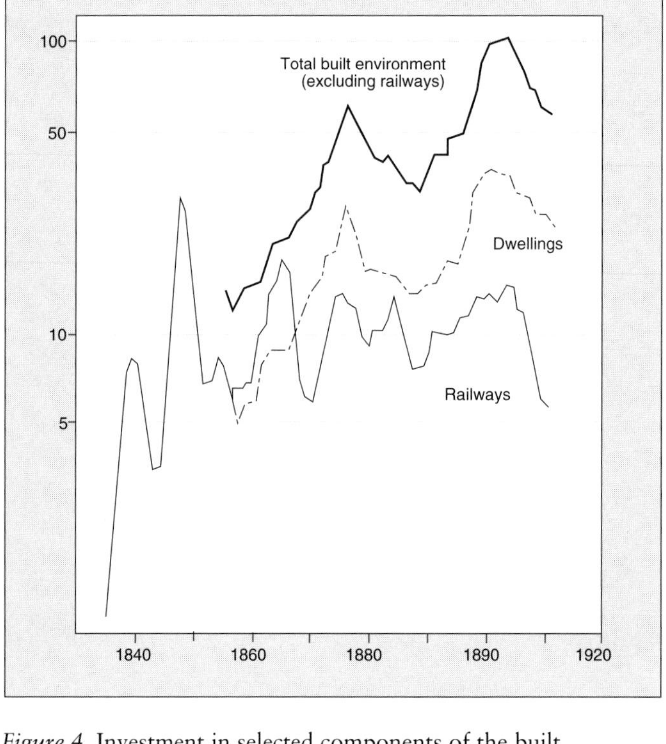

Figure 4 Investment in selected components of the built environment in Britain (£million at current prices)

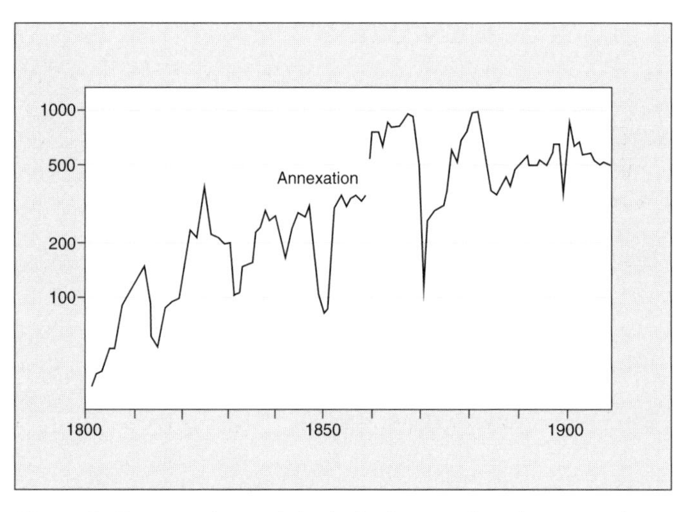

Figure 5 Construction activity in Paris – entries of construction materials into the city (millions of cubic metres)

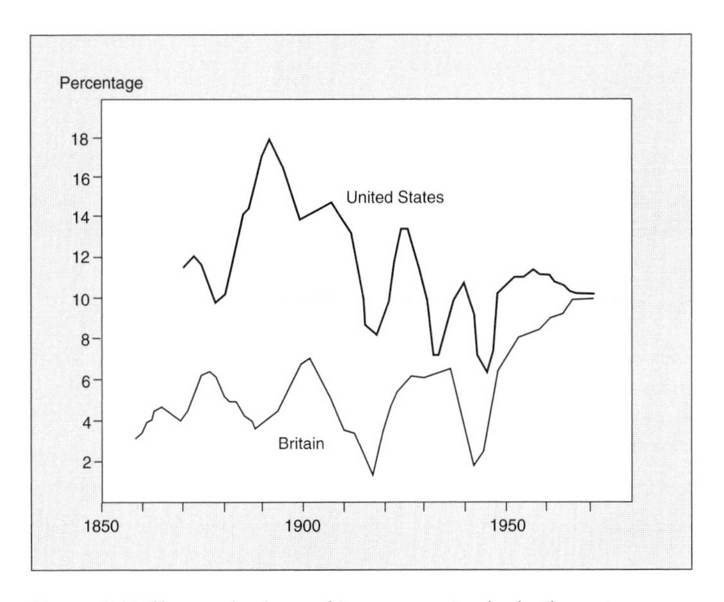

Figure 6 Different rhythms of investment in the built environment in Britain and the United States (percentage of GNP (USA) and GDP (Britain) going to investment in the built environment – 5-year moving averages)

Source: D. Harvey (1978) 'The urban process under capitalism: a framework for analysis', in M. Dear and A.J. Scott (eds) (1981) *Urbanization and Urban Planning in Capitalist Society*, London: Methuen, pp. 107 and 110.

g
g
g

•GRAVEYARDS•

by Alastair Bonnett

The German prince Hermann von Puckler-Muskau designed a vast park within which to set first his wife's and then his own tomb. Each mausoleum was fashioned in the shape of a pyramid, and constructed side-by-side on an island at the centre of the park's principal lake. On each, the words 'Graves are the peaks of a far new world' were chiselled.

Death is traditionally a site of fantasy, of strange and pleasing gestures that both affirm and ironise the conceits of the living. One of the most troubling aspects of the modern city is that this complexity has been expunged. As urban populations have grown, and corpses have come to fill the city's earth or be blown as dust along its pavements, the subject of death has become unacceptable. Any evidence that the metropolis is built on a pile of bones is swept away and removed from sight. Moreover, the globalisation of this attitude – what we may call the globalisation of 'western' death – is making inroads into diverse urban cultures across the world. The principle of economic rationality, of corpses as an inconvenience, is worming its way through traditions and practices that have long taken a very different attitude to burial. Graveyards are banished to ever more distant suburbs and turned into places where the human imagination is actively discouraged from operating:

> Walking through a cemetery or churchyard today is like walking through the saddest Sixties development. . . . Great tracts of land have become grim, modernistic cities in miniature. With rules and regulations as to the size, height and material of the memorials – resulting in an unremitting uniformity. (Lambton, cited by Worpole, 1997, p. 25)

Many contemporary cities have no place, intellectually, emotionally or physically, for the dead. Indeed, British local authorities' leisure departments (oddly it is the people at 'leisure' who look after the largest proportion of modern dead Britons), having decided that graveyards are a financial burden, actively encourage cremation (Jupp, 1993). Thus 70% of Britons are now burnt, leaving no useless trace, no pointless bump in the ground to intrude upon the ever-busy, ever-productive activities of the living.

For the Victorians the urban cemetery was a place of contemplation, a place to wander through and picnic in. When Cardiff's Cathays Cemetery opened in 1859 the *Cardiff Times* opined that it 'would form the principal walk of the inhabitants of Cardiff'. In *On the Laying Out, Planting, and Management of Cemeteries*, John Claudius Loudon's influential text published in 1843, a new, rationally planned approach to graveyards was formulated. Yet Loudon still envisaged cemeteries as municipal gardens, places of pleasure and moral instruction that were fully integrated into city life. Such sentiments appear to be incompatible with today's cities. The Victorian cemeteries remain, but there is no available ideology to integrate them comfortably into the rhythms of the metropolis.

Consider the following extract from the *Friends of Nunhead Cemetery Newsletter*. A teacher is explaining why she has taken her class to explore this English inner city cemetery. 'Nunhead has much to offer schools,' he notes, namely:

> social history, monument design and symbolism, sketching, photography, creative writing, woodland ecology, tree growth and measurement, fruits and seeds, birds, insects, art from leaves, geology of monuments, architecture, Victoriana, the two World Wars. (Cited by Walter, 1994, p. 177)

If these children were taken to an art gallery and asked only to consider the fire hydrants and security attendants' caps, few would not consider it bizarre. Yet here is a trip to a graveyard in which the subject of death, of mortality, of bodies and what happens to them, is completely overlooked. Only in the modern city could such behaviour appear normal. Only in a city where the corpse is banished, considered more disgusting than sewage, more upsetting than violence, could a refusal to acknowledge the dead be taught and learned as an ordinary, acceptable part of children's education.

The teacher explains, 'Nunhead is atmospheric, and children react strongly to death, burial and mortal remains. . . . So my briefings to the class emphasised the *living* aspects of the site, rather than the cemetery as such.' The modern imagination has learned to be unnerved by physical signs of mortality. They are 'creepy', 'eerie', the stuff of 'horror'. People prefer their houses not to look over graveyards. People prefer to ignore cemeteries. Or destroy them. One of the few signs of an active contemporary engagement with older graveyards comes in the form of vandalism. Headstones are knocked over. Monuments are graffitied. Graveyards seem to have the power to unleash a particular anger, a particular contempt, among vandalistic youth. It is an extreme response. But it is entirely in keeping with current attitudes towards the dead, attitudes that go to the heart of what it is to be an urbanite at the turn of the twentieth century.

Dying is not a modern thing to do. The compartmentalising logic of modernity finds death an anachronism, a medieval relic. Today, people 'succumb' to particular 'killer diseases'. They 'struggle' with cancer, they 'battle' with heart complaints, there are 'winners' and 'losers'. Walter observes, 'The human being is no longer shadowed by a single skeleton personifying Death, but by any number of germs and diseases which attack medically identifiable organs of the body' (1994, p. 12). Hence the corpse has come to represent an inability to 'fight off' disease. It represents failure. Graveyards offend and refute this absurd logic. They refuse and mock the city's pretensions to rationality and control. Remorselessly they remind visitors that each generation dies, that bodies must be piled on bodies, that the most modern and healthy of folk rot down just as well as the rest of us.

References

Jupp, P. (1993) *The Development of Cremation in England 1820–1990: A Sociological Analysis*, unpublished Ph.D. thesis, London School of Economics.

Walter, T. (1994) *The Revival of Death*, London: Routledge.

Worpole, K. (1997) *The Cemetery in the City*, Stroud: Comedia.

• GROUND •

by Iain Chambers

The ground upon which buildings, roads and cities are constructed is invariably conceived as empty, awaiting cultural inscription. This invokes a violent amnesia: for what ground has not already and always been culturally inscribed by someone, by some body? Nevertheless, and to continue, the ground itself, unless particularly resistant or unsuitable, is invariably ignored in the subsequent appropriation and exploitation of the terrain. Yet the land, although seemingly passive and dumb, replies. In climatic and telluric terms the abstract site witnesses the persistent, sometimes dramatic and irruptive, insistence of its terrestrial coordinates – from inclement weather (storms, blizzards, floods and drought) to geological instability (earthquakes and volcanic activity). But a deliberate ignorance of the ground is exposed not only in turning away from the materiality of the site to becoming engrossed in the imposed prospects of the plan; it also lies in the direct subordination of earthly forces to societal desire and confirmation.

Here the ground, whether flattened and cropped by earth-moving equipment in contemporary southern California, or else hewn, hacked and extracted in the form of building stone directly from the bowels of the city (in order to evade attempts at urban planning and control that forbade the importation of building materials), as in sixteenth-century Naples, is subjected to the same unilateral and exploitative grasp. To exploit, to extract from the land in this manner, seeking an asymmetrical benefit, is to impose an architecture whose essential violence, no matter how ancient and weathered it has subsequently become, is an enduring testimony not only to political and economic power, but also to the profounder dominance of plan and projection embodied in the imperious reach of the architectural gaze. In this field of vision (and realisation), the twelfth-century Norman castle of Castel dell'Ovo in Naples and the contemporary business tower at London's Canary Wharf (and its equivalents in New York, Shanghai, Singapore and elsewhere) are far more proximate than is usually presumed.

Floods, storms, landslides, earthquakes and volcanic activity are only the most immediate signals of the constant mutability of the ground; the perpetual interrogation of its assumed constancy. Before this uncertainty there opens up a forking of the path between a life lived in the sought consistencies of buildings, towns, cities and their infrastructures, and one experienced as if in transit: the choice between inhabiting a monument, a permanent site, and experiencing a shelter. To dwell in a place on a ground (in German, *Grund* also signifies 'reason') that is secured in the relative stability of property laws, local governance and architectural design also evokes the idea of to tarry, to delay, to desist from action, to stop, and eventually arrives at the signification of to stun and to stupefy. To dwell in this manner is to move into a place, not over it. It is to occupy rather than pass through. It is to transform the terrain and its sedimented forms and forces into the mere mirror of the self. It is to negate alterity. Cities have been historically elaborated upon these foundations. The ontological and political import of such a manner of dwelling has been extensively and elegantly rehearsed in recent years in both Bruce Chatwin's *Songlines* (1987) and Paul Carter's *The Lie of the Land* (1996). It is perhaps no accident that these particular reflections matured in the *Terra Nullis* that European settlers subsequently constructed as Australia.

To move over, rather than move in, implies a lighter yet more extensive involvement with the ground that supports our bodies, our lives. It is a relationship that is constantly, even daily, renewed. Contrast this with a historicist understanding of dwelling that seeks in the continuity of ownership a continual confirmation of identity; an identity in whose name a murderous tax in blood, repression and death is regularly extorted.

Clearly it would be impossible for most of us to detach ourselves from the domestication of Earth that the accumulation and identification with property and the seeming permanency of urban life has bequeathed. Without choice I, and those who read these lines, inevitably find ourselves thrown into these metropolitan, increasingly global, calculations. Still, rendering the history of building, dwelling, thinking a little less secure can serve to register, and interrogate, the limits of certain ways of dwelling. This would be to consign ourselves to a more vulnerable, hence more responsible, state; that is, to the constraints of a metropolitan becoming that reveals in the locality of our lives a responsibility for our terrestrial frame.

g
g
g

•GYPSY SITES•

by David Sibley

Gypsy sites in British cities are rarely easy to find. One approach to a typical site in a northern city is by a footpath through a semi-derelict industrial district. The path takes you across the weed-strewn remains of an old cement factory and past mounds of gravel to a busy road fronted by a cocoa mill and a chemical works. Across the road, behind a high wall, is the home of twenty-five Gypsy families, the site itself hemmed in by another chemical plant, an incinerator and a refuse tip. The cocktail of smells includes the sickly sweet smell of cocoa and the smell of decay from the refuse.

The locale conveys in a tangible way the meaning of 'residual space' and 'space of exclusion': a group of people who were considered to be a polluting presence elsewhere in the city is consigned to a space marked by its serious air pollution and the pollution of material residues.

At the entrance to the site, down a side street cleared of its terrace houses since the area was declared to be unsuitable for residential use, is a house trailer which provides a base for the *gaje* (non-Gypsy) warden. It is positioned so that its large rear window looks out over 'Gypsy space'. But this Gypsy space is not a space shaped by Gypsies. It is a sterile, concrete grid filled in with standings for trailers, identical breeze-block buildings with toilets, washrooms and some storage space, and a patch of ground next to each trailer, sometimes grassed and one or two with the edges softened by flowers and shrubs. This is a disciplining grid on which families have had to modify their traditional interactions and social relationships to avoid appearing deviant or disruptive, to be marked as trouble. 'Johnny', a great grandfather in his seventies, once put a shed on his patch of grass. This broke the rules, so he was evicted.

Surveillance by the warden and a book of rules tend to keep the adults indoors. Breaking up cars, sorting scrap and lighting fires, activities that provided a focus for collective experience, are strictly prohibited, so most adults seem to have concluded that it is better to live on the social, watch television or drive off the site to visit friends, or go to a pub where their Gypsyness is not a problem. On the site the inside of the trailer is all that is left of Gypsy space, a space where Gypsies decide what is clean and what is polluting. Only the children move freely from one space on the grid to another, oblivious to its disciplinary purpose. There is one concession to Gypsy culture. When someone dies, the warden retreats. It is an occasion that demonstrates the power of the nomad, when large numbers of relations congregate to celebrate a life, when a men's fire and a women's fire burn all night on the site and the space is crowded with cars and vans. It is a carnivalesque inversion, confirmed by the common provision of a police escort for the long procession of vehicles following the hearse. Gypsy death as spectacle.

In 1994, the British government determined that Gypsies and other travellers should no longer be given special treatment. According to the perverse logic of the New Right, sites provided by the state under an Act of 1968 channelled resources to Gypsies which the rest of the population were denied. Let us forget centuries of oppression, the galleys, transportation and the Holocaust. In future, Gypsies could buy land and apply for planning permission for a residential site, like any other members of society. So, the sites secluded in heavy industrial zones, wedged between railway viaducts or appended to sewage works on the urban fringe, and regulated by rules which penalised transgressions of their Cartesian designs, were really spaces of privilege. These Gypsy sites are now to become a part of the archaeology of the welfare state.

Gypsy sites are but one small manifestation of the racialisation of space. In the popular imagination, race and space in Europe and the United States are more usually conjoined in large-scale enclosures and exclusions – notably, inner cities in Britain and the United States or the *banlieues du cauchemar*, nightmare suburbs of large French cities, as *Le Figaro* put it. These spaces express the accommodation of core states to major diasporas and they resonate with histories of slavery and colonialism. Importantly, they register in mainstream politics. They are troubling reminders of the failure of pluralist ideals. The systematic exclusion of Gypsies from residential areas and their relegation to scattered pockets of left-over land in British cities, by contrast, has gone largely unnoticed. Once the eruptions of hostility preceding proposals to build sites have subsided, Gypsy spaces are lost from public view, although not from the disciplining eyes of the control agencies, from the police and officers of the Department of Social Security. Gypsy sites serve to remind us of the privileging of a sedentary life and the imputed deviancy of nomadism, but for most of the population these sites mean little. For them, the conflict between nomad and house dweller has been won and the Gypsy settlement behind high walls signifies no more than a refuse tip, something arguably necessary but undesirable and best located elsewhere, in a place where it does not disturb the comfortable order of suburban life.

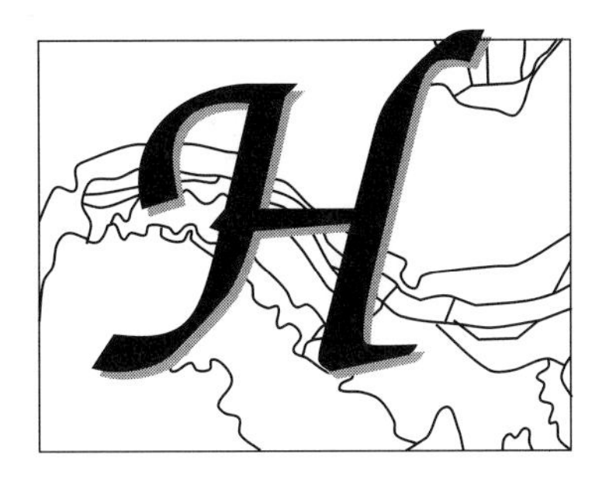

•HABITATION•

by Ross King

> That it is a sad trade to be a lovely woman,
> That it is the banal labor
> Of a dancer who entertains, manic, cold,
> and fainting
> With her mechanical smile.
>
> (Baudelaire, 1962)

The street – even more the grand boulevard – the urban masses passing by, new leisure, new money and the idle stroller (the flâneur), the voyeur and his leer, the dandy and the embourgeoisement of fashion, the streetwalker plying her trade (the 'banal labor[er]'), constitute the city of the new industrial world against which Baudelaire wrote his poetry. In the modern metropolis, the oldest profession also moves from the brothel to the boulevard – democratised as urban display, presenting as mass article in a wonderfully expanding world of mass consumption. So every woman that walks (in) the street will be imagined as the (at least imminent) whore.

The street, and even more the grand boulevard, promises freedom: anonymity, drifting, the cornucopia of displayed commodities, limitless contingency, boundless choice. For women – and for many men – the promise is a chimera. The carceral street masquerades in the gorgeous guise of liberation.

There was a small-scale parallel to Baudelaire's Paris in the Sydney of my childhood and early youth, now only dimly recalled. The Rowe Street Coffee Shop, off a long-gone lane in the city's centre, was venue of the 1950s 'Sydney Push' – coffee, philosophy, nonconformism in confrontation with a stiflingly rigid, conformist society, and the iconoclasm of philosophy professor John Anderson as antipodean and anticipatory Foucauldian. For so many nice 'middle-class' girls it was indeed freedom and escape – to a scene of drugs, libertinage rather than liberty, deflowering in the cloak of unmasking, the inevitable abandonment, and abortions in an age before the pill.

From both the boulevard and the coffee shop let us protect good women. The home is reversal (inversion) of the streets of Paris (Sydney) – off the streets and into the suburbs, and women are safe at last from the leering eye and the temptation of ideas. Instead, the physical street itself becomes the watcher, the new panopticon – the nonconforming dress, the unexplained visit, the incontinent dog, might be observed by a dozen prying eyes from tastefully curtained windows, or it might be unobserved – there is never certainty. The fear of observation and of the censuring judgement makes the protected woman her own gaoler. Even more effective is the self-surveillance of the watching home – an anti-erotic realm of self-repression and hidden violence. But the suburb is still something more.

Watching, and the privileging of sight and the visual over the other senses, are widely argued to characterise the masculine (flâneur, voyeur). The turn to all the senses, and 'real' immersion, might mark the feminine – an experience of being, based not 'on the gaze which objectifies but on the touch which unites' (Suleiman, quoted by Garb, 1990). The discourse of architecture has been one focused on the visual – on images and their critique, and thereby (so it seems) on the male experience of space. The purposive, architectural representation of the female experience of space – a feminist architecture and urban design and their critique in discourse – is problematic. But a feminist spatial representation of the female experience of space does exist, and probably ubiquitously. Throughout the present age, women's making and unmaking of space has been the stuff of jokes and fun – endlessly rearranging the furniture, selecting or else imagining new curtains or carpets, fiddling with flowers, crocheting doilies, or doing other pretty things. What men do has been far more serious. Men's typical aesthetic production (in architecture, urban design, painting, sculpture, music composition) has been embraced in 'serious', reflective, critical, 'theoretical' discourse; and the discourse has ensured that the voices will remain male. The same is true of a few areas shared by men *and* women (the novel, some musical performance). But not those more typically female. And we get the false, ideological distinction between 'art' and 'craft'; more catastrophically, we get the denigration of the real genius of women's purposive spatial production – the achievement of habitability, and that remarkable variety and freshness that flows from subtle change, adaptability, textures, new smells and surfaces. The making of ground surfaces, walls and ceilings is called architecture and is critically explored; the occupation and personalisation of those surfaces is not. Yet both are representation; the latter, moreover, is representation of personal emancipatory visions – carceral space always carries, though imminently, emancipatory space. I have explored these issues in King (1996).

A feminine spatial design will be very different from the male. It will break things down; make them accessible; break the façades with flowers and scented fruits; reduce the scale; reinstate the tactile, the sounds of water and birds, the place of children's play (Nietzsche's 'play that calls new worlds to life'), the impermanent and the appropriable – Bataille's 'architecture against architecture'. Its venue, overwhelmingly, will be the suburb and the home – as the carceral is imminently the emancipatory, so the anti-erotic is imminently the sensual.

References

Baudelaire, C. (1962) 'Confession', in C. Baudelaire, *The Flowers of Evil* [*Les Fleurs du Mal*], revised edn edited by M. Mathews and J. Mathews, New York: New Directions.

Garb, Y. J. (1990) 'Perspective or escape? Ecofeminist musings on contemporary earth imagery', in I. Diamond and G. F. Orenstein (eds), *Reweaving the World: The Emergence of Ecofeminism*, San Francisco, CA: Sierra Club Books, pp. 264–278.

King, R. J. (1996) *Emancipating Space: Geography, Architecture and Urban Design*, New York: Guilford.

Suleiman, S. R. (ed.) (1986) *The Female Body in Western Culture: Contemporary Perspectives*, Cambridge, MA: Harvard University Press.

•Hell•

by Jane Rendell

One of the most popular forms of social activity for men in the early nineteenth century was gambling. This took place both in exclusive clubs and low gaming houses. In London, both were located in the same part of the city – St James'. Although gambling between members of the same class was not seen as a problem, social hierarchies were transgressed when those at the top of the hierarchy lost money to those below them. Gambling houses were considered a social problem connected with immorality, commonly referred to as 'hells', places of sin and evil, where men made pacts with the devil. They were also represented as 'temple[s] of ruin, indolence, and guile' (Badcock, 1823, p. 291), and the gambling room itself was described as the 'Sanctum Sanctorum', a sacred and special place. Gambling was represented as removed from everyday life, located at the transgressive extremities of good and evil, heaven and hell.

The frequenters of hells were known as 'greeks' and 'pigeons' (see G. Cruikshank's illustrations to verses called *The Greeks and the Pigeons* 1817). These terms were defined by profit and loss; greeks made money while pigeons lost money. Although the term 'greek' often represented desirable qualities of political democracy in the eighteenth and early nineteenth centuries, in this instance the reference is specifically to paganism and scenes of debauchery and cunning. Pigeons on the other hand, suggest naiveté, a homing instinct and lack of autonomy, where unhappy gamblers have no choice but to return to hell, in the hope of recouping the debts they have incurred. The representation of gambling as heaven and hell, sanctuary and trap, represented social fears concerning both the heavenly and enticing promise of instant profit through gambling and the subsequent hellish entrapment of debt repayment.

The gambling system was often compared to a spider's web; with clear differentiations between those who benefited, gaming reptiles, sporting spiders, and those who lost, bleeding victims. The representation of the gambling hell as an entrapping and web-like place was intensified by upper-class fears that the profits made by working-class proprietors were at the expense of members of their own class. A number of texts told stories of young aristocrats losing their inheritances to duplicitous gambling hell entrepreneurs. One such man was Crockford, originally a fishmonger, who turned to betting and made a huge profit out of the gambling habits of the upper classes at his club on St James' Street. Crockford was caricatured by Robert Cruikshank in 1824 as a shark, located at the centre of his web, trying to 'attract, entrap, and ruin the unwary'. Aspects of entrapment were associated with entry to the building.

> These dens have the appearance of private dwellings, with the exception, that the hall door of each is left ajar, during hours of play, like those of trap-cadgers, to catch the passing pigeons (Weare, 1824, p. 354)

Separated from the street, the gambling clubs were treacherous territories, difficult to enter and to leave. The gambling room was usually located at the end of a series of locked doors, where those guarding the doors could survey the visitor through circular peep-holes. The journey to the gambling room was important in providing a dislocating effect from the world outside, in order to create the atmosphere of suspense and excitement conducive to gambling. On entry to the hell, the gambler had to surrender control, to give himself up to the unexpected, to the devil.

> There seemed to be great difficulty in getting in; and we had to pass through several doors strongly barricaded before we came to the gambling room, which was in the front room upstairs. (Weare, 1824, pp. 409–410)

In clubs, gambling rooms were usually placed on the first floor along the front elevation. This location, on the principal floor at the front of the building, gave these rooms a special status within the spatial hierarchy of the club. Given its illegal status, gambling required a location in a hidden and protected place. Temporally as well as spatially, separation from everyday life was essential to sever connections with diurnal cycles that could potentially shorten long gambling sessions. This was achieved by permanently drawing the curtains of the gambling room and lighting the rooms artificially with chandeliers and lamps. The careful arrangement of this lighting was indicative of the desire to seduce: 'a profusion of chandeliers and candelabras were tastefully arranged to light the victim to the alter of seduction' (Weare, 1824, p. 354).

To place such an internalised and highly private activity at the front of the building raised above the street in an exposed position was paradoxical. Closed curtains during the day would have indicated subversive activity, and one can only suppose that this was intentional. Through power, prestige and wealth, the members of the most exclusive clubs were immune from legal interference, and it was this privileged status that was displayed via closed curtains to the street.

The surface of the building was associated with deceit, both externally and internally. In the neighbourhood of St James', the physical proximity of the respectable male club and the disreputable gambling hell meant that they might be mistaken for one another. Here the fear that working-class gambling hell proprietors might sap the upper classes of money resulted in the representation of the hell as a place of false pretence. Low gambling houses were considered to 'masquerade' as respectable male clubs, in order to create an aura of respectability. By 'imitating' social forms of club organisation in terms of committees, membership, balloting, entrance fees, annual subscriptions, and its listing in the *Court Guide*, Crockford's was considered to be 'disguised' as a club:

> in order to keep out those whom they have already plundered out of their last shilling, their houses assume every specious appearance. Thus the great 'hell' in St. James's Street is called 'Crockford's Club!', the 'hell' in Park Place is called the 'Melton Mowbray Club!' and the 'hell' in Waterloo Place, the 'Fox Hunting Club!!!' (Deale, 1828, vol. 1, p. 65)

In clubs, the gambling took place in the saloon or subscription room, part of an élite suite of first-floor rooms, including an antechamber, eating room and smoking room. These rooms were the most ostentatious in the club, often the only ones to be architect designed. At Crockford's, these apartments, designed by Benjamin and Phillip Wyatt in 1828, consisted of a dining-room, drawing-room and gambling room, furnished with a writing table for loans and demands for winnings, chairs, dice bowls, small hand rakes, and the hazard table, oblong with a green cloth. These rooms were lavish and highly ornate:

> The drawing rooms, or real hell, consisted of four chambers: the first an anteroom, opening to a saloon embellished to a degree which baffles description; thence to a small curiously-formed cabinet or boudoir, which opens to the supper room. All these rooms are panelled in the most gorgeous manner; spaces are left to be filled up with murals and silks or gold ornaments; while the ceilings are as superb as the walls. (Waddy, 1919, p. 121)

The extravagance of the interior decoration in Crockford's was considered falsely seductive. The use of glass in mirrors and chandeliers was intended to create an air of illusion and a dazzling atmosphere which would produce a state of fantasy and delusion in the gambler, encouraging the play of unrealistically large sums of money. The lavish display of exotic foods and the free wine enticed and intoxicated, providing the stimuli to promote excessive behaviour in gambling.

> The beautiful chandeliers, large pier glasses, in superb gilt frames, with curious designs, and the handsome side boards, loaded, with the costly glass and plate . . . his sight was quite dazzled by looking glasses, chandeliers, platecut glasses and decanters, all glittering with the glare of light emitted from an abundance of wax candles . . . some choice paintings, rich curtains, rare fruits, every delicacy in abundance . . . wine sparkling in the decanters – the whole formed a coup d'oeil of the most fascinating, dazzling and intoxicating appearance. (Deale, 1828, vol. 2, p. 253)

Concerns with the immorality of gambling were voiced by both the upper classes, who feared losing money to working-class entrepreneurs, and the rising middle classes, who saw gambling as a lazy and extravagant upper-class vice, where the easy profits which could be accrued without labour contradicted the work ethic of capitalism. For the middle classes gambling opposed values of stability, property, domesticity, family life and religion. As upper-class men lost their inheritances and aspiring middle-class men accumulated fortunes, cultural representations of gambling shifted. For the upper classes, the wealthy male gambler whose virile masculinity represented aristocratic power became increasingly associated with the feminine, with the immorality of female gamblers, and with the weak and deceitful qualities of effeminate men, such as dandies. It was the duplicitous appearance of the gambling hell which was held responsible for the social problems posed by the extravagant and deceitful nature of gambling. It was the element of 'masquerade' or 'disguise' in the confusion of low gaming hells and respectable clubs which was held responsible for encouraging the entrapment of gamblers. It was the allure of the interior through decorative embellishment and the illusions of glass and light which allowed the intoxication of the gambler. The hell was a duplicitous surface of seduction.

References

Badcock, J. (1823) *Slang. A dictionary of the Turf, the ring, the chase, the pit, the bon-ton, and the varieties of life, forming the completest and most authentic lexicon balatronicum . . . of the sporting world*, London: T. Hughes.

Deale, I. (1828) *Crockfords or Life in the West*, London: Saunders and Otley.

Waddy, H. T. (1919) *The Devonshire Club and Crockfords*, London: Eveleigh Nash.

Weare, W. M. (1824) *The Fatal Effects of Gambling*, London: T. Kelly.

•HEROIC MONUMENTS•

by Allan Pred

AT A POINT SITUATED IN THE VERY CENTER OF THE CITY,
at a site located at the southern end of the Royal Gardens
(*Kungsträdgården*),
at a place that long served as a venue for Stockholm's upper bourgeoisie

to mark their social location, to parade and strut in the most recent and fashionable of finery so as to publicly define their difference,

at a location within a park whose daytime spaces are (weather permitting) now alive with activities of a more clearly popular nature: small children scampering up and down various climbing devices; the seriously faced as they slowly pace an over-sized chessboard simultaneously playing to casually clad passers-by and playing each other – pondering the next repositioning of their meter-high pieces; the hundreds seated before a raised stage, absorbing the performance of a rock group or some other free entertainment; the coffee sippers and salad eaters conversing at outdoor serving tables;

at a dwarfing statue, a heroic monument which, in the popular geographical imagination of much of the city's youth, serves as a boundary indicator heavily laden with symbolism, as an eveningtime geographical marker between that territory – to its south – which is the stomping ground of skinheads and extremist nationalist groups and that territory – to its north – which is the province of young migrant gangs and anti-racist groups,

at a towering statue which is ideologically and historically supercharged because it is a representation of King Karl XII, a representation of that absolute monarch who reigned over Sweden between 1697 and 1718, a representation of that warrior ruler whose efforts to defend the country's role as a 'Great Power' led at first to martial successes against Denmark, Saxony-Poland and Russia but eventually to both costly defeat en route to Moscow and battlefield death in Norway,

at an imposing statue of a king whose death – by enemy or Swedish bullets? – is as controversial as his historical record,

at a regal statue whose material form was originally designed to promote national identity, to co(m)-memorate, to (re)construct collective historical memory and patriotically enthuse, through having its eyes gaze at the Royal Palace a few hundred meters away across the water, through having its right hand firmly grasp a sword handle, through having its fully extended right arm point with

'He was a great king. He was Great-Power Sweden. He was the Will of the People. He made the decisions, there was law and order in the country. . . . In any case, I would like to have a somewhat similar society. Where there's one person who decides and there's law and order.' (Anonymous celebrant/demonstrator (this and subsequent anonymous quotes from Anna Lundström, "Vi äger gatorna i kväll!' Om hyllandet av Karl XII i Stockholm den 30 November 1991', in Barbro Klein (ed.), *Gatan är vår! Ritualer på offentliga platser*, Stockholm: Carlsson Bokförlag, 1995, pp. 134–161)

Newspaper headline, December 1, 1991: BATTLE OVER THE WARRIOR KING

unflinching determination
in the direction of Russia,
AMIDST THE ROUTINE PRACTICES AND OCCURRENCES OF EVERYDAY AND EVERYNIGHT LIFE,
amidst the chill and damp of the dark fall evening,
amidst what otherwise would be relative quiet,
amidst a leafless-tree dreary tranquillity normally only broken intermittently – either by the rumble of buses and other vehicles trafficking the nearby park perimeter, or by the raised voices of a 'yuppie' clientele entering and leaving the popular *Café Opera* nightclub, but 30 meters away,
there occurs, every November 30th, on the anniversary of Karl XII's death, a dramatic event throughout which

Demonstration chant: 'Sweden for the Swedes, Sweden for the Swedes.'

VIOLENCE MAY ERUPT (UN)EXPECTEDLY,
a particular conjuncture of not so ordinary circumstances in which there is the ever present risk that metaphoric bad blood between opposing groups may sometimes metamorphose into spilled blood, red and real,

'Both of the combating sides regard themselves as representatives of goodness struggling against evil.' (Anna Lundström)

a symbolic struggle over past historical meaning, current cultural categorization and collective identity that may quickly degenerate into a concrete, real-blows-exchanged struggle,
as it did in 1991,

'Of course, one stands for his country. I am absolutely 100 percent prepared to die for my country. And preferably in battle.' (22 year-old anonymous celebrant/ demonstrator)

when about 700 ultra-nationalistic Swedish Democrats, skinheads and others who would celebrate Karl XII gathered less than one kilometer away in order to march toward the statue and place a wreath at its base,
when anti-racist counterdemonstrators assembled in the Royal Gardens with most intent upon preventing any wreath-laying and some intent upon placing a hammer-and-sickle banner instead,
when police efforts to detour the march resulted in some scuffling,

'For those paying homage to Karl XII the counterdemonstrators are non-Swedes and traitors. Their actions are seen as the equivalent of occupation by a foreign power. . . . The immigrants and anti-racists are regarded as trespassers upon the historical and geographical domains of the patriotic.' (Anna Lundström)

when, despite riot fencing, a few made it through to the statue area only to be bruised and kicked,
when, as a consequence of the entire

'We want a Sweden that is culturally exactly like it was before. . . . That's

episode, the mass media were able, yet again, to focus inordinately on the activities of a relatively small number of 'Nazis', 'fascists' and 'skinheads', to label them as 'racists' and, thereby yet again, to contribute to the widespread practice of discursive and social/ economic violence, to do so by enabling people to ignore or deny the burgeoning of racist discrimination and segregation in everyday life throughout the country,

when, long before

November 30th, 1992

when,

what we're fighting for. Its become awfully kind of multicultural in this country, that's the thing we're fighting – Coca Cola, Levis and all that other stuff there is now. We don't want a kind of USA here, a big mudhole where you toss all kinds of trash. Then you get increased criminality and violence – nobody is safe here in the city. . . . We want a Sweden that's fine and pure, not one of those concrete ghettos.' (Anonymous celebrant/demonstrator)

AT ANOTHER TIME AND PLACE IN THE CITY

•HOARDINGS•

by Iain Borden

h
h
h

Hoardings are those impermanent constructions, those two-dimensional panels which line the streets and spaces of the city. Not really buildings, nor really here, they are simply a temporary covering, a mask across the face of the city at its most leprous. They hide a multitude of sins.

any deviance from its utopian image, an erasure of space in depth, time as flux, and social being as difference. Nor do such hoardings have to be physical – Barcelona erected hoardings just as effectively when it scrubbed its pavements and displaced the homeless for its hosting of the Olympics.

Hoarding 1

Hoardings have, of course, been around since the birth of the western city, used to line the political and processional routes in Ancient Rome, Renaissance Florence and Revolutionary Paris, transforming the city through false buildings and impossibly impressive perspectives into a magnificent theatrical stage-set. By extension, such hoardings became the surface performance of ritual, the means by which the city could be an idealised version of itself, a denial of

Hoarding 2

Around wasteland and building sites hoardings are the blank walls of tired plywood, tattered posters and invisible architect boards. By extension, these hoardings are the exclusion from the 'real' city of dirt and disorder, keeping danger, noise, massive girders of steel and exposed builders' buttocks from the normalcy of the city streets. The occasional peep-hole is provided not so much to let you see what is different as to reassure you that soon all will be back to normal, all construction will be complete, and the city as finished object, unrealistically permanent and clean, will once again reappear. In the meantime, the hoarding is the fire-screen, the safety curtain, the modesty veil that allows the reproductive cycle of construction to pass unnoticed.

Hoarding 3

Space is nothing if not an opportunity, and the hoarding is nothing if not an available surface. So along urban streets, posted on the sides of buildings, across bridges and on all manner of fantastical metal structures the advertising hoarding multiplies. These hoardings layer buildings and urban space with another filter, another varnish, projecting images over brick, photography over smells, gloss over depth. Some of them even move every few seconds (lest we realise the inanity of their suggestions). They offer us cigarettes, cars, politicians, telephone services, clothes, air travel, chocolates, computers, alcohol, holidays, newspapers, movies, jewellery and, by implication, body hygiene, a winning personality, job fulfilment, satisfying sex and a truly democratic world in which to live. All human life is there, not down on the gritty, filthy street but up there, 4 metres above your head, 8 metres beyond the edge of the road, out of reach but clearly visible. All you have to do is get there. And you cannot.

Hoarding 4

Other hoardings are within reach, however, at least physically. As capitalism searches for ever more space to conquer, as the cosmos, ocean depths and Antarctica all start to fall to the outward reach of commodifying forces, so too do the micro-sites, those interstitial and downright *small* spaces still left vacant in the city. London has now become the most extreme site of these miniature proliferations, with micro-advertising appearing everywhere with a flat surface to spare: taxis and buses dematerialise into advertisements for the *Financial Times* and Levi jeans, tube tickets offer alternatively license to travel (obverse) or an invitation to the new Giorgio Armani store (reverse), while the Yuletide pavements of Covent Garden are strafed with the fire of neon projectors.

Most invasive of all, travellers leaving tube stations across the city are confronted with the commodified hamburger staircase, a set of steps on which the risers have been transposed into a wall of McDonald's advertisements, indicating both the presence and direction of the nearby fast-food restaurant. Unlike conventional advertisements these are unavoidable, as they must be looked at in order to judge and negotiate the steps safely. The micro-hoarding steps recognise this movement, and exploit its open dialogue with architecture. At this level they insert themselves directly into the line of sight, working through the eye/body co-ordination of the traveller as they move around the city. These micro-hoardings are also directly instructive, the less than subtle aura of the inevitable Golden Arches being accompanied by a wall of arrows telling travellers where to go as soon as they leave the station.

These are temporal as well as spatial insertions, for the exact moment of intrusion is precisely judged in time as well as space, invading the psychology of the traveller at the very moment of decision-making. Just as they leave the semi-somnambulant passivity of the tube station and escalator journey and make the transition to the decisive activity of the busy urban street, the Golden Arch steps recognise this move, selecting this space and this moment of awakening awareness into which to inject their message.

Hoarding 5

Lest we forget, we too can become hoardings. With Jean-Paul Gaultier spectacles, Ted Baker shirt and Kenzo boots we are moving advertisements, spectacular versions of ourselves on display to our fellow hoarding-citizens. The human body becomes a mechanism for symbolic exchange, part of Baudrillard's 'genetic code, an unchanging radiating disk of which we are no more than interpretive cells' (Baudrillard, 1988, p. 140). And if this is not (im)personal enough, we can now go beyond hairstyle and cosmetics as proclamations of our class and cultural status – fitness is now a social issue. To be slim and trim is to be not just attractive but thinking, moneyed, urbane. My body is not just my temple, it is an advertisement for myself.

Further reading

Baudrillard, J. (1988) 'Symbolic exchange and death', in M. Poster (ed.), *Jean Baudrillard: Selected Writings*, Cambridge: Polity Press.

Borden, I. (2000) 'Thick edge: architectural boundaries in the postmodern metropolis', in I. Borden and J. Rendell (eds), *InterSections: Architectural Histories and Critical Theories*.

Venturi, R., Scott-Brown, D. and Izenour, S. (1972) *Learning from Las Vegas*, Cambridge, MA: MIT Press.

•HOME (1)•

by Adrian Passmore

h
h
h

13, Acacia Avenue

The estate agent who sells the thing wears a poorly fitting suit. It doesn't fit because his frame is not right for it, or because the cloth is cheap, or because both he and it have changed shape since first joining. He puts this suit on each morning with the air of a condemned man, but at least there's a pension in it somewhere. He hates his shirts too; nobody knows this. He leaves to go to work. A professional man.

Plot 3, Somewhere Close

Quiet, cosy, comfortable, isolated, warm, haven – these are words for this house. It's the place for the nuclear family, for the rosy-cheeked child, and for the shared decision about what comes next. You can have fun inviting friends around and cooking them exotic dinners. You can enjoy stocking the cupboards for Christmas and having the family over. You can match the wallpaper and the curtains. You can put a fine car on the drive. Alternatively you can invite your neighbours for dinner, but slaughter them in the double garage and bury them under a new patio with integral conservatory – just a thought.

69, Empire Drive

The number of people experiencing violence in their own home is uncountable. Whilst an open fear can sometimes drive discourses on public space it is bad form to own up to your own domestic beatings – receiver or giver. They cross a dodgy line between aberration and culpability. Very few people exact violence on their guests. It seems that beating up cohabitants is far more acceptable. And since the home is the privileged terrain of the family, why not take it out on them, or on a lover? They say people often do: I heard next door at it once or twice, but didn't like to say anything in case those responsible opted for some neighbourly advice.

16, Ford Avenue

Instruments of domination: microwave, blender, whip, mower, chip fryer, electric blanket, unfurled belt, hi-fi, patio, alarm clock, children, pressure cooker, bills, duster, scalextric, flower border, lawn, telephone number, remote control.

2b, Mash Street

A woman with a withered hand does the shopping. The withering was a gift from her husband who shoved her good fingers into the blender for Christmas. It was something to do with chippolata sausages and a turkey dinner; unfortunately the discussion was underwritten by the brewing industry. She would punch him with it but she is scared it will disintegrate with the impact she anticipates as just. She spends her time tending her youngish son, who lost his leg when kicking a sheep carcass into the mincer at the abattoir where he used to work. She is shopping for his dinner; she buys chops without irony.

111, Someplace Avenue

Row after row of boring houses where nothing ever happens. A dull street, they say.

Flat 4, 2, Needle Street

This junkie home is the stripped down thing. It reveals a neat housekeeping logic. Modern through and through. There is no superfluous decoration and nothing mock, just a battered dralon sofa where they sit muted in pleasure. The kitchen has no food in it, like the inventory of the latest factory: just-in-time calories always keep them from starving. The house is cold, but their drugs are warming. A simple world in another dimension, not the scheming den which some might think. The shopping list is shorter than any of the neighbours'. Coincidentally, they are all environmentalists, one of them is a nurse, and two of them knit.

64, Bit Street

They love to watch the telly here. Their favourite is anything with pictures and sound. Going along with the clichés, the children in particular, are sponges to its relentless flow, although the parents are much longer term addicts. Meals are often eaten in a mute junkie row. If they ever argue about the television, tempers can really fray: this is strange considering that three more televisions are dotted around the house. Since supply is no problem, it seems that where you watch your programme is the important issue at number 64, with the front room being the number one broadcast slot. If they made a soap opera about these people, television would not be the ruination of the family but its last glue.

23, Love Lane

When I am a burglar in your house, I can't see what you see. All the habitual fondnesses are absent. I survey things and tally them with a freight logic that puts size against value. I rummage carelessly through your preciousness. I can snigger about this because I once sold you the house, but left estate agency as a profession because the suit no longer fitted. Since you are not likely to return for a while, I slip on your underwear for a bit. Good fit. I have always loved it in other people's homes.

37, Newman Street

That it has come to this is a distressing thing. He copes outwardly, but has a million words for the minor angers that stand him where he is. It ruins him to see his wife leaving to go to work; he hates the shops full of decisions and arithmetic; he curses the low height of the pushchair handles; he can't take washing up, putting the dishes away and then taking them out to dirty them again; daytime television makes him mournful and he cannot abide turning the central heating off during the day. He dwells on these things as he repeats his routines, and believes he comes up with a twist: he can't deal with the fact that he believes he was meant for better things.

The Old Mill

Although this is a dream home, with lawns more carefully manicured than the estate agent's nose hair, there is a chill in the house. A couple of accountants live here. The skins of their white faces are both grained with the compulsion of numbers, although hers is slightly more masked by ten minutes in front of a makeup mirror each morning. Over dinner their urges never get the better of their manners, but underneath the veneer he has thrown wine in her face and she has skewered him with a carving fork, which she rightly anticipates will rip him more than the knife will. The ice air is unthawable even with the best central heating and a fake log fire, so they often both drink steadily. One day it will erupt and the straight logic of accountancy will be overrun by a righteous violence. She will compulsively murder her man: no surreptitiousness, it will take place in front of the warm grate and the dinner guests. Her defence? It wasn't really personal or particular – he has had it coming for years.

The Boarded-up House

Empty houses are uncountable.

•HOME (2)•

by Jane Rendell

h
h
h

> In a love affair most seek an eternal homeland. Others, but very few, eternal voyaging. These latter are melancholy, for whom contact with mother earth is to be shunned. They seek the person who will keep far from them the homeland's sadness. To that person they remain faithful. (Benjamin, 1979, p. 75)

On a leafy street in Clapham, minutes from the common, is a terraced house which was my home for two years. Scattered all over London, all over the world, are other homes, houses where I have lived at one time. Some stand still, others are erased, physically in military *coups*, or from conscious memory, only to be revisited in dreams. In all the places I have lived I recognise myself, but this particular house represents something very special to me. Its neglected and decaying fabric, its disparate and drifting occupants, offered a way of living which had nothing to do with comfort, security, safety, permanence. This home, and the friend I shared it with, showed me what I can only call 'the rhetorics of architectural misuse' (Bourdieu, 1984).

Refusing rent

> The gift has no goal. No for. And no object. The gift – is given. Before any division into donor and recipient. Before any separate identities of giver and receiver. Even before that gift. (Irigaray, 1992, p. 73)

Squatting resists ownership, occupies without permission, without wishing, or being able, to pay rent. Squatters question issues of purchase, property, occupation, and challenge design intention. The woman who owned the house refused rent. Although her home was large, five stories, she lived frugally off her pension, in two first-floor rooms. She occupied a world beyond the everyday, inhabited by spirits – 'the powers that be'. The 'powers' were not adept in the material world, their decisions were unreasonable and random. Large pieces of furniture moved nightly; plumbing, electrics and general household maintenance followed erratic management systems. The 'powers' refused council money for repairs – this disturbed the karma of decay.

Secretive display

> At the baths, a very different kind of temperament tends towards dangerous daydreams: a twofold mythical feeling that is quite inexpressible comes to the surface. First, there is the sense of intimacy in the very centre of a very public place, a powerful contrast that remains effective for any one who has once experienced it; secondly, there is this taste for confusion which is a characteristic of the sense, and which leads them to divert every object from its accepted usage, to pervert it as the saying goes. (Aragon, 1994, p. 53)

Living space is mapped and defining according to ideologies of domesticity, where sleeping is divided from playing, from working, from cooking, from eating, from cleaning and so on. Every activity has its compartment. In my home the boundaries which usually control and contain were intentionally blurred and transgressed. The bath sat in the centre of the roof – bedworklivingspace. From the bath you could talk to the person lying next to you in bed, look up into the sky, down on to the stove, beyond to those eating, and further, through the window on to the street. From the bath you could see into the toilet – a place where we traditionally demand privacy from prying eyes, ears, noses. The doors to this tiny blue room were spliced open like a swing saloon. Barebottomed, in an intimate space, to flush, you placed your hand through a smooth circular hole out into the public void of the stairwell, where you grabbed a wooden spoon hanging from the ceiling on a rope.

Form follows . . .

> space is broad, teeming with possibilities, positions, intersections, passages, detours, U-turns, dead ends, one-way streets. Too many possibilities indeed. (Sontag, 1979, p. 13)

Doing architecture, we play by certain institutionalised codes – planning and building regulations regarding services and construction detailing; and principles regarding form and function, structure and decoration. Our house was resistant to such ideologies. Services challenged ideals of low maintenance and opted for a high degree of strenuous user involvement. No room for complacency, every moment of occupation was *écriture feminine* – a writing from, and on, the body. The ladder to the upper floor, far too short, had missing rungs, and in one place a piece of sharp cold iron. Vertical movement, especially at night, took place as a series of jolts and slipped footings. The soil pipe gushed diagonally through the stairwell and out of the rear wall of the house; a proud dado rail. The removal of structural members, the stripping back of partition walls, asserted the fabric of the building as a living component of the space. Cracking brickwork and rubble, revealed between the splintering stud partitions, formed a decorative skin. Metal rivets holding the decrepit ceiling plaster together shone at night like stars.

h
h
h

De-stabilising structures

> She may go anywhere and everywhere, gaining entrance wherever she chooses; she sails through walls as easily as through treetrunks or the piers of bridges. No material is an obstacle for her, neither stones, nor iron,

> nor wood, nor steel can impede her progress or hold back her step. For her, all matter has the fluidity of water. (Germain, 1993, p. 27)

To occupy the roof as a habitable space, a truss was removed, the central one of only three. To connect the roof space to the floor below, a huge hole was cut out of the ceiling. There were structural implications; the roof space had not been designed for occupation. Challenging the propriety of structure rejects the ordered comforts of domestic routine, comfort and laziness, questions the spatial relationships we take for granted, but also tips the balance of safety and danger. One morning I awoke to a horrible crash and a scream; a friend, unfamiliar with the intricacies of the household, had missed her step and fallen three metres to the kitchen floor below. Her head narrowly missed the cast-iron stove. Existing in the time of decay, architecture, here, was no longer solid and dependable, but transient, as fragile as human life.

Wandering objects

> In any case, what is delightful here is the dissimilarity itself between the object wished for and the object found. Thus trouvaille, whether it be artistic, scientific, philosophic, or as useless as anything, is enough to undo the beauty of everything beside it. In it alone we recognize the marvellous precipitate of desire. (Breton, 1987, pp. 14–15)

When we do architecture, we adhere to all kinds of codes. We buy and use spaces and objects for certain purposes, in the ways for which they were designed. The bricoleur is a home-maker who finds new uses for found objects and collages them in space. Collecting, scavenging, recycling – bizarre hybrids of junk-shop and designer pieces replace the logic of commodity consumption and design for use. A limited number of possessions provides a catalyst to achieve flexibility through transformation. Within one life a table was the crowded focus of a drunken evening, several café tables, frames for candle-lit icons, and a hot blaze on a cold night.

Wandering subjects

> If we have retained an element of dream in our memories, if we have gone beyond merely assembling exact recollections, bit by bit the house that was lost in the mists of time will reappear from out of the shadow. We do nothing to reorganize it; with intimacy it recovers its entity, in the mellowness and imprecision of the inner life. It is as though something fluid had collected our memories and we ourselves were dissolved in this fluid of the past. (Bachelard, 1969, p. 57)

Recognising the shifting relation between an object and potential utilities produces doubt and uncertainty, a heightened awareness of the ever-changing nature of static objects. Placing objects and subjects in unusual combinations positions us in uncharted territory. Lost, our cognitive mapping devices de-stabilised, we can imagine a new poetics of space and time, unlimited, no longer stagnant with the inscription of specific and expected responses. Such potentiality opposes the autocratic architect's pompous regimes of mono-functionality but also rejects the banality of highly flexible multi-purpose spaces designed for anything (but nothing) to happen in. The accidental and continual juxtaposition of apparently unconnected things produces a density of potential interpretation. There is no moment of completion, rather you are aware that dreams are lived and lives are dreamt.

> A border is an undefined margin between two things, sanity and insanity, for example. It is an edge. To be marginal is to be not fully defined. (Levy, 1993, p. 73)

> I was your house. And, when you leave, abandoning this dwelling place. I do not

> know what to do with these walls of mine. Have I ever had a body other than the one which you contracted according to your idea of it? Have I ever experienced a skin other than the one which you wanted me to dwell within? (Irigaray, 1992, p. 49)

References

Aragon, L. (1994) *Paris Peasant*, Boston, MA: Exact Change.
Bachelard, G. (1969) *The Poetics of Space*, Boston, MA: Beacon Press.
Benjamin, W. (1979) *One Way Street and Other Writings*, London: Verso.
Bourdieu, P. (1984) *Distinction: A Social Critique of the Judgement of Taste*, Cambridge, MA: Harvard University Press.
Breton, A. (1987) *Mad Love*, Lincoln, University of Nebraska Press.
Germain, S. (1993) *The Weeping Woman on the Streets of Prague*, Sawtree, Cambridgeshire: Dedalus.
Irigaray, L. (1992) *Elemental Passions*, London: The Athlone Press.
Levy, D. (1993) *Swallowing Geography*, London: Jonathan Cape.
Sontag, S. (1979) 'Introduction', in W. Benjamin, *One Way Street and Other Writings*, London: Verso, pp. 7–28.

•HOTELS•

by Ross King

Definition: Hotels (lobbies): simultaneously hyperspace and heterotopia, the non-places that link all places.

> This long lane behind us: it goes on for an eternity. And that long lane ahead of us – that is another eternity.
>
> They are in opposition to one another, these paths; they abut on one another: and it is here at this gateway that they come together. The name of the gateway is written above it: 'Moment'. (Nietzsche, 1961, p. 178)

We travel in undifferentiated hyperspace: the Boeing 747 – the international terminal and the unvarying displays of the same duty-free enticements – the airport to hotel transfer bus (limo, taxi) – the hotel lobby (perhaps some plastic pretences of local difference) – the hotel room. But the lobby is also the point of intensest anxiety, the excitement of the new, the sudden experience of *place* and *difference* (seemingly in spite of the miniatured parodies) – simultaneously gateway to the abyss of the unknown (Nietzsche), and the cornucopia of wonders – the heterotopia (Foucault). Who can ever forget those fabulous first moments of walking out into the babble, noise and gibberish of a new, hitherto unimagined place! (I once came down to the lobby at the old Hotel Baghdad – a little piece of England – to emerge from my colonial survival breakfast to a cool desert morning, and to hail a taxi from the turmoil. My driver asked where I wanted to go. 'Babylon!' We had to go via Kerbala, a place I had never heard of – the Battle of Kerbala, in the year 680, might have been one of the terrible, defining moments of humanity, but it was not a moment for Australian or English history. The road was jammed with a million pilgrims, brown-robed and waving shimmering silk banners of brilliant green – 'ignorant people', declared my taxi driver, a Sunni. At Babylon I climbed the Tower of Babel. Late in the day we drove across the desert again towards Kerbala, the gold domes of its shrines hovering from afar in the silver mirage, to dematerialise to a dream city floating in a steely sky, and suddenly the magic of the 1001 Nights became as real – or as unreal – as any other reality. The end of the day saw the end of the dreams, back through the lobby, to the hotel restaurant and an English dinner of roast beef and Yorkshire pudding – one of my favourites. Though a day in a million, there are a million others like it.)

The hotel lobby is the non-space (universally undifferentiated) that links all hotels, all cities (all, at least potentially, different), and thereby all places. The hotel lobby door, in turn, is the gateway to the modern world. In one direction extends the infinite nightmare of the hyperspace, everywhere on Earth the same, monotony as Hell (Benjamin). In the other direction is the path of the road to Kerbala: the babble (Babel) of unknown languages, banners and flags whose meaning is lost to us, histories to which we have no access, and the delusion that the stories we have half heard (the 1001 Nights) are suddenly comprehended.

Jakarta is one of my favourite cities. It is more a city of paths than of places; there is no 'city centre' (a western cultural construct, as is 'city' – and I start to wonder if, indeed, it is a city?). But it does have hotel lobbies, the new venue of trade in a globalising world, thrown haphazardly on to the landscape and giving seemingly random access to the pathways. You look out, uncomfortably chilled, to the fabulous turmoil of the most energised place (surely!) on Earth. But when you pass through the revolving doors there is also the sound of it, and the air itself grabs you, thick, clammy, soon drenching and clinging to you; and it carries that stench of the wet tropics, raw sewage, poorly tuned trucks burning an abundance of cheap gasoline, those aromatic clove cigarettes that 'they' all smoke, and the wonderful smells of the street stalls – oh, the street stalls! You also move out into a world where you are no longer able to pay people to be civilised and speak English. For language (Babel) is also resistance.

Nowhere is place and difference more sharply defined than in those street stalls – the roadside edge is 'living room', 'town centre', 'community centre' (the western constructs again just do not work!). The sun goes down early in the tropics, and under dim lights the roadsides come to life as food is cooked, debate begins, and smells, food and language are foreign to those in the hotel restaurants.

The crowds on the pathways of magic also observe us, the tourists, the packaged, the mobile phone connected, the camera bedecked, and we are all so obviously rich. If only they too would write essays about *their* experience of the revolving doors and of the hyperspace (the 747, the international terminal . . .) beyond. Sometimes the resentment is palpable, but it is more frightening when it is masked as politeness or, even worse, deference.

In my student days, before the hyperspace, I once stayed at the Raffles Hotel in Singapore – a run-down establishment, in a run-down colony bursting with energy for all that, and feeling and smelling not unlike the Jakarta of today. But the Raffles was an essential part of the place itself, symbol of foundation and conquest, and like a fading dream, and sunset; somehow it could be nowhere else. I revisited again, after nearly forty years. It is no longer run-down, but it is also no longer part of Singapore. It is a sanitised, gorgeous trace of a past in a city that would have only a present (conformist, regimented) and a future. And it does replicate deference. That gateway called moment is still there, but the pathways have been reversed, and the hyperspace is outside. There was indeed a time when hotels were part of the place itself: Raffles, certainly, but also Shepherds (colonialism by insinuation), The Algonquin (cultural exclusion), The Ritz (self-mockery of the ruling class) . . . and down to the humblest hostelry in the tiniest town. Most are obliterated; or, if they survive, the hyperspace has reversed the pathways for them too, as one enters for the delusion of going back. Hotel lobbies present the moments of supreme excitement and expectation (the road to Babylon, Babel and Kerbala, or to the hyperspace and its disillusionment masked as comfort); but they also focus the distortions – increasingly the reversals – of time and the political.

Reference

Nietzsche, F. (1961) *Thus Spoke Zarathustra*, Harmondsworth: Penguin.

h
h
h

•INNER CITY•

by Fran Tonkiss

A sign on a disused commercial building in a triangle of back streets marks the beginning of a transformation: ANOTHER PRESTIGIOUS LUXURY INNER CITY LIFESTYLE DEVELOPMENT. After thirty years the front line of gentrification has reached this pocket of railway arches in inner south-east London. There is no longer anything remarkable about these small economic miracles. Similar transfigurations happen all over the city, in buildings whose industrial proportions conduce to an attractive arrangement of domestic light and space; whose unyielding contours lend themselves to twenty-four-hour porterage and the other maximum-security accoutrements of 'lifestyle developments' whose imaginative situations can bring with them forms of unwanted adjacency. If such packaging of wharves, warehouses, offices and factories represents a cut-rate variant on the bespoke conversions of the original gentrifiers, it now has its own imitators – in new-build developments which promise 'loft-style' living or affect the look of deconsecrated churches and shut-down hospitals.

Only the developer's sign is of interest. This is a new sense in which to understand the term 'inner city'. Such an expression has rarely been associated with any concept of lifestyle, much less one that might be considered prestigious or be set about with promises of luxury. If a language of urban pathology invented the 'inner city' as a problem, a different sort of language is now seeking to valorise it as an investment prospect. Urban sociologists once mapped the inner city as a 'zone of transition'; between industry and joblessness, between the urban middle classes and the urban poor. In this contemporary reversal, the inner city of the advertising hoarding marks another transition – one which takes hold at the level of meaning.

Imported – like so much else of British urban policy – from the United States, in the 1960s the 'inner city' referred to the exhausted zones of North American cities. The flight of capital and attendant 'white flight' out of inner urban areas left spaces which were characterised in terms of a tenantry of 'deprivation': the old and the very young, the unemployed, the poorly educated and badly housed, the criminal – and, more or less explicitly, the black. Marked out between boundaries conceived as both physical and psycho-social, such spaces posed special problems of rationalisation. Needy, impacted, impoverished, deprived, racialised – within urban policy programmes the inner city was imagined as a space which must be pacified, whose commerce must be regulated, and whose utility must be maximised.

Gentrification, it might be said, is only a new variation on this theme – the apache zones of the inner city tamed by what Sharon Zukin has called 'pacification by cappuccino' (Zukin, 1995, p. 236). These twin versions of the inner city – one an object of government, the other an object of speculative desire – say something about the shifting balance between state and market as 'engines' of urban renewal. They also say something about the way that material spaces can be re-imagined, empty space partly filled up with meaning. In spite of a recent convergence, the language of urban policy and the language of private development still talk up the inner city in rather different ways. The new jargon names a new space. To think about space as being produced, in this context, is to think about how imagined geographies ('another prestigious luxury inner city lifestyle development') are realised in concrete forms, how spatial artefacts are made with resources which are ideal as well as material.

This new development is not in that real and imaginary place which has been called the 'inner city'. That other place is invoked as a kind of perverse picturesque; near enough to add an edge to the lifestyle, but held back by the certainties of garage parking and high prices. In these kinds of transformation, not only buildings but urban perceptions are altered. Language gets put to work alongside the engineers, architects and developers, in ways which reinvent both physical spaces and cultural meanings.

There is a pathos to buildings like this one – the dejection of empty libraries, municipal baths, factories. It is something like a residue of meaning and of utility, left over as material traces, inscribed in cornerstones, spelled out as faded lettering. It is a quality which contrasts starkly with the peculiarly bland housing compounds which these buildings become. The remnants of redundant functions, dead labour – a crane attachment, roll of honour, an inscription about the value of book-learning, a gated entrance for Infants – may all be incorporated as grace notes in the jejune rhythm of the new structure. The historical sense of the lifestyle development is that of a fictive nostalgia and quotation which is sometimes called 'postmodern': a *mélange* of bastardised façade, mezzanine, retro-classical column and portico. Alongside these gestures, the closed-circuit cameras and the entry-phone system recall a different kind of inner city 'lifestyle'.

For over thirty years, social statistics on inner city populations in Britain showed them to be disproportionately inclined to poverty, joblessness, crime, victimisation, mental illness, substance abuse, premature death, and the eating of free school meals. Doubtless market researchers, pulp novelists or other futurists have identified the new social classes that will be drawn to the conjunction of 'inner city' with 'prestige' living – although such groupings do not always materialise in the images held out to them by cultural experts. Whatever, an older version of the 'inner city' recedes behind the developer's hoarding.

Reference

Zukin, S. (1995) 'Cultural strategies of economic development and the hegemony of vision', in A. Merrifield and E. Swyngedouw (eds), *The Urbanization of Injustice*, London: Lawrence & Wishart, pp. 223–243.

•INTERSECTIONS•

by Allan Pred

Sveavägen

AT THE CORE OF THE CITY,

at a point along *Sveavägen* (Mother Svea Road), along a street named after the mythic mother of all Swedes, along a street thereby personifying the entire country, along a street renamed in 1884 so as to inscribe patriotism upon the urban landscape (and thereby – hopefully – so as to help quell the burgeoning of class antagonisms and social tensions), along central Stockholm's principal north–south thoroughfare,

at a location but a few feet from the intersection with Tunnelgatan (Tunnel Street), where the simultaneous hellish and pleasurable qualities of the modern city are given concrete metaphor, where one is faced with the choice of moving forward and descending a stairway into the Hades of the subway system, or going to the left and ascending the heavenward-directed steps which flank *Brunkebergs Tunnel*,

at a site occupied by *Dekorima*, by yet another retail establishment emblematic of Stockholm's thorough enmeshment in the daily workings of globally interdependent capitalisms, by yet another retailing establishment catering to narrowly defined markets through offering the niche-tailored products of multinational corporations, by a retail establishment specializing in up-scale house paints and art and hobby supplies,

A decal on the Dekorima entrance consists of a masked man pointing a gun at the viewer and a text reading 'STOP! Protected against robbery by time-lock.'

Word-play jumping out from a Dekorima window-display ad: 'Next time its your turn (luck).'

at a spot now marked by a bronze plaque embedded in the sidewalk, by a plaque super-charged with uncertainty and ambiguity about the past, by a plaque that resonates with the confusion and discontent with which so many Stockholmers – and Swedes in general – regard both present circumstances and future possibilities, by a plaque congruent with the widespread inability of Stockholmers – and many Swedes elsewhere – to recognize either the world in which they live or themselves, by a plaque that is a painful reminder, a distilled disabuser of any lingering sentiment that Sweden is the best place in the world to live, an exceptional country immune to the crises affecting the world at large,

AMIDST THE ROUTINE PRACTICES AND OCCURRENCES OF EVERYDAY AND EVERYNIGHT LIFE,
amidst the whiz-by, wheeling-on of taxis, heavy trucks, delivery vans, buses and cars of German, U.S. and Japanese – as well as Swedish – origins,
amidst the heterogeneous foot traffic,
amidst the flow of randomly mixed passers by, of migrants, refugees and natives, of the pale-complexioned and the olive-skinned, of the impeccably dressed and the paint spattered, of people conversing not only in Swedish, but in Turkish, Arabic, Finnish, Spanish and the otherwise foreign,
amidst the stream of embodied difference and the not usually discernible eddies of personal discomfort and social conflict accompanying it,

About 18.5 percent of Sweden's population consists of migrants, refugees, and their first-generation offspring.

Post-office poster across Sveavägen from Dekorima: 'Everybody loves a postcard, who do you love?'

amidst the movement of those who stop and ponder at the sidewalk marker, of those who give it no more than a quick glance, of those who nonchalantly roll their baby carriages over it, of those who are totally oblivious to it,
VIOLENCE MAY ERUPT WITHOUT PRIOR WARNING,
a particular conjuncture of seemingly ordinary circumstances may, in the passing of a single second, become extraordinary in the bloodiest of ways,
the pent-up may – with or without prior intent and scheming – be suddenly and furiously released,
the current moment of danger may become a site-specific corporeal reality rather than a general state of the hypermodern present,
as on a late winter evening, more than a decade ago,
when an unaccompanied middle-aged couple – strolling subwayward and chatting casually as any husband and wife might upon leaving a movie – were approached from the rear by a stranger,
when that unidentified man, pushed over the top by an invisible flash of adrenaline,
pushed a trigger to produce a highly visible flash and then rapidly disappeared up the steps of Tunnel Street,

Plaque text: At this spot Sweden's Prime Minister, OLOF PALME, was murdered, February 28, 1986.

when what was immediately apparent thereafter was a frantic, anguished woman and a fast dying national leader,
when what was not immediately apparent, and was still not apparent in mid-1997, was who had done it and for what purposes, was a question of what had actually

'Many politicians describe the unsolved murder of the

i
i
i

come into conjunction at this particular node of time and space, was a question of which were the more widely operating structures or processes – if any – that had been brought to bear at this precise point, was a question of whether it had been a random or planned act of violence, whether it had been carried out by a petty criminal (such as Christer Petterson, who was tried and found not guilty, but is once again the prime suspect), or a loner whipped into action by the virulent excesses of neoliberal/anti-Palme rhetoric, or a right-wing 'madman' acting on his own or in conspiracy with suspiciously behaving policemen, or unassisted police officers with neoFascist leanings, or someone somehow linked up with one of Palme's extramarital affairs, or an agent of one of the parties displeased with his mediating role in the war between Iran and Iraq (the Irani secret police, the Iraqi secret police, or the revolutionary Kurdish group, the PKK), or a clandestine operative for some other government threatened less by his overt criticisms than by his various covert maneuvers (a hit-man in the service of the KGB, the CIA, the Chilean secret police, or – what briefly appeared most probable – South African 'security' agencies bristling at the enormous sums of money he was having secretly channelled to the ANC),

prime minister as 'an open wound in the nation''. (*Dagens Nyheter*, Aug. 11, 1996)

In the fall of 1996, with neoliberalism still thriving in Sweden, some were seriously proposing that crime solution become subject to private competition, be opened up to 'market forces.'

when, as at any other site of ordinary or extraordinary urban activity, there was multiple intersection, a conjoining of numerous local and non-local influences, a juxtaposing of numerous interrelations and interdependencies,

when

AT ANOTHER TIME AND PLACE IN THE CITY

i
i
i

•INVISIBILITY•

by John Law
and
Ivan da Costa Marques

First Observation

A car manoeuvres to pull away from the curb. And there is a child. How old is she, 5 or 6 maybe? He or she, the child is both attractive and androgynous, is dark, almost black, with dark curly hair and coffee coloured skin. S/he isn't wearing much. A pair of underpants. That's it. Nothing on her feet. S/he's jumping up to the passenger window of the car, talking into it. The car pulls away, leaving the child who walks off.

Does she belong to the car? Answer. No. She's a street child. The censuses of people sleeping on the streets of Rio show (if you believe them) that there are perhaps a hundred children under ten living by themselves on the streets. And many more live on the streets as part of some family group – though 'family' does not necessarily mean blood ties.

Second observation

The Linha Vermelha is the motorway from the airport to downtown. It suffers from frequent traffic jams in the evening. The cars and the buses come to a complete halt as all the middle classes try to pack themselves into the tunnels in order to flee the desperation of North Rio in favour of the relative domestication of South Rio. And where the cars come to a halt, high up on an elevated part of this motorway there are people, mostly teenagers, mostly black, trying to sell the motorists packets of biscuits and canned drinks. Carrying big bags, they weave between the cars, finding narrow spaces to stand between the railings and the traffic.

The first time the visitor sees this he is a little alarmed. Partly for himself. Does he want large (and not so large) teenagers bearing down on him trying to sell refreshments? And what will happen if he doesn't buy any? (The answer is: absolutely nothing.) And partly for them: good God, people must be desperate if they stand on motorways, inches from the traffic, trying to sell drinks and biscuits at R$1 a time. For the traffic is very erratic, not to say delinquent. At one moment it is standing still, and the next moment it is moving at 60 miles an hour. Do these people not get run over?

That is the first time. But after just a few journeys they have become invisible. Not in the sense that you do not see them. You see them with your eyes. Still. Somewhere or other they are registered. But the conversation does not stop in the car. It carries on. There is no reason to divert from the topic at hand.

On denial

The city makes denials in many ways. But here in at least two senses. There is that which is denied because it is not recognised. And then there is that which is denied because it is recognised too well. Sometimes it is possible to peep with fascination through the narrow gap between the two. Through the gap between what is not yet known, and what is known too well. Which is a narrow gap that moves, displaces itself, with a rapidity that is both interesting and disturbing.

It is something like this. At first we do not see an awful lot that we will later see. We do not have the context, the background, the knowledge to make sense of it. For instance, it may be that if we are from the affluent North we do not immediately see that the street child is indeed a street child who is begging. We do not see her that way. 'Does she belong to the car?' The question presupposes something else. That children talking to adults in cars perhaps belong to those adults? That those adults are their parents? At any rate it certainly doesn't imagine that the children belong to no one. Which is a lack of recognition, an example of the first form of denial: that which we have not yet seen at all which is still, of course, invisible.

But then we have many of the categories that we *do* need to see. Coming from the North we can see, for instance, the people sleeping in their rags in the doorways. We can see them perhaps because we have been trained in nearly twenty years of Thatcherism, to see people sleeping in cardboard boxes above the grills of the underground in London. Or, indeed, in places which are much closer to home than that. So we can see the homeless young and not so young people of Rio de Janeiro. And the beggars. And the impoverished people who are selling things in the streets. But we don't yet *see* the street child.

One kind of denial. And then another other kind: the denial of familiarity. Perhaps that of 'civil' inattention? Not attending to those in the street out of deference? No, not that. But rather that of 'civic' inattention. Not attending to those who do not belong. For instance, those who stand on the motorway selling biscuits to passing motorists. Those, in other words, who do not belong to the world of the citizen.

Two forms of invisibility. That which is not seen because it is unknown. And that which is not seen because it is known too well but does not belong. And that rare place, somewhere in between, where vision is possible for a moment.

•JOB CENTRES•

by Fran Tonkiss

'Spatial images,' the journalist and cultural theorist Siegfried Kracauer once wrote, 'are the dreams of society.' Just as the interpretation of dreams could reveal the workings of the unconscious, so different spaces might yield up the social realities that lay behind them. Kracauer was thinking of the labour exchange – the space given to those who had been 'compulsorily liberated from society' (Kracauer, 1997, p. 60). He believed such places might tell something of Berlin's unemployed that couldn't be found in bloodless statistics or in the echo of the parliamentary chamber: how this disparate group stood in relation to other social strata; their joblessness to a wider system of production; the labour exchange to other sites of commerce and value.

To think about the geography of unemployment is to conceive the city as fractured; to trace the ill fit between contours of economic movement and places of economic stasis. Where his teacher Georg Simmel had seen the modern city as a space of circulation – of money, people and things – Kracauer saw the grim inertia of the labour exchange. It was the underside of the time-and-motion order of the modern factory – a space where empty time existed in the absence of motion:

> In the employment agency, the unemployed occupy themselves with waiting . . . the activity of waiting almost becomes an end in itself. . . . I do not know of a spatial location where the activity of waiting is so demoralizing. (Kracauer, 1997, p. 62)

The passive space of the labour exchange gave the lie to modernist images of the city as a growth machine. Flows of labour, of money and commodities, the exchange of energy and value, get stuck at this point. There is an ironic echo of other modern spaces of commerce when Kracauer imagines the labour exchange as a sort of 'arcade', a passage which should conduct the traffic of bodies back into work, but one which only grew more crowded. Circuits of capital mark out blank spaces even as they animate others. Unemployment had then, as now, a baleful tendency to attach itself to certain places, and to certain bodies.

Kracauer reads the labour exchange as a kind of bad dream; bleakly *unheimlich*, it is 'the opposite of a home and certainly not a living space' (p. 59). A double to the factory, it is not properly an economic space. Although the men might talk or play cards, it is not a social space. Neither one nor the other, it is yet a place where the economic and the social run together. While they do the work of waiting, the unemployed became objects of hygiene and moral regulation – the walls of the labour exchange showing posters for venereal health and birth control, as well as admonitions against smoking, taking milk without food, and wrecking the furniture.

The link from social and moral to economic order has a long pedigree in the government of the modern city. From workhouse to labour exchange, from dole office to job centre, the ranks of the unemployed have been marshalled in special sites where the demands of the market encounter the patronage of the state. The changing nature, even the changing names of these spaces, tell the story of unemployment in different ways. By the end of the twentieth century, the problem had come to be understood primarily as one of *skill*. In the view of liberal economists, it was necessary for individuals to develop their stocks of human capital, to become entrepreneurial with regard to themselves, to cultivate flexibility. Blameless waiting was no longer enough. A mongrel notion of 'human capital' – this most recent conjunction between social management and economic good – marks the completion of a modern project to render people's labour simply in market terms. It encourages you to think about yourself in particular economic ways; as a type of investment, as something to be capitalised in the market. What's more, this language and its logic spills over the boundaries of any real labour market into the spaces of everyday life, now re-imagined as a kind of social factory. Quite where the imperatives of the market end and the freedoms of everyday life – even those of 'compulsory liberation' – begin is hard to say when productivity and efficiency become values to be realised as much in the gym or the kitchen – or the dole queue – as in the office or on the shop-floor.

Kracauer read the labour exchange in terms of dead time, forced out of the virtuous circuits of capital in the modern city. Now, the work of being jobless is no longer one of waiting but a burlesque of efficiency. In one recent version of Kracauer's arcade, ambitiously named the job centre, the passive language of unemployment is translated into the quest narrative of 'jobseeking'. Parody of an office, mint of human capital, in this anodyne space it's all go. As people perform the motions of actively seeking work, unemployment as blank time no longer exists. You have to work at it.

It's a nice euphemism. Job centres – peopled by those who have no job on one side, and those unhappily in the business of unemployment on the other – still tell the bad dreams of cities that are only good places to live if they are also places to work.

Reference

Kracauer, S. (1997) 'On employment agencies: the construction of a space', in N. Leach (ed.), *Rethinking Architecture: A Reader in Cultural Theory*, London: Routledge, pp. 59–64.

•JUNGLES•

by Stephen Cairns

The 'urban jungle' is a potent yet troubling term in the West. It refers to the congestion which urban forms and jungles sometimes seem to share, but taken literally there could be no more obvious contradiction. The weighty pronouncements of Hegel, for instance, remind us that it is precisely through urban form and architecture that we can 'disentangle' ourselves 'from the jungle of finitude and the abortiveness of chance' (Hegel, 1993, p. 90). While the more profane observations of Henry Miller warn that when such entanglement is allowed to happen a kind of 'muddled' madness results – New York, for him, is a 'mad stone forest' (Miller, 1961, p. 69). So the rhetorical force of the urban–jungle coupling comes from the way it makes intimate such traditional opposites as predictability and chance, culture and nature, civilization and savagery. The urban jungle, then, is not simply about congestion, but names a congestion in which the very wildness the city was supposed to guard against spontaneously wells up from within.

Rem Koolhaas calls this phenomenon the culture of congestion and, like Miller, exemplifies it with New York, in particular the Manhattan of the pre-1940s. In the early part of the century on Manhattan – this island nestling against the vast land mass of the New World – a strange experiment spontaneously and unconsciously evolved. Beginning in the fairgrounds and side-shows of Coney Island under the direction of a rag-tag band of assorted developers, entrepreneurs and speculators, the experiment produced unprecedented juxtapositions of functions, figural excesses and urban congestion. Soon the experiment was transferred to the larger laboratory of Manhattan itself. Here a full-blown 'vernacular of capitalism' emerged. All of this, Koolhaas adds, flourished outside the authority of formal architectural culture.

This was a relatively brief phenomenon however, and for Koolhaas it was Le Corbusier's modernism which contributed to its end. Le Corbusier proposed a fully aerated urbanism which burst Manhattan's fantastic bubble. In place of the urban jungle, he planned a cultivated urbanism in which architectural forms faithfully mirrored their functions, where breezes circulated around free-standing towers, and where an uninterrupted ground plane was given over to rolling parkland and speeding traffic. Manhattan's unconscious drives are exorcised by this machinic programme, its vernacular congestion was relieved and the law of form reinstated. The wildness was rationalized and the savagery was civilized; jungle and city are made discrete.

Le Corbusier's cultivated urbanism was shipped off around the globe. A particularly direct route was taken to Singapore where at first glance it seems to have arrived intact. Speeding along the freeway from Changi Airport to the city, it is impossible not to notice the functionally homogenous, free-standing towers which seem to sprout from rolling picturesque parklands. It is as though the old anxieties about congestion and urban wildness have indeed been sorted out. But Singapore unlike Manhattan is a lush tropical island where nature has already provided its own jungle. Consequently, to follow the fortunes of modernist urbanism to Singapore is to find jungle and urban form entangled again – this time literally.

Jungle anxieties find a new and literal potency in this part of the world, so generating the impression of an intact civilized urbanism is no small matter in Singapore. In a city built on a jungle effecting civilization meant domesticating vegetation: greenery needed to be tended, thickets manicured, tendrils trimmed, herbage curtailed. The jungle needed to be kept at bay. As Stephen Yeh makes clear, in

Singapore State, investment in landscape aesthetics is extraordinary, particularly around high-rise public housing estates. He describes in great detail the various manicured themed gardens and the parks filled with trees from around the world. This landscaping demarcates territory, decongests space and articulates the broader values of civilization – there is not a hint of irony, for instance, in the official goal 'of turning Singapore into a fully developed country inhabited by a cultivated society' (Yeh, 1989, p. 831). And in a cultivated society the jungle and the city cannot mix.

The twentieth-century anxiety around the urban jungle is already pre-figured in this part of the world. In North Sumatra, across the Malacca Strait from where Singapore now stands, was the city of Aceh. Sixteenth-century English traveller Captain John Davis was confused by his first sighting of the city: it was 'very spacious' yet 'built in a Wood', so it was not visible from a distance and seemed to leak out across the countryside. He reports

> that we could not see a house till we were upon it. Neither could we go into any place, but we found houses and great concourse of people: so that I think the town spreadeth over the whole land. (Captain John Davis, cited in Reid, 1980, p. 240)

French Jesuit S. J. de Premare writing to a colleague in Canton in 1699, is likewise puzzled by the entanglement of dwellings and jungle. So much so that he could not see the city at all:

> [i]magine a forest of coconut trees, bamboos, pineapples and bananas, through the midst of which passes quite a beautiful river all covered with boats; put in this forest an incredible number of houses made of canes, reeds and bark, and arrange them in such a manner that they sometimes form streets, sometimes separate quarters; divide these various quarters by meadows and woods: spread throughout this forest as many people as you see in your towns, when they are well populated; you will form a pretty accurate idea of Achen [Aceh] and you will agree that a city of this new style can give pleasure to passing strangers . . . . Everything is neglected and natural, rustic and even a little wild. When one is at anchor one sees not a single vestige or appearance of a city, because the great trees along the shore hide all its houses. (S. J. de Premare, cited in Reid, 1980, pp. 240–241)

To western eyes Aceh is a conundrum. It has no figure, it has no ground, it is not legible, it seems to obey no territorial rules. Yet it accommodates a population as substantial as anywhere else in the world at the time. The convolution of civilization and savagery presented here forecloses on the possibility of thinking this place as anything like a city. Yet, at the same time, these reports open the way to imagining an urban formation otherwise. This is to allow the idea of the urban jungle full reign. In this place urban pleasures are not necessarily generated by recourse to the law of form. This is a city in which wildness flourishes, where civilization is thought savagely.

References and further reading

Hegel, G. (1993) *Introductory Lectures on Aesthetics*, London: Penguin.

Koolhaas, R. (1978) *Delirious New York: A Retroactive Manifesto for Manhattan*, London: Thames and Hudson.

Le Corbusier (1967) *The Radiant City: Elements of a Doctrine of Urbanism to be Used as a Basis of our Machine-age Civilization*, New York: Orion.

Miller, H. (1961) *The Tropic of Capricorn*, New York: Grove Press.

Reid, A. (1980) 'The structure of cities in Southeast Asia, fifteenth to seventeenth centuries', *Journal of Southeast Asian Studies*, 11(2) (September).

Willis, C. (1995) *Form Follows Finance: Skyscrapers and Skylines in New York and Chicago*, New York: Princeton Architectural Press.

Yeh, S. (1989) 'The idea of the garden city', in K. S. Sandhu and P. Wheatley (eds), *Management of Success: The Moulding of Modern Singapore*, Singapore: Institute of Southeast Asian Studies.

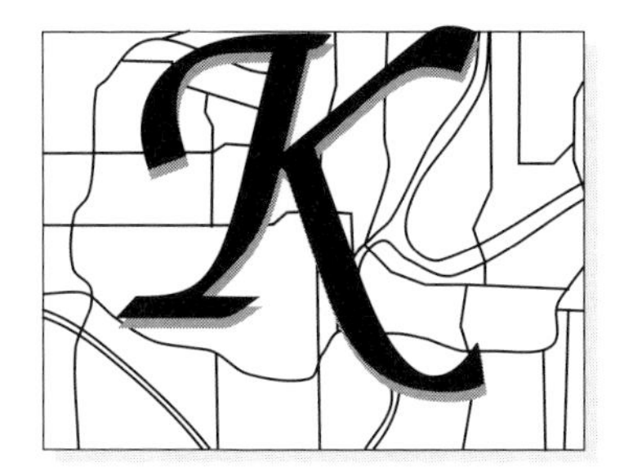

•KNOWLEDGE•

by Jane Rendell

In wide arcs of wandering through the city
I saw to either side of what is seen,
and noticed treasures where it was thought
there were none.
I passed through a more fluid city.
I broke up the imprint of all familiar places,
shutting my eyes to the boredom of modern
contours

(Dunn, 1995, p. 9)

In contemporary urban and architectural discourse we are increasingly obsessed by figures which traverse space: the flâneur, the spy, the detective, the prostitute, the rambler. These are all spatial metaphors, representing urban explorations, passages of revelation, journeys of discovery – they are 'spatial stories' (de Certeau, 1984, pp. 115–122). Through the personal and the political, the theoretical and the historical, spatial stories are told, narratives exchanged in, and of, the city.

[This spatial story is for Deborah Miller and our late night story-telling. See also Borden *et al.*, 1995; 2000.]

Inspired to 'know' the city, our fragile understanding of urban and architectural space is framed by desire. Our attempts to reveal unknown aspects of known cities create new cities – as we desire them to be.

> Unlike promises we make to each other, the promise of the city can never be broken. But like the promise we hold for each other, neither can it be fulfilled. (Burgin, 1996, p. 7)

As an historian, I tell spatial stories somewhat differently from other story-tellers; my stories are inspired by a desire to 'know' the past. Historical epistemology is a complex area: forcing the evidence, teleological argument, drifting into fiction, are well-known crimes. Being female complicates it further. Historical knowledge may be characterised as a masculinist pursuit. To make purposeful decisions about historical lines of enquiry and interpretative strategies assumes clear and certain knowledge, 'knowing' without doubt. This kind of knowing does not call the self into question, indeed it assumes it is possible to 'know', to know one's own mind. But this female subject knows doubt.

> For the master's tools will never dismantle the master's house. They may allow us to temporarily beat him at his own game, but they will never enable us to bring about genuine change. (Lorde, 1979/1996, p. 160)

This female subject places herself in complex relation to her subject matter. Historical knowledge is not objective, it is founded in subjectivity:

> why not then continue to look at it all as a child would, as if you were looking at something unfamiliar, out of the depths of your own world. (Rilke, 1986, p. 55)

She desires to know the city, but, for her, 'knowing' the city invites, and invokes, a need to know the self, the one who seeks knowledge. The two are in dialectic relation. The (hi)stories we tell of cities are also (hi)stories of ourselves. For feminists, the personal is an important epistemological site. (Who I am makes a difference to what I know, conversely what I know makes a difference to who I am.) For a feminist historian the personal forms a threshold between past and present. Negotiating a meaningful relation between the two is critical terrain.

> Critical work is made to fare on interstitial ground. Every realization of such work is a renewal and a different contextualization of its cutting edge. One cannot come back to it as to an object; for it always bursts forth on frontiers. . . . Instead, critical strategies must be developed within a range of diversely occupied territories where the temptation to grant any single territory transcendent status is continually resisted. (Cixous, 1976, p. 229)

In the pursuit of historical knowledge, in the desire to know, to reunite with the past, poignant forms of exchange are entered – (re)reading, (re)searching and (re)writing – between the theoretical and the historical, between past and present, between city and self, are sites of methodological struggle – places where difficult questions of spatial and historical knowledge are raised but also where tantalising glimpses of the relation between outer and inner worlds are offered. 'Knowing' and 'unknowing', the politic, poetic, personal, pragmatic, seduce at every turn, promising the unrealisable, reunion between my desirous self and the city, the object of my desire.

> When desire takes over, the body gets the upper hand. In our intense contemplation of the beloved – as if to discover the secret of that which binds and confuses – we are looking for our past. (Carotenuto, 1989, p. 17)

In her second diary, Anaïs Nin wrote of Fez that the image of the interior of the city was an image of her inner self. Backwards and forwards, forwards and backwards, what we have is not an after-image of what has gone before, but a veiled view of what we are to become. Inside to outside, outside to inside, all we can know is that what we seek outside ourselves is already within:

> Myself, woman, womb, with grilled windows, veiled eyes. Tortuous streets, secret cells, labyrinths and more labyrinths. (Black Koltuv, 1990, p. 7)

Knowledge is labyrinthine. In writing the city, I am writing myself. . . .

References

Black Koltuv, B. (1990) *Weaving Woman: Essays in Feminine Psychology from the Notebooks of a Jungian Analyst*, Maine: Nicolas-Hays.

Borden, I., Kerr, J., Pivaro, A. and Rendell, J. (eds) (1995) *Strangely Familiar: Narratives of Architecture in the City*, London: Routledge.

Borden, I., Kerr, J., Pivaro, A. and Rendell, J. (eds) (2000) *Unknown City: Contesting Architecture and Social Space*, Cambridge, MA: MIT Press.

k
k
k

Burgin, V. (1996) *Some Cities*, London: Reaktion Books.

Carotenuto, A. (1989) *Eros and Pathos: Shades of Love and Suffering*, Toronto: Inner City Books.

Cixous, H. (1976) 'The laugh of the Medusa', in E. Marks and I. de Courtivon (eds) (1981) *New French Feminisms*, Sussex: Harvester.

de Certeau, M. (1984) 'Spatial stories', in *The Practice of Everyday Life*, Berkeley: University of California Press.

Dunn, A. A. (1995) *Vale Royal*, Uppingham: Goldmark.

Lorde, A. (1979) 'The master's tools will never dismantle the master's house', reprinted in *The Audre Lorde Compendium: Essays, Speeches and Journals* (1996), London: Pandora Press.

Rilke, R. M. (1986) *Letters to a Young Poet*, New York: Vintage Books.

k
k
k

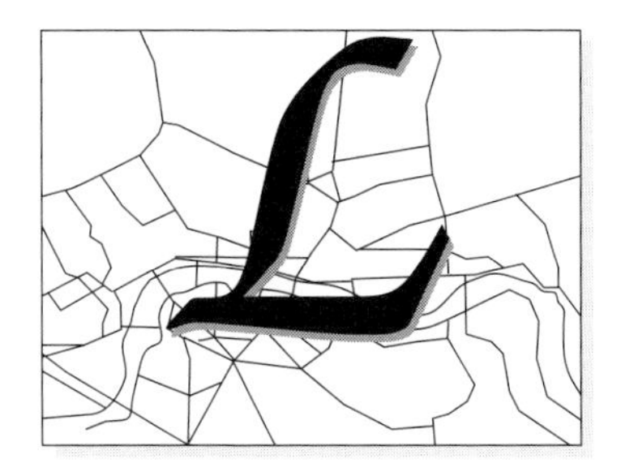

•LABYRINTHS•

by Ross King

All architecture (the representation of another, alternative space of everyday life – the promise of redemption or is it the prison?) is translation – the graveyard of dreams made concrete. This is a story about (the impossibility of ever telling the story about) a place. A theme of this essay is that all originality is chimerical: the site chosen is Parc de la Villette, Paris, already picked over by many authors, and dwelt upon in many texts (a starting point is Tschumi, 1988).

The story starts with Le Corbusier, who designed a hospital (unbuilt) for Venice, to 'define' that place (as it were) by asserting its opposite. On a geography of labyrinthine contingency and Byzantine decay, there would be imposed a Cartesian grid of columns to support a new, Modern Movement icon.

Peter Eisenman takes Le Corbusier's Venice grid and imposes it on his urban design for the Cannaregio, an amorphously ill-defined remnant of open space, also in Venice. It would memorialise what Le Corbusier had failed to achieve. But Eisenman also translates his own (unbuilt) House 11a into the design, at three distinctively different scales. So what is House 11a? It is architecture reduced to an abstract mathematical puzzle that, by challenging its contract with habitability, throws architecture itself into question. When it becomes sculpture – scaleless objects in a landscape such as the Cannaregio – the very promise of architecture shatters, and we have another memorial to a lost dream (this time Eisenman's) woven into the first. The design becomes a play on grids, mathematical puzzles and distortions of scale.

Bernard Tschumi allegedly appropriated Eisenman's Cannaregio into his own prize-winning design for Paris' Parc de la Villette – well, it does have a grid, and it does have scaleless plays in not so solid geometry. It is built, great fun, and wonderfully successful.

Eisenman is invited (by Tschumi) to prepare garden designs to be incorporated into the Parc de la Villette, and duly appropriates Tschumi's appropriation. However, la Villette, like everywhere else, has a genealogy: the ancient walls of Paris crossed the site, and it was for these that Claude-Nicholas Ledoux, the most rationalist of Enlightenment architects, designed his

Porte de la Villette. In another age the site became the Abattoir de la Villette – to which, Dadaist Georges Bataille informs us, we banish bloody slaughter and our own unseemliness – the graveyard of the real, the purchase of the fantasy of cleanliness and respectability, the price of forgetting. So Eisenman imports into the garden designs the false memory of the ramparts of the city (false because they are in the wrong place and at the wrong scale), and the labyrinth of the abattoir (again transformed in scale, and now far distant from that earlier, generating labyrinth of Venice) (see Bédard, 1994). Ledoux, it seems, has been missed – too close a rival for the radical mantle, no doubt.

Eisenman's collaborator at la Villette is Jacques Derrida, for whom the question of architecture and rationality would come to the fore. In his *Fifty-two Aphorisms for a Foreword* (1989), we get aphorism 29:

> To say that architecture must be withdrawn from the ends that are assigned to it and first of all from the value of habitation, is not to prescribe uninhabitable constructions, but to take an interest in the genealogy of an ageless contract between architecture and habitation. (Derrida, 1989, pp. 67–69)

Reality itself is problematised; but so is representation, and we are back to something very ancient.

Derrida, for his part, translates into the design an idea from Plato's *Chora* – from before that other contract between language and rationality (so certainly no place for Ledoux). A gold, gridded cube would be inserted into the mishmash of traces that would constitute a garden for Eisenman – each, presumably, to stand as a Derridean (deconstructive?) preface calling the other into question – which memory, which history? What is place, what is memory, what is authenticity?

And in turn, Tschumi: the points on the la Villette grid are not columns (Le Corbusier) or fake excavations (Eisenman), but fire-engine red, steel constructions serving a variety of functions (kiosks, cafés, exhibitions, etc.), but in their iconography evoking the Constructivist dreams, forever unfulfilled, of the era of the Russian Revolution – lost hopes indeed! So the ideals of a better space of everyday life, from Leonidov, Chernikhov, Rodchenko, become fun and public toilets, in the city of the Bourgeois Revolution and of the events of 1871 and of 1968. Revolution overlays revolution, epoque interweaves with epoque.

Great cities are labyrinths of time (past lives, hopes, dreams, despairs), lost in the incomprehensible labyrinths of space. All architecture, for its part, is (mere) translation – from the chaotic, jumbled, forever unreadable 'text' of the city (those labyrinths of space, quintessentially two-dimensional) and from the architect's half-recalled memory of a million other texts (other labyrinths of time, the confusion of all things ever seen or ever imagined). Sometimes it is a very arid translation, as if nothing of the city as labyrinth has been observed, nothing of life as layered events remembered; but it is also sometimes very self-consciously, even self-indulgently recollected and paraded, and that is the danger of the la Villettes of this world. If la Villette succeeds it will be first because it is also fun, a game, play – 'play that calls new worlds to life' (Nietzsche).

Architecture is also translation in another sense: of dreams whose potential was never realised (Benjamin's 'mythic past'); for that great promise of architecture – an alternative, more fulfilling, emancipatory space of everyday life – is never achieved. Whatever we design, it seems, we memorialise the ruins of past dreams; and whatever we build are graveyards, the depthless gatherings of past ages and their hopes – wonderful, still able to motivate and show us glimpses of those past dreams and perhaps of future hopes, but graveyards for all that.

References

Bédard, J-F. (ed.) (1994) *Cities of Artificial Excavation: The Work of Peter Eisenman, 1978–1988*, New York: Rizzoli.

Derrida, J. (1989) 'Fifty-two aphorisms for a foreword', in A. Papadakis, C. Cooke and A. Benjamin (eds), *Deconstruction Omnibus Volume*, New York: Rizzoli.

Tschumi, B. (1988) 'La Case Vide', *AA Folio*, 8.

•LIFTS•

by Jenny Robinson

It's a lazy Sunday and the whole city is sleepy with the heat, so I take my towel and suntan lotion and make the five-minute drive down to the beach. I lay down my towel, spread my body out on the sand, and feel the warm sun soaking into my back. My hat starts to slip off my head. I wriggle, replacing my hips in the indentations they've made in the sand, and slide my hat back over my vulnerable (pink) nose.

Later, I stand up. I run into the sea for a swim, returning wet to my spot. The sun cools down as I sit, lost in the sea, my whole self drifting into its endless moving warmth while it softly reflects the sun setting behind me. The sea catches my fears, my sorrows, fills me with its endless flow, pulls me into its infinity. Losing myself, I feel finally at home.

As I stand up and walk, slowly, reluctantly, towards my car, I pass a hotel where I used to go for school dances as a teenager. The best fun was catching the glass lift on the outside of the building up to the top floor where there was an expensive night-club and bar, sneaking a look inside as we were told we couldn't go in. The ride down was magnificent. Laughing as we covered our embarrassment about being too young and scruffy for the night-club, we stared out of the windows of the lift, mesmerised by the sea. Lost again, at home in the sea's dark endlessness.

I decide to take the lift once more, nervous in my costume and towel, but drawn to re-live the excitement I had felt before. No one else is around, and as the doors close I relax a little. Leaning against the red rail, nose close to the glass, I stare as the street gets smaller and the cars moving slowly along the Marine Parade start to disappear from view. As the lift goes higher, the sea fills the glass and I look across the horizon, count the ships waiting to go into the harbour and slowly lose myself in the sea's depths. The surface is choppy with waves and white horses as the wind is coming up. At the top I feel as if I had never left the beach – sea is everywhere.

de Certeau writes about his trip in the lift to the top of the World Trade Centre:

> Beneath the haze stirred up by the winds, the urban island, a sea in the middle of the sea, lifts up the skyscrapers over Wall Street, sinks down at Greenwich, then rises again to the crests of Midtown, quietly passes over Central Park and finally undulates off into the distance beyond Harlem. A wave of verticals. (1984, p. 91)

I really enjoy the way he portrays the city as a sea. When I feel homesick now that I live in a city without a sea, I go at night to a high point – up in a lift or to the top of a hill, to a viewing point. I stand and stare across the dark sea of the city, watching as the sparkling flashes of street lights and headlights set the darkness in motion. The endless waves of the city draw me to the edges of myself, stretch my sense of who I am across the whole city. I am lost again in the sea as if I had never left home.

de Certeau sees this differently, though. He stands at the top of his lift, rigid, cut off, distanced, finding only the vanishing place of the geometric, map-making viewpoint and the guilty pleasure of the voyeur, desperate to be found out in this incriminating position. Rather than melting into the endlessness of the city-sea, he retreats into the nowhere point of the objectifying viewer.

How curious then that his desire to go back down to the street, down again in the lift, is in order to find the pleasures of losing himself as a practitioner of space, as a streetwalker. For him, space always leaves an excess, experiences which can't be captured completely in language. He wants to experience the spaces of the city somatically, 'blind as . . . lovers in each others arms' (p. 93). Introducing her *Tales of Love*, Julia Kristeva writes: 'in the rapture of love, the limits of one's own identity vanish, at the same time that the precision of reference and meaning becomes blurred in love's discourse' (1987, p. 2). A little further on she talks of love as a 'vertigo of identity, vertigo of words' (p. 3). de Certeau's image of the lovers, blindly wrapped in the ecstasy of each other, points to the same desire as the sea evoked in me – a longing to abandon oneself, to be transported beyond oneself into the infinite space of the inexpressible. de Certeau longs to lose himself in the spaces of the street, just as I found pleasure losing myself in the sea, in the city-sea.

To heal myself, to lose myself in endlessness, to smooth the damaged and difficult edges of my self: I take a lift to stare at the sea, at the city-sea.

References

de Certeau, M. (1984) *The Practice of Everyday Life*, Berkeley: University of California Press.

Kristeva, J. (1987) *Tales of Love*, New York: Columbia University Press.

•MADNESS•

by Roy Porter

An early Georgian squib, *Hell Upon Earth* (1729), depicted London as a

> great, wicked, unweildy [*sic*] overgrown Town, one continued hurry of Vice and Pleasure; where nothing dwells but Absurdities, Abuses, Accidents, Accusations, Admirations, Adventures, Adversities, Advertisements, Adulteries, Affidavits, Affectations, Affirmations, Afflictions, Affronts, Aggravations, Agitations, Agonies, Airs, Alarms, Ailments, Allurements, Alterations, Ambitions, Amours, Amphitheatres, Anathemas, Animosities, Anxieties, Appointments, Apprehensions, Assemblies, Assessments, Assurances, Assignations, Attainders, Audacities, Aversions, &c.

The list continued. Intriguingly, this alphabet of urban outrages did not specifically refer to insanity or its subdivisions by name: doubtless when everything metropolitan was mad, the pamphleteer thought that quite redundant.

Others, however, were more explicit. Ned Ward, one of the first great urban journalists, enjoyed eliding the city of London as a whole with Bedlam as its epitome. Bedlam (or more properly Bethlem Hospital) had become one of the sights of London; most tourists paid it a visit. In 1610, Lord Percy, his wife and her two sisters 'saw the lions [in the Tower], the shew of Bethlem, the places where the prince was created, and the fireworks at the Artillery Gardens'. Ward played upon the ambiguity of what was actually seen – and who it was who was actually crazy. In *The London Spy* (1698), he drolly depicted two sightseers standing outside the new and palatial Bethlem building just outside the city walls in Moorfields and showed them assuming that this must be the residence of 'Quality'. The implication was that men of quality who chose to live in town were all off their heads.

Ward of course was one of that new breed, the 'Grub Street' hack, and it is no accident that the actual Grub Street was a mere stone's throw from Bethlem Hospital. The insinuation was clear: it was not only the Quality who were crazy: it was also the new breed of 'writers for hire'. The writing being spewed out by the Grub Street hacks was symptomatic of delirious or distracted brains.

And that satirical connection may have been a bit near the bone, because no small number of literary figures in Ned Ward's London ended up crazy: the Restoration playwright, Nathaniel Lee, was confined to Bethlem itself, while the sublime religious poet, Kit Smart (whom Johnson befriended) had to be lodged in St Luke's Asylum, founded in 1751 just down the road from Bethlem.

The trope of metropolitan madness was of course not new to the eighteenth century. If Samuel Johnson's satirical poem *London* (1738) offered a vision of a mad city, demoralized, disordered and destructive . . .

> Here malice, rapine, accident, conspire,
> And now a rabble rages, now a fire;
> Their ambush here relentless ruffians lay,
> And here the fell attorney prowls for prey;
> Here falling houses thunder on your head,
> And here a female atheist talks you dead.

. . . that was in part because he could draw upon a tradition of anti-urban satires going back at least as far as Juvenal in Imperial Rome, exposing urban psychopathology. For its part, the Christian tradition had Babel and Babylon ready to hand to add to the identification of the town with all that was sinful and infernal, even before the Papacy doubled the demonization of the Eternal City.

But what made cities manufactories of madness? A long tradition presented them as the works not of God but of man, and hence the products of pride and pomp. In William Cowper's pithy if sentimental phrase, 'God made the country, and man made the town', and generations of pastoral poets and painters drew the obvious conclusion that rural life was sane and sanitive whereas in town all was chaos and cacophony – or, in Cowper's words:

> Such London is, by taste and wealth proclaim'd
> The fairest capital of all the world,
> By riot and incontinence the worst.

William Blake sang the same themes in more apocalpytic phrases.

One practical consequence of these views lay in decisions regarding the actual siting of lunatic asylums. As, from the late eighteenth century onwards, these institutions increasingly became purpose-built, opinion strongly favoured locating them out of town, indeed deep in the countryside, on the supposition that rural views, open skies, peace and the prospects of agricultural labour would all prove recuperative to the patients. (The fact that the out-of-mind, thus exiled, would also be out-of-sight doubtless also had its attractions.) Bethlem Hospital itself finally moved out of town in the twentieth century, from its building in Southwark (destined, appropriately, to become the home of the Imperial War Museum) to Beckenham in Kent, which, if not exactly rural was at least leafily suburban.

By then, however, the prospects of emptying the mad out of cities had become futile, since an enormous range of commentators – sociologists, journalists, moralists and novelists, to say nothing of psychiatrists themselves – had concluded that cities were the inevitable breeding-grounds of derangement. *Fin-de-siècle* culture espoused the paradigm of degeneration and saw city life as deterioration incarnate. According to major psychiatrists – Moreau de Tours in France, Henry Maudsley in Britain, Cesare Lombroso in Italy, and Max Nordau in German-speaking parts – the tensions of modern living were eroding basic constitutional stability and sanity. The factory production-line and its division of labour; the *anomie* accompanying vast towns and the lonely crowd; the dissolution of once stable household units, employment, rural communities, and organized religion; the vapid and vicarious excitements of Megalopolis, and its accompanying destitution, alcoholism, ubiquitous prostitution and venereal disease epidemics – all these were inherently psychopathological manifestations, that were spawning a degenerate *classe dangereuse* of drunks, syphilitics, paralytics, defectives and atavists. Worse still, the interbreeding of misfits and profligates – and it was widely assumed that the degenerate, driven by perverted sexual appetites and lacking self-control, would breed disproportionately – would lead, over the generations, to the swamping of the healthy by the residuum. Society would grow ever more disgenic, eventually disintegrating not, as Marx predicted, through revolutionary class conflict, but through internal psycho-physiological decay.

The psychiatry associated with the 'degenerationist' movement has long been discredited. But the close association through the twentieth century of city life in, for instance, Vienna and Manhattan with the psychoanalytic movement shows that many have continued to love their poison.

•Maids•

by Ivan da Costa Marques
and
John Law

One

The girl is 20. This modern young woman is independent minded. Like her mother, she is studying law and wants to be a judge. She says that it is time to leave home.

> 'I can't wait to move to a place of my own. To live by myself.'
> 'Will you have a maid?'
> 'Oh yes . . . my own maid.'

Yes, she is modern and independent. But the terms carry a special twist in Rio. Here middle-class modernity at the beginning of a new millennium does not necessarily mean that professionals do their own housework. Yes, the girl is independent minded but she still wants to have her own maid. She will still be living by herself – living with no-body – because the maid belongs to the infrastructure. Is it that she does not see the maid as someone with whom she might necessarily, if unconsciously, share her life? Would such a shift point to the final abolition of slavery in Rio? Are maids or domestic servants being denied the status of full human beings, free people? Is this how a young middle-class woman can be modern and independent minded and still expect to have her own private maid?

Two

Why does Ivan *not* have a maid, an *empregada*, when most middle class people in Rio do?

> 'I did have a maid when the boys were younger. Someone to look after the flat, and be there when they came home from school. But then I gave up. I didn't want a maid.'
> 'Why not?'
> 'It was awkward, because I didn't really like having someone in the house the whole time. There are all these negotiations and agreements that you have to make. What the maid does, how she relates to you.'
> 'But doesn't she stay in her own part of the flat?'

'Yes, unless she is cleaning and tidying the rest of the flat. Most maids don't want to come into your part. It would be very embarrassing for the maid if you wanted her to eat with you, for instance. She would rather eat by herself. But, basically, it wasn't that. I didn't want to share the space, and didn't like the negotiations involved in sharing. And now the boys are grown up, I don't need a maid. Not at all.'

The flat is not particularly large, but it is built on architectural principles that would have made Robert Moses proud. Living room. Bedroom. A second bedroom. And a third. A bathroom with a lavatory. A kitchen. And then, beyond the kitchen, the articulation, in concrete, of a divided way of living. Or, perhaps better, two ways of living that are partially connected. At any rate, a little room, or a passage to a little room. The little room of the maid. And her lavatory. Her basin. Her shower.

Three

The changing conditions of work:

'Your mother has a maid?'

'Yes. She does. She didn't have one for a time, but she is old now. She really needs someone to look after her. And she is used to it. Though sometimes there are problems.'

'How so?'

'Well, things have changed. At least in theory the conditions of work for maids are tighter. More regulated. They are supposed to have time off. And if a day is a holiday, they are not supposed to work. My mother doesn't like that. She thinks, 'Well if I've got a maid, why shouldn't she work if I need her?' And then I tell her, 'No mother, times have changed, things aren't how they were. Maids have certain kinds of rights now that they didn't in the past.' But then I don't want you to think that my mother is harsh. Not at all. She certainly didn't expect the maid to work all the time she was there. Just to *be* there if she needed her. I have to tell her it isn't like that any more. Times have changed. But my mother isn't hard. That's not what it is about. Because she is just as likely to help out if something goes wrong. For instance, she would probably pay her maid's medical bills – or those of her sister or her children. And she has no need to do that at all.

'So there is a conflict between my own imagined adherence to the rights of the maid (a modern First World pattern of behaviour, that of being politically correct) and my mother's "realistic", embodied, and situated approach to having a maid. Which is, no doubt, a heritage of slavery. Because my mother was brought up on a farm. She only moved to Rio when she was 20. And her grandfather had owned slaves – and the traces of slavery were very alive on the farm when she was a girl. A division, then, between country and city. Which is, at the same time, also a partial but strong connection. Which is embodied yet again in the modern city girl who wants to become a judge.'

Four

Quite a few maids don't like being maids. They don't like to go to the supermarket, for this reveals that they are maids.

'I once had a maid who would use a backpack to bring home food and other items from the market, when there was no alternative to shopping. Using a backpack made her look like a middle-class young woman or teenager. In her case this worked well because she was young, good looking, slim, the colour of her skin was light, and she had light green eyes. She did not have the typical body of a maid. There was another way in which she was not like a typical maid, because all through the years she kept on studying. Though she has now got a secondary school certificate she could not find a better job. She is still a cleaning woman.'

Ironies

A place of one's own. With a maid? Without a maid? A maid who belongs? Or doesn't belong? A maid who is? Or who isn't? The stories tell about the partial intersections of two worlds. Two worlds that are changing. Two worlds that articulate themselves together, or don't. Which depend upon one another, or don't. Which deny one another, or don't. Complement one another, or don't. Include one another, or don't. Recognise one another, or don't. Patches of feudalism and modernity. All mixed up. Or not.

• MALLS •

by Mary King

From the East, the mall looms like a concrete island set in a vast sea of cars. Brightly illuminated signs beckon toward the street, and flowers decorate the contours of the parking lots. The entrances are genially lit and hung with ornamental banners. From the road the initial impression is one of welcome and delight. By way of the West exit, however, an entirely different scene materializes. Here low-paid work crews maintain the back alleys and rubbish shoots that weave their way behind the stores unseen to mall patrons. The mall operates much like a self-enclosed city with its own regulated flows, apportioned spaces and socially sanctioned separations, and, within the narrowly demarcated perimeters of the mall's interior, these in-built divisions become conspicuously pronounced.

In the twenty or so years since the Westfarms Mall on the outskirts of the city of Hartford, Connecticut, has expanded, and donned all the cumbersome trappings of pretense, some significant spatial displacements have materialized. The initial stores, a chaotic scramble of shops where the inexpensive and the trite were recklessly proffered, have been replaced. One by one windows festooned with winking lights and hanging beads have been removed. Draped in subdued shades of mauve and beige and embellished with the dreadful masks of mannequins, the mall has unfolded through calculated design from an early consumer experiment to one of the largest income-generating shopping centers in the United States, thus marking a trend toward an improved form of lavish consumption. Like a global bazaar of refined privilege, the mall offers garments from the world over and products of a specialty, one-of-a-kind variety. Such items signal the fashion change of industry from a marked interest in merchandise for the masses toward a focus on a reinvigorated elite.

Through this process of social sorting the mall now caters exclusively for those who can afford its extravagant wares – a teacup for forty dollars, a pair of brocaded socks for ninety, silver cuff-links for thousands. Those who cannot buy serve food in the restaurants and use long poles to pluck trash from the carpeted atrium. It has been said that in this meritorious system you get the job you deserve; however, these changes represent deliberate plans and disinvestments, not just unintended consequences. The mall has a distinctive racial complexion unraveling in a segregated atmosphere in which the better shops have all-white personnel who serve an almost exclusively white clientele, while the maintenance and garbage details are administered almost entirely by African-Americans. Such workers must travel a distance from the city to procure the minimum-wage jobs in those shops which require a constant supply of cheap labor.

At the same time the mall has thrived, the city of Hartford has endured a remarkable depreciation. Many prosperous businesses and established department stores have folded as local commerce has been wrung out of the city and into the more prestigious residential enclave. Merchants and markets caught in the reversal of traffic and fortunes faced the decision to either watch their business shrink and close or fabricate the transition from the public to the private sphere. Hartford, once a congregation, now occupies a dark place in the suburban imagination. The city has become emblematic of widespread urban degeneration and decay in a transition developed along intrinsically racist themes. Almost any patron at the mall can recount tales of criminality and terror in the city, no matter how fictitious. Incidents that occur even thousands of miles away are woven into narratives that elaborate on the dangers of Hartford. These accounts translate into a discriminate avoidance of the city.

By contrast, the mall presents itself as a mini-city of perfection where the climate is controlled, the products are familiar, and safety is artificially insured. The consumer experience is hyper-real, for everything is as it should be, all items in their expected locations, each shop and employee hierarchically arranged. Customers of identical complexion and income bracket walk past mirrored walls reflecting more of the same. Credit card information serves as an electronic double so that everything, each product and patron, is effectively coded to include a refraction of itself, a simulacrum image. All is recorded on surveillance videos where the vertiginous hum of security systems produces a dizzying condition in which the shoppers are actually imprisoned by their own watchful eyes.

Still, shopping at the mall represents the possibility to try on many articles and simultaneously experiment with many selves. It appears to be a site of adventure and novelty, of freshly cut dollars, and shiny change. Strolling through the stores is like an exploration, a quest, a place to seek out and also to be seen. People are compelled there by the anticipation of something still unrevealed, a discovery of unknown surprises.

Yet each purchase signals an ending, for wants once satisfied have already reached completion. The acts of trying on, lacing up, and fastening signify closure more than beginning. Jackets zipped, garments hung and put away, packages wrapped and sealed – these gestures speak resolutely about finitude. The mall is the end of novelty; in its essence it is an escape from freedom and choice, a virtual confinement. Shopping at the mall is a dubious chore because cumulative purchase after purchase cannot abate satiation indefinitely. As a result, desire, the most fundamental of motivations, is reduced to a banal preoccupation. The mall represents not only a world of monstrous inequality but one of perpetual boredom. For though propelled by motion and pursuit, one can only go round and round the same space over and over again. What is realizable there is what is already known and has already been experienced, more of the same, more of the same.

•MAPS (1)•

Lunch with Giambattista Nolli

by Dolores Hayden

Giambattista Nolli, architect and surveyor of Pianta Grande di Roma, *the first modern city map, 1748. 'Despite praise heaped on the work by everyone from the pope on down, it was not a financial success. . . . After two years, most engraved copies were still unsold' (Allan Ceen).*

Remember when we two young architects
measured a street with a dozen crooked houses?
I draw all Rome now, every way-out quarter,
the Pope himself signed me a pass, I enter
everywhere, yes, even cloistered convents.

Rolling and clanking my iron chain, I slice
at space, cut ground and figure, figure and ground.
The riverbanks and cypresses, you'll know,
Nature I've sketched in trunks and twigs and leaves,
the plan's all new, stretched flat on twelve wide sheets.

'Lacks charm,' a colleague carps, he can't see grid
as science. 'No taste, no style,' a rival sneers.
'Buy it,' the pilgrim friars beg their abbots.
Craving the martyrs' tombs and holy relics,
they swear the saints themselves guide my *bussola*!

No one has ever drawn a map like mine,
or understood its mathematic power,
or counted up its thousand uses – taxing,
policing, buying, selling, spying, wooing –
that's not to mention ordinary viewing.

You build, my friend, you know our art is urban.
Just four *zecchini*. No? I wager you –
surely I'll sweep away your handful of coins –
one day we'll all own city maps in Rome.
So please, be one of the first, put down your cash!

•MAPS (2)•

by Harvey Molotch

City maps are supposed to help people get where they want to go, but because the map-makers ignore so many differences among cities, their maps are often much less useful than they ought to be. 'One size fits all' is not a good policy for city maps.

Maps of San Francisco, for example, show the neat grid of straight-aways criss-crossing the city as laid out by the original city fathers who paid no attention to natural topography. This can bring some big surprises for pedestrians trusting their map – climbs the likes of which are more common to an Andes trek than, say, a visit to the podiatrist. For the frail, trusting the map could be fatal, even without a famously windy San Francisco day; for the able-bodied it can mean being tardy or arriving in a scum of sweat. While wilderness hikers know their maps must show topography to be of any use, city maps are silent on the issue.

Even for those in cars, maps have their flaws. Sydney maps aren't very good about telling you that certain roads are actually flyovers, making it more than a leap of faith to turn off them. Sydney maps also use small circles to indicate traffic-lights, and not roundabouts as they would in countries where drains flow (properly) counter-clockwise.

Of more general importance to drivers, maps do not provide clues about what is often the most important piece of information one can have: traffic. The shortest distance on road paper is not necessarily the shortest distance in road life. Combined with the fact that the dimension of *time* is a map absence, there are no clues of when during the day particular routes have more or less congestion. In tourist towns, all changes by season. Thickness of lines tells the quality of roads in terms of traffic engineers' professional classification schema, but not the quality of the road as it will be experienced by those who use it. For travels through deserted towns in Kansas, none of this matters, but knowing something about best times and patterns of congestion is the only way to get through Bangkok.

Maps don't tell about crime and danger, including psychological anxiety. Certain parts of some cities are not places Mom wants us to go (and again especially at particular times and seasons). Nor are there clues of cultural distance: who should be where and when in Jerusalem, say – the best route for Jews may not be the best for Arabs. The route which blacks might prefer in Johannesburg may be different from the one preferred by whites. Even where there is no security issue or racial divide, movements through 'uncomfortable' terrain feel like it takes longer than traveling a greater distance where one feels at home. Map-makers do sometimes show 'scenic routes' but they have not gone much further to demarcate route subjectivities.

Standardization allows diverse kinds of people to use maps they have never before seen for places where they have not been. But this holds back the information that may be the most crucial in getting around quickly, safely, and comfortably – what maps are supposed to be all about. Without the all-important background knowledge of topography, society, and culture, maps can't provide a good sense of direction.

•MARATHONS•

by David Bell

Starting line . . .

Running twenty-six miles through a city – one amidst thousands upon thousands of runners – makes the marathon runner a strange symbol of urban life; he or she is pitted against the city, and against the self, and victory comes from defeating both city and self – from out-running not the other runners, but out-running the body, out-running the streets: here are urban men and women running *against the city* in 'a communal ritual of shared embodiment, constituted in moments of shared intimacy of a sort which urban life rarely allows' (Frank, 1991, p. 65); the community of long-distance runners, panting and straining in a surreal spectacle which borders on the hysterical – a kind of freakshow born out of the fitness craze (the jogger is the suburban cousin of the marathon runner) and packaged as a celebration of individual endeavour, buffed up by mediatization and poeticized by TV commentators for armchair consumption; runners against the city, but not running *away from* the city – if the marathon can be read, as Frank suggests, as an act of 'resistance' to an 'urban space which denies embodiment' (p. 65), then the marathon might be seen as a kind of anti-commuting; a loop through the city, carried by the body, bringing its own distinct form of communion, traced by people who usually follow the route by bus, taxi, car, train – who travel in isolation, the only point of the journey being arrival at their destination, whether the site of production (work) or the site of reproduction (home); the marathon, then, as the commuters' version of *Reclaim The Streets* – beating the streets with their bodies (or being beaten – not everyone makes the finish), asserting their individual and collective presence in the urban fabric which ordinarily invisibilizes them; in the same way that protest marches achieve their fullest potential through the very act of being on the city streets – using those streets to articulate a kind of politics of visibility – so the marathon runner takes to the streets to make a statement, to make a difference – by re-enacting the daily commuter flow of people through the city, but recasting it as a spectacle of urban bodies in urban space; of course, there's another side to the contemporary urban marathon, and one which is particularly focused upon by media commentators as providing an alternative rationalization and legitimation for the spectacle of urban bodies enduring limit experiences: charity – if it's done for a good cause, almost any seemingly mindless self-abusing act can be invested with a suitable, upstanding, collective civic and moral meaning; this charity legitimation is especially interesting, perhaps, if we view the urban marathon as a (middle-)class-specific statement about the relationships between the self, the city and charity – about selfless benevolence as class responsibility, and about rendering hyper-visible the charitable body – making a spectacle of charitable acts in the same way that past generations of benevolents aided the urban poor through very public rituals of philanthropy (the difference being that rather than dip into one's own pockets, the marathon runner deploys his or her body as a way to get non-marathon runners to dip into theirs – charity here is not about the giving of one's own money, but the giving of one's will and one's body, and *other people's money*, which the marathon runner symbolically launders); of course, money enters the marathon through a second route, too: through corporatization – urban marathons all bear the mark of corporate sponsorship, emblazoned on the runners' bodies, bringing to the streets a very different agenda, of business image-making and relentless promotionism: the urban marathon, then, as individual yet collective, benevolent yet commoditized – as a distillation of the late-modern urban human condition: irrational recreation offered the rationales of health, charity and business . . .

> In driving rain, with helicopters circling overhead and the crowd cheering, wearing aluminium foil capes and squinting at their stop-watches, or bare-chested, their eyes rolling skywards, they are all seeking death, that death by exhaustion that was the fate of

the first Marathon man some two thousand years ago. . . . The marathon is a form of demonstrative suicide, suicide as advertising: it is running to show you are capable of getting every last drop of energy out of yourself, to prove it . . . to prove what? That you are capable of finishing. (Baudrillard, 1988, p. 20)

References

Baudrillard, J. (1988) *America*, London: Verso.

Frank, A. (1991) 'For a sociology of the body: an analytical review', in M. Featherstone, M. Hepworth and B. Turner (eds), *The Body: Social Process and Cultural Theory*, London: Sage.

Finishing line

•MEETING PLACES•

by Allan Pred

Stureplan

AT THE CORE OF THE CITY,
at a square of simultaneous convergence and divergence,
at an open space that serves as a pivot between Stockholm's most
fashionable residential area(*Östermalm*) and its downtown
shopping and office district,
at a hub where the on-the-ground
circulation of local consumers comes
into conjunction with the global
circulation of commodities and
capital (facing in toward what is
popularly known as 'The Mushroom' –
a low concrete column with a wide
circular cap at the center of
Stureplan – are neon signs flashing
the representations of multinational
corporations, retail outlets offering
foreign as well as domestic goods, a
major bank, a SAS tourist bureau,
a Burger King outlet, and an
assortment of restaurants and night
clubs much of whose fare is imported),
at a site where the largest of
the surrounding buildings from the 1880s –
with its century-old postmodern
pastiche of classical and
neorenaissance elements – houses an
arcade of up-scale establishments
which cater to the niche markets so
favored by the agents of flexible

Between 1989 and 1994
the disposable income of
80 percent of Sweden's
population declined.

Since 1993 the unemployment
rate in Sweden for
those of non-European
background has fluctuated
between 37 and 45 percent.

'Our youths are on the
lowest rung of the ladder
when they look for work.
They have almost no

production and flexible accumulation,
at a location where contradictions and
compatibilities are juxtaposed with
one another,
AMIDST THE ROUTINE PRACTICES
AND OCCURRENCES OF
EVERYDAY AND EVERYNIGHT
LIFE,
amidst the cacophony of vehicular traffic,
amidst the slow blur of pedestrian
movement – the shuffling and the rushed, the stylishly attired and
the casually clad, the plastic-bag toting shoppers and the white-
shirt-and-tie financial service workers, the backpack bearers and
the attaché case carriers, the well-oriented locals and the direction-
asking visitors,
amidst those standing beneath 'The Mushroom' – the foot tapper and
the head swiveller, the yawner and the distant-horizon squinter,
the face scratcher and the glazed-over gazer, the ceaseless smiler
and the unrelenting frowner, waiting, waiting, waiting,
patiently or nervously, for a lover,
spouse or friend,
amidst those chattering at outdoor
serving tables,
amidst the cheerful and the discontent,
amidst those who walk at ease and those
who are unable to release the pain
of economic uncertainty or
(un)employment anxiety for a single
step,
VIOLENCE MAY ERUPT WITHOUT
PRIOR WARNING,
a particular conjuncture of seemingly
ordinary circumstances may
prove explosive,
the long simmering may boil over,
the current moment of danger may
become a site-specific concrete
reality rather than a general
state of the hypermodern present,
as during the late evening hours of
December 3 and the wee hours of
December 4, 1994,
when, as during any ordinary Saturday
night, the teeming life of
Stureplan is at its most teeming,
when throngs of young women and men
are making their ordinary club,
restaurant and disco rounds, often
forming long outdoor queues as
they await admission to the most

chance to enter the labor market. . . . It's a ticking bomb.' (Anneli Ward, a youth social worker in a segregated Stockholm suburb (*Dagens Nyheter*, December 10, 1994))

'Racism is worse here than there. . . . In France I never experience the same feeling of inferiority as here. For you we are nobody. We share no history and you have had terribly little contact with black people. . . . the longer one lives as a black in Sweden, the more one realizes how widespread racism is. How deep it goes.' (Papa Sow, Senegalese migrant studying in Stockholm to be a teacher (*Dagens Nyheter, June* 1, 1996))

'Roughly speaking, here is how people are assessed in nightclub environments: Swedes are atop the pyramid, then second – generation migrants, and lastly, 'the rabble', e.g., most blacks.' (Inti Cecil Sondlo, African migrant (*Dagens Nyheter*, December 8, 1994))

m
m
m

popular establishments,
when, in nothing out of the ordinary fashion, those of apparent non-European background, those of dark skin or dark hair – those who are racially slurred in everyday Swedish as 'blackheads' – are repeatedly denied entry by door-guards who provide no due cause,
when, in particular, three men in their early twenties – one of them the son of Chilean political refugees and another of them of Irani background – are not allowed into a currently 'in' club, *Sturecompagniet*,
when, in the unfolding of that increasingly heated moment, in the midst of that petty demonstration of power, racist insults as well as a brutish blow are dealt out by one of the 'entrance hosts' before the eyes of two patrolling policemen who shortly require the Chilean to identify himself,
when in the aftermath of that joint debasement and public humiliation, in the one-hour-later sequel to that shared loss of face, the three men return from the suburbs, armed with an automatic weapon, determined – according to court testimony – to demonstrate to one another that 'they are men and not cunts', determined to get back at the door-guards (and the countless other degraders and epithet hurlers lingering in their memory?), determined to scare the shit out of them, determined – under weapon protection – to beat any remaining shit out of them, intending nothing more,
when the situation grows out of control,
when hell is quickly reached via a road paved with bad intentions,
when first the psychological trigger goes off and then the AK4 trigger goes off,
when first there is scurrying and confusion and then there are four dead and twenty wounded,
when

'Apartheid can assume many forms, also via systematic queue selection.' (Aleksandra Ålund, sociologist ('Lilla Aktuellts förkrympta världsbild – Brott och bruk av kulturella koder,' *Kulturella perspektiv, 1995, no. 4, 2–17,* quote from p. 9))

According to a 1996 survey conducted by the Center for Migrant Research at Stockholm University, two out of three African men had been turned away from a restaurant or club during the past year. Middle Easterners and Latin Americans resident in Sweden also readily account such discriminatory experiences.

'To be a migrant [of non-European origins] *in Sweden in the year 1996 is obviously not easy.'* (Frank Orton, *diskriminersombudsman*, or head of the Anti-Discrimination Board (*Dagens Nyheter*, August 29, 1996))

•MEMORY•

by James Donald

Why did Atget's photographs of deserted Paris streets so unsettle Walter Benjamin? The absence of people or actions suggested to him the detachment of an official record of the scene of a crime. They provoke the question: What happened? Or rather: What happened here?

This *frisson* of urban space being haunted by unknown past events may be a clue to the way that memory is inscribed in a city. Of course, this is quite different from the official version of a collective memory set in place by the nineteenth-century rebuilding of the great European capitals. There, a peremptory and apparently confident architecture told citizens what the past meant and where they fitted into its narrative of the nation. Against that pedagogic architecture, though, there can be found a more performative, less disciplined way of projecting meaning on to the symbolic space of the city through the act of recollection.

What sort of act is it, then, to remember a city, to make the past city present?

I discovered while thinking about that question that the novelist Virginia Woolf grew up in the same London street as I did, in Hyde Park Gate – 'that little irregular cul-de-sac which lies next to Queen's Gate and opposite to Kensington Gardens', as Woolf described it. She was there sixty years before me, around the turn of the century, and her family was much grander than mine. Even so, there is a shared sense that living in that place in some way shaped what happened later. 'Though Hyde Park Gate seems now so distant from Bloomsbury, its shadow falls across it,' Woolf recalled. '46 Gordon Square could never have meant what it did had not 22 Hyde Park Gate preceded it.' Woolf's memoir also reveals the structure of memory: a dramatization of the past with yourself as both actor and spectator.

> I felt as a tramp or a gypsy must feel who stands at the flap of a tent and sees the circus going on inside. Victorian society was in full swing – Vanessa [her sister] and I beheld the spectacle. We had good seats at the show, but we were not allowed to take part in it. We applauded, we obeyed – that was all.

In this theatrical spirit, I recall that 8B, our rented flat, was at the top of a building on the corner of Hyde Park Gate. At the front, it looked over Kensington Gardens. At the back, the cul-de-sac was the stage for a rich cast of characters (even if at the time the drama seemed generally unremarkable). Number 20 was the home of Winston Churchill, then in the dog days of his final term as Prime Minister and later in resentful retirement. Opposite him, at number 18, lived the sculptor Jacob Epstein, who represented a face of modernism acceptable to the English at least to the degree that he had been the butt of endless philistine humour. An interesting conjuncture, from the perspective of my present interests, and one that Woolf might have appreciated: the embodiment of a patriarchal myth of Empire over the road from a cosmopolitan and Bohemian artist.

In recalling the London of my childhood – remembered as London, even though just one street – what is it that I am acting out? It is not just the bringing to mind of past facts. When I look closely at photographs of the place at that time, there is a dissonance between the visual record and what the memory feels like. This cannot be ascribed solely to the datedness of the cars and the clothes. That is not a place I can get back to; I can only imagine recollected events taking place. Notoriously, of course, the vivid events of childhood may or may not have taken place. Yet remembering them, however inventively, remains a way of working through current desires and anxieties. In my case, the new knowledge about Virginia Woolf becomes part of my narrativization of Hyde Park Gate – a street I have not visited for twenty years or more, and a home demolished in the 1970s, but still a place that is central to my experience of myself as formed in and by London.

Two things about memory and the city are clear: the narrative quality of memory, and the essential temporality of the city. There is a city formed from strata of recollected events and ascribed stories. In her novel *The Four-Gated City* (1969), Doris Lessing describes how a Londoner sees her neighbourhood not as socially deprived and ugly, but as textured and animated by such layers of memory.

> Iris, Joe's mother . . . knew everything about this area, half a dozen streets for about half a mile or a mile of their length; and she knew it all in such detail that when with her, Martha walked in a double vision, as if she were two people: herself and Iris, one eye stating, denying, warding off the total hideousness of the whole area, the other, with Iris, knowing it in love. With Iris, one moved here, in a state of love, if love is the delicate but total acknowledgement of what is . . . Iris, Joe's mother, had lived in this street since she was born. Put her brain, together with the other million brains, women's brains, that recorded in such tiny loving anxious detail the histories of windowsills, skins of paint, replaced curtains and salvaged baulks of timber, there would be a recording instrument, a sort of six-dimensional map which included the histories and lives and loves of people, London – a section map in depth. This is where London exists.

This is the city as palimpsest: a space on to which meaning is inscribed, and then obliterated as new meanings are inscribed on top of them. It is that opaque and enigmatic surface which Atget captures in his Paris photographs. Sedimented into the city's very fabric, these meanings can be recovered only through a symbolic archaeology; the act or art of memory.

•MONEY•

by John Allen

Not so long ago the streets of many a city – if not exactly paved with gold – were dotted with pawnbroker shops. In such places you could deposit often the most intimate of possessions as security for money borrowed: that prized wrist-watch, the family heirloom, or even father's best suit. Crucially, such brokers provided the breathing space of credit; that is, the space between what you lack financially and what you want to purchase, or rather possess. Much of the urban – its production and circulation – turns on this space of tomorrow. Most obviously, the built environment where anything from flats, studios, houses, restaurants and shops to the ubiquitous monumental office blocks and more, are realized through imaginative financial arrangements ranging from mortgages, loans, rights issues, derivatives and other such instruments. Perhaps David Harvey (1982, 1985) was right when he intimated that if it were possible to trace the circuits of capital through the built environment, their twists and turns as well as their fixture in land and property, we would know much about how cities work to reproduce themselves, often in contradictory and restless ways.

But Harvey would have been only half right, if that – for the movement of money and capital around the city involves more than simply land and property. Consumption is a key element in the turnover of city life too, in all its many and varied forms. Necessity takes people out to shop in the instrumental pursuit of basic household goods, whether it be to the local markets, discount stores, or supermarkets. Coins and notes, cash in various denominations remains the mainstay of routine monetary exchange, but it is credit, principally in the form of 'plastic cash', which provides the means to indulge in the creative act of consumption. The contemplation of a new fashion accessory, the desire to shop for the latest in household design, the craving to possess that little item of splendour in the department store, all arguably are encouraged by the sense of deferred cost which credit makes possible.

It was George Simmel, however, who first put his finger on the liberating effects of a modern money economy and, indeed, stressed the congruence between urban lifestyles and the machinations of money. In *The Metropolis and Mental Life* (1903), his best-known essay on the city, he drew a parallel between the objectification of urban relationships and the pure abstraction of monetary exchange. The resemblance between the two sets of relationships is critical for Simmel, for it is through the cold, calculating nature of market indifference that the individual is freed from the bonds of trust and obligation associated with traditional, communal ties. Put another way, the city has no claims on the city dweller who is now free to indulge – pursue means over ends – without fear of the social consequences. Indeed, the latter is made possible by the fleeting patterns of abstract social interaction which push people to distance themselves from others, to become blasé about the pressures of urban living, even if it is at the expense of others or, perhaps today, at the risk of a state of permanent indebtedness.

Since Simmel's time, much has been made of the heightened pace and intensity of city life, if not always related to the speed-up in monetary circulation, then at least to the quickening pace of late capitalism. Even if transport in the city is not going any faster than it has in the past, urban life for many seems to have moved up a gear. It seems more rushed. The gap between event and experience seems to have diminished further as what goes on around the city intensifies. Perhaps more than anything today this is a script written for those who work in the financial districts of the major, global cities – the dealers and brokers of the big finance houses who try to make sense of the bewildering array of facts and information about the swirls of money which oscillate around the globe. The symbolic landscapes at the heart of New York, London, Tokyo, among others, signify the prestige and dynamism associated with such fast monies; monies which connect loosely or otherwise to everyday patterns of consumption, style and fashion, as well as to Harvey's circuits of capital.

It is not as if money binds the city as a whole, however – as if it were a ball of string which, once unravelled, would reveal the many different facets of city life. There are interconnections to be drawn, but in cities as far apart as Manchester, Mexico City, Moscow or Manila, there are disconnections as well as discontinuities. Within cities of the developed world as well as in the less developed world, there are those excluded from the moneyed circuits of the formal economies. Credit, in such instances, may come in the form of a 'backhander', a loan shark, a promise, an exchange in kind, or, if lucky, a local credit union, little of which may enter into the formal world of banking and finance. This is not a world of cheques and credit cards, nor of mortgages and overdrafts, but rather one bypassed by most of Simmel's modern and more recent transactional styles.

Even if the coded 'hole-in-the-wall' cash dispenser has replaced the pawnbroker's shop as the symbol of money on the high street (in the post-industrial city at least), this does not mean to say that everyone subscribes to the same colour of money. It is superficially green for those 'high-powered' individuals engaged in dollar-denominated transactions, pink for those trading in gay retail spaces, or even black (is that a colour?) for those caught up in the shadowy world of illicit markets. Yet, for the most part, its colour has been drained by the commodification of almost everything, including urban relationships. Simmel was right, perhaps. Money is a frightful leveller. Everything it seems has a price.

References

Harvey, D. (1982) *The Limits to Capital*, Oxford: Blackwell.

Harvey, D. (1985) *The Urbanization of Capital*, Oxford: Blackwell.

Simmel, G. (1903 [1950]) 'The metropolis and mental life', in K. H. Wolff (ed.), *The Sociology of Georg Simmel*, New York: The Free Press.

•MUSEUMS•

by Iain Chambers

The great Renaissance scholar and magician Giordano Bruno was born in the shadow of Vesuvius. He was burnt at the stake as a heretic in Rome in Campo dei Fiori in 1600. But here in downtown San José, amongst palm-treed streets and convertible cars, his presence is very much alive. There is here a hole in time that connects the capital of Silicon Valley directly to sixteenth-century Europe. For San José is also the home of the Rosicrucian Museum and the largest collection of Egyptology west of the Rockies.

The ghosts of Pharaoh Thutmosis III, Hermes Trismegistus, Bruno, and others who sustained the lore of Egyptian metaphysics are discreetly fostered by the Rosicrucian Order through the display of the ancient Egyptian way of life and death. The museum itself is located in the Rosicrucian Park which occupies a whole city block. Its entrance is flanked on either side by four ram-sphinxes. In the park there is a hieroglyphic-covered red obelisk, capped in copper: 'a three-quarter size replica of the original, which stood before the House of the Sun at Heliopolis.' Elsewhere there is the Grand Temple, a modified reproduction of the Temple of Hathor at Dendera; a statue of the falcon god Horus associated with the sun and the heavens, son of Osiris and Isis; a statue of Thutmosis III (1505–1450 BC), the inspirational source for much of Rosicrucian thought; and the Pylon gateway, similar to the one leading to the temple of Medinet Habu (which serves as a model for the Administration Building), that is adorned by a baboon who 'sacred to Thoth – the Egyptian god of wisdom and judgement – sits atop a pair of scales balancing the ib, or heart, of a deceased person, against Maat, the feather of truth.' Inside the museum, after wandering amongst the glass case exhibits, sarcophagi, canopic jars and a replica of the Rosetta Stone, you can descend into a tomb from Luxor – a full-scale reproduction of a 4,000-year-old Egyptian noble's resting place.

Such appropriations are immediately susceptible to facile accusations of cultural imperialism, or, in a lighter vein, exotic kitsch. Yet there is a deeper, more fascinating current that throws light on this scene. From the desert to the desert, from one heliocentric culture to another, from ancient, feudal Egypt to modern, multifarious California, there courses a metaphysics of the spirit, and a simulation of life, that yesterday was embodied in a pyramid and today in a silicon chip.

(All quotes are from the self-guided tour book, *A Walking Tour of Rosicrucian Park*.)

•MUSIC (1)•

by Arturo Escobar

> Music is prophecy. Its styles and economic organization are ahead of the rest of society because it explores, much faster than material reality can, the entire range of possibilities in a given code. It makes audible the new world that will gradually become visible, that will impose itself and regulate the order of things; it is not only the image of things, but the transcending of the everyday, the herald of the future. (Attali, 1977, p. 11)

Cities, like music, carve up spaces. The musical space, however, as some theorists tell us – such as Boulez or Xenakis, the musician-architect – while marking out the materiality of objects, is not primarily a geometrical but an existential space. Contemporary electronic musicians have made an art of the creation of new aural possibilities of being-in-sound by combining space and technique. Something of the sort is always going on in many urban quarters where popular music blasts its sounds in the heat of the night. Mention Papa Wemba or Ismael Lo or Salif Keita, and the streets and lives of people in Bamako, Kinshasa or Dakar will reverberate with meanings and pleasures. Something similar can be said of New Orleans during Mardi Gras or Rio, Kingston or Havana, or of well-demarcated spaces here and there in New York or Paris and in so many other cities of the world. Music partakes of the creation and practice of space. In dance, music becomes a practiced space.

One may say that there are distinct musical cities in this sense, particularly those where popular music regularly moves a significant proportion of its inhabitants to the kind of enactment of body-in-space that is peculiar to dance. Cali, Colombia is a city of this sort; the city's passion for music has been vividly celebrated by its young fiction writers and, more recently, by communications scholars working under the paradigm of reception and consumption of popular culture. With strong roots in the popular classes and the city's significant and growing black presence, for the past fifty years tropical music has enveloped this city, a central element in the struggle for urban space. It is as if, with rapid industrialization and urban expansion, music had always been there, presaging and effecting at the same time the class and gender contest over culture and space. And while spatially contained for some time, tropical music today seems to have reterritorialized the city in unambiguous ways, as a powerful contribution of the subaltern classes to the urban scape.

Cuban music – particularly that which is generically known as salsa, made up of many distinct Afro-Cuban rhythms and popularized in New York in the 1960s – became well known in Cali since the 1950s, certainly after the first visit of the most famous Afro-Cuban orchestra of all time, the Sonora Matancera, direct from Cuba with the then-young but already renowned singer, Celia Cruz, the Reina of the Rumba. That legendary concert took place at a working-class theater in the early 1950s, and since then salsa has spread throughout the city, slowly transgressing class boundaries and becoming 'hegemonic' in the 1960s when the best Puerto Rican, Nuyorican and Cuban orchestras started to frequent the city and to compose songs to it – from Richie Ray and Bobby Cruz in the 1960s to Eddie Palmieri and Willie Colón in the 1970s and 1980s, to La India and new and old Cuban groups in the 1990s. While the city exhibits a more heterogeneous music landscape today – the result of an increasingly complex globalization of media industry, culture and identity – salsa remains the dominant musical genre. The best international salsa groups tour the city throughout the year, alternating with the growing number of highly accomplished and celebrated local groups. Many of them converge at the International Salsa Festival that takes place each year at the soccer stadium – as one of the highlights of the city's peculiar brand of carnival that takes place between Christmas and January 3 – when 60,000 people vibrate to the music of their favorite local or foreign groups non-stop for fourteen hours, forgetting about the Colombia outside which each year seems an inch closer to the verge of falling apart.

It is true that if you go out at night you may find people at most if not all the hundreds of dance places – of all prices, styles and types, revealing a complex geography of class, gender and race – that abound within the city limits and beyond. It is true that in any of them you may find couples that have made an art of their dance, to which they devote not one but two or three nights every week; it is true that in a few of them there are salsa shows of the most talented dancers – always working class, often black – sometimes also staged for tourists to see. You may also come across the most radical dancers who believe that music is the only redeeming force in a fragmented world, that musicians have in their hands the reins of the universe, and that one should give oneself to the night in a mixture of alcohol, music, pleasure and dance, as if their lives depended on the ability to perform to perfection the last step which they have just improvised, and one more after that, in a perpetual and sensual embrace with their partner of choice. Dancers, in fact, seem to craft a space of their own in the dance floor throughout the night.

The 1980s came, and with them – a reflection of the influx of drug money and the steady deterioration of the social and political situation of the country as a whole – grew insecurity at night and increased the prices of liquor at night-clubs and bars. The people of Cali could not resign themselves to staying at home on weekend nights, and so there appeared in 1994, unnoticed, the first *viejoteca* (literally, old people's discotheque) as an answer to both lack of safety and soaring prices for everything, from rentals to food and liquor. This first *viejoteca* – which began with a strict age restriction, for people older than 45, and with a music selection of tropical and Afro-Cuban music from the 1940s to the 1970s, which older couples would dance as in the days of their youth – soon gave way to many more (about twenty or thirty today) with the distinct mark of providing a safe environment for people to dance, moderate prices, and the most wonderful music from the 'classic' periods of Afro-Cuban and salsa music.

Take the *viejoteca* Séptimo Cielo (The Seventh Heaven), which took its name from one of the oldest working-class dance places of bygone days. It opens its doors at 4 p.m. (another innovative feature as compared with conventional night-clubs which open much later at night) to women and men from both working and middle classes between the ages of 25 and 45, and it goes on until the mandatory closing time of 4 a.m., couples dancing to 1960s and 1970s salsa, with some newer rhythms interspersed in between (fox, charanga, rumba, cumbia, merengue, vallenato, newer salsa and rock read the sequence pre-prepared by the disk-jockey on that night). On a typical night it might sell – to the approximately 700 people who come – 180 half-bottles and 100 bottles of *aguardiente* (local liquor made from sugar cane alcohol), forty half-bottles and twenty bottles of rum, and smaller quantities of Finlandia vodka, local beer and whisky. You would see men and women drinking at roughly the same pace, dancing with the same passion and agility, equally enjoying themselves; it is rare to see people drunk. You do not see women dancing together, but many women do come alone, perhaps after work, to have a drink and enjoy a dance. Male groups are not admitted without female partners, because 'they might get drunk and their macho egos might surface in troublesome ways,' although the place prides itself on total safety (people are frisked at the entrance, and those few carrying hand-guns are asked to leave them at the office), insured if needs be by twelve big security guards who immediately take out to the street trouble-makers or those perhaps engaged in a disruptive marital dispute.

This well-known *viejoteca* – not really for old people, although families are encouraged to attend – exemplifies this city trend, with its lower prices (about 50% for drinks of what it would cost at a regular place), welcoming ambience (not even a watch has been lost since its opening three years ago, the manager takes pride in saying), great music, and a large and well-thought-out dance floor. And so the city seems to have found a way to hold on to its musical vocation, to make of music and body and space one single and intense ludic experience perhaps giving some meaning to life in a country that seems to have lost all sense of direction. 'Que viva la música,' the city seems to say every night, echoing the voice of one of its young writers that described long ago the whirlwind that swept over Cali when salsa music first arrived, the same moment that old and young people alike re-enact every weekend in their *viejoteca* of choice.

Indubitably a contest over culture and space, can we also say that this re-enactment/production contains the possibility of a world to come? Explaining his notion of the prophetic value of music, Attali (1977) adumbrates the emergence of a new kind of musical practice, which he labels composition and which heralds new social relations. Perhaps in the body-knowledge (also a form of power) of the most committed salsa dancers we might see the reverberations of the open order – an urban order? – Attali talks about, one that is not totally subjected to the economy of the lack and the commodity:

> [Composition is] becoming a real potential for relationship. It gives voice to the fact that rhythms and sounds are the supreme mode of relation between bodies once the screen of the symbolic, usage and exchange, are shattered. In composition, therefore, music emerges as a relation to the body and as transcendence. . . . Composition thus leads to a staggering conception of history, a history that is open, unstable, in which labor no longer advances accumulation, in which the object is no longer a stockpiling of lack, in which music effects a reappropriation of time and space. (pp. 143 and 147)

Sources

Conversation with managers and disk-jockeys at Viejoteca Séptimo Cielo (Carrera 44 con Calle 14) and Viejoteca del Parque de la Caña (Avenida 2 Norte con Calle 32A); the novels of Andrés Caicedo and Humberto Valverde; works on popular music by communications scholars at the Universidad del Valle in Cali (Alejandro Ulloa, Sonia Muñoz, Jesús Martín Barbero).

Reference

Attali, J. (1977) *Noise: The Political Economy of Music*, Minneapolis: University of Minnesota Press.

•MUSIC (2)•

by Susan J. Smith

Music is the last of the arts to be scrutinised through the critical lens of cultural politics.

Thanks to Edward Said we have little difficulty accepting that the novel speaks of colonialism and imperialism; Denis Cosgrove and Stephen Daniels show how landscape painting is a statement about how powerful people think the world ought to be. Cindy Sherman and Ingrid Pollard make us think twice about how women are (and are not) depicted in photography; and bell hooks confirms how much art matters if we are to engage with the politics of the visible world.

But music!

Music has been above all this. An ethereal art in a world apart from the skirmishes of material life. Perhaps we hoped that at least one art, even if it were 'only' music, could be above worldy things, beyond politics, apart from the economy.

But it isn't.

Enlightenment principles and modernist practices were so pre-occupied with what could be seen, that we forgot, or thought irrelevant, the study of sound. 'For twenty-five centuries, Western knowledge has tried to look upon the world. It has failed to understand that the world is not for beholding. It is for hearing.' Now we know that there is more to cultural politics than the visible world. So here is a story which is unthinkable without music. It is one story among millions rooted in the strains of the city; one among many which rely on the sounds of the streets.

It is the story and the song of the birth of the US metropolis.

The practicalities are documented in the standard (if terminologically dated) texts:

> The 'Great Migration' of 1915 marked the beginning of a significant shift in the distribution of black people from the rural South to the urban North. . . .
>
> Forced to migrate to the northern cities, they found adjustment hard, and life in the slum ghetto far from pleasant; but a move to a northern city was the only realistic means of improving income, obtaining an education, and gaining political power. . . .
>
> Like the immigrants from abroad, the Negro migrants to Northern cities filled the lowest occupational niches and rapidly developed highly segregated patterns of residence within the central cities. . . .
>
> As the poorest and newest migrants, the Negroes were forced to double up in the slums that had already been created on the periphery of business and industrial districts. The pattern has never been broken.

This is what some scholars call 'American Apartheid' – the practice of segregation, the process of exclusion, the coupling of spatial separation with 'racial' inequality. Yet, material facts are only part of the tale. The picture seems black and white, cut and dried. But oppression and exclusion are not the whole story.

'Critics are often at a loss as to how to articulate exactly what it was . . . that made black music central to a description of these times.'

The city provides the clues:

> The valorization of black rhythm, spontaneity, laughter and sensuality . . . contrasted starkly with Harlem's squalor.

Attali is right: 'All music, any organisation of sounds is a bid for the creation or consolidation of a community.'

Duke Ellington heard it loud and clear:

> That's the Negro's life. Hear that chord! That's us. Dissonance is our way of life in America. We are something apart, yet an integral part.'

> It was the power of the music, after all, that had named the epoch 'The Jazz Age'.

And as Theodor Adorno knows, 'Every musical phenomenon points to something beyond itself by reminding us of something, contrasting itself with something or arousing our expectations . . .'

Toni Morrison captures the spirit:

> The music bends, falls to its knees to embrace them all, encourage them all to live a little, why don't you? since this is the it you've been looking for.

Music is the voice of those who are otherwise silenced or struggle to be heard.

Music is a position from which to say the unspeakable, think the unthinkable, tackle the undoable. Music is a way of being, a space of identity. It is a powerful place. Here life can be improvised and elaborated, reworked and renewed.

> The desire to make music into the medium of cultural rebirth and to hear in it the characteristic signature of racial genius has been a recurrent feature of thinking about black culture.

And the place to work this magic? The spaces of the city. And the space of one city in particular, New York. And one place in New York in particular. Between 130th and 145th, North of Central Park. The space and the symbol of Harlem.

> Stories are legion of African-American and African pilgrims progressing to Manhattan then plunging headlong into the ultimate symbolic black cultural space – the city within a city . . . Harlem was not so much a place as a state of mind, the cultural metaphor for Black America itself.

In these days when Harlem said it all

> the uneducated blues-loving poor . . . the hedonistic jazzers . . . all proceeded from the same remarkable assumption. All saw a new place for music as the central source of cultural value.

Up there, in that part of the City – which is the part they came for – the right tune whistled in a doorway or lifting up from the circles and grooves of a record can change the weather. From freezing to hot to cool.

Harlem, the great black metropolis, could nurture the life of an emergent people, provide a crucible for their cosmopolitan consciousness . . . form the majestic kernel of a novel modern enterprise, bringing new life to the race after . . . the catastrophic shock of adaptation to impoverished life in America's cities.

And so it was that 'What institutional politics . . . could not do, what violence, crushed by counterviolence, could not achieve, free jazz had to bring about.'

There were struggles of course:

'Who would be the custodians of the racial spirit, the guardians of the distinctive heritage that would be the awe-inducing gift of black folk, black Americans, to the modern world?'

The *Negro Music Journal* missed the point:

We shall endeavour to get the majority of our people interested in that class of music which will purify their minds. . . . The day of low, trivial, popular music should be cast aside forever.

Toni Morrison didn't:

[she] knew from sermons and editorials that it wasn't real music – just colored folks' stuff: harmful, certainly; embarrassing, of course; but not real, not serious.

Yet Alice Manfred swore she heard a complicated anger in it . . . it made her hold her hand in the pocket of her apron to keep from smashing it through the glass pane to snatch the world in her fist and squeeze the life out of it for doing what it did.'

. . . And if that's not enough, doors to speakeasies stand ajar and in that cool dark place a clarinet coughs and clears its throat . . . the City is smart at this: smelling good and looking raunchy; sending secret messages disguised as public signs.

Tales of the city are always more than they seem. 'The limit is your own imagination, really.' (Courtney Pine, *Modern Day Jazz Stories*).

Acknowledgement

'Music' is a montage, cut and mixed from *Rhapsodies in Black* (1997, The South Bank Centre, several authors), *Jazz* (1992, by Toni Morrison), *Noise* (1977, trans (1985) by J. Attali), and various standard social geography texts.

•NECROPOLIS•

by Jill Forbes

Cities of the dead

The Necropolis is the gateway to Glasgow, a monumental *memento mori* to the motorist driving too fast along the M74. Perched on a drumlin (mound) overlooking the Merchant City district, its granite obelisks and urns outlined against the racing clouds are relics of the city's Victorian splendour, articulating a powerful corrective to Glasgow's modernising ambition. The great cemeteries of nineteenth-century London, designed to accommodate an exploding population and its attendant epidemics, are full of neo-Gothic exuberance unquelled by Marxist rationalism, even, and especially, in Highgate. Not so Glasgow which, like its model the Père Lachaise cemetery in Paris, aimed to treat death hygienically and scientifically. Its site ignored the contemporary trend to locate burial grounds on what were then the outskirts of town. Instead, the triangle formed by the Royal Infirmary, where the burghers died, the Cathedral where their funeral rites were conducted, and the Necropolis, where they are buried, proudly proclaims economy of effort since it is contained within little more than a square kilometre. Indeed, if scientific rationalism inspired its creation, neo-classicism presided over its design. Just as Père Lachaise is marked out (*quadrillé*) in 'divisions' of a quasi-military precision, so the Necropolis (avoiding the banality of 'cemetery') is divided into sectors named for the letters of the Greek alphabet (*alpha*, *beta*, etc.) with recourse to Latin (*primus*, *secundus*) when the former were exhausted, and embellished with a funerary architecture of Doric temples, Corinthian columns and Roman sarcophagi, much of it designed by Alexander 'Greek' Thompson (Black, 1992).

Here it's hard to believe that 'Glasgow smiles better'. Not only is it splendidly windswept and gloomy, it is also neglected. The visitor to Père Lachaise is met with a notice explaining the history of the cemetery and is given a leaflet listing the names of the great and the good who are buried there (Mairie de Paris, *Cimetière du Père Lachaise: Les Personnalités*). According to taste, one can leave a scrawled message on the plinth of Oscar Wilde's art

n
n
n

deco monument, examine the expansive marble of the dignitaries of the French Communist Party who, even in death, 'occupent le terrain', question the wisdom of unearthing Yves Montand to settle a paternity suit, and still be home for an apéritif. But in the Necropolis you can wander for hours before finding the tomb of the rather frail-looking Charles Tennant whose patent bleaching process poisoned generations of Glaswegians and allowed his descendants to retire to Mustique. The Necropolis has remained untouched by the successful revival of the East End of the city and even the admirable ecumenicism, which invited members of all faiths to be buried here, is belied by the Jewish sector which is disused and almost completely overgrown.

If it is true that all good Glaswegians are in Australia, Canada, South Africa, the United States, England – anywhere, in fact, except their native plot – this may explain why the Necropolis is dominated by a 'personality' who isn't even buried there: high on his column, John Knox casts a baleful, Presbyterian eye over this essentially Catholic city and appears to wave a monitory finger at the loose-living dons whose own tower, on the city's western side, responds to Knox's provocation with neo-Gothic fantasy.

Graveyards, in general, are either moving or impressive. St Clement's, by the beach in Aberdeen, brings tears to my eyes with its litany of local patronymics – Anderson, Donaldson, Fraser, Forbes – of local occupations – rope-maker, shipmaster – and its almost universal cause of death 'drowned at sea'. Its structural equivalent, Sète's 'cimetière marin', for all the celebrity Valéry's poem has given it, and for all its sunny Mediterranean setting, is in essence another burial ground for fishermen. Military cemeteries – Arlington as much as Ypres – impress by the sheer quantity of the carnage that first caused their creation. Conversely, in the great metropolitan cemeteries we admire the architecture and seek out the tombs of those whose fame lives beyond the grave. But Glasgow's Necropolis is neither small enough to be intimate, nor large and well-known enough to attract more than the odd tourist. Yet it was, perhaps appropriately, the landmark that first told me I had arrived in Glasgow. To an English outsider, or even to a Scot from the other side of the country, it exemplifies how Glasgow's seemingly eternal and highly parochial conflicts are both self-contradictory and paradoxical. Otherwise, why would the neo-Gothic embody the home of rational enquiry while fanaticism is dressed in neo-classical forms, and how else could the traditions of European art and culture boil down to a religious version of the ice-cream wars?

Reference

Black, J. (1992) *The Glasgow Graveyard Guide*, Edinburgh: St Andrew Press.

•NETWORKS (1)•

by Andrew Barry

In the early part of this century the display of electricity was associated with both wealth and pleasure. As David Nye reminds us, the first electrified places in America were 'wealthy residences, hotels, theaters, department stores and clubs' (1990, p. 382). Served by electrified trolley lines, the population flocked in to see and be seen in the newly illuminated city centres. Urban electrification had a spectacular and attractive effect. It drew in the crowds.

But if the new technology of electricity played such an important part in early twentieth-century consumer culture and urban life, the city itself was primarily visualised in terms of organic metaphors. It made sense to talk about urban ecology. The city was organised functionally in terms of neighbourhoods and communities, industrial areas and financial districts. Certainly, electrical technology played a key role in the visual and commercial culture of the city, but it did not, at this time, provide the conceptual resources with which the city could be imagined.

Today, electrical and electronic technologies perhaps have a more central and complex role in the urban imaginary; for above all, the city has become a place of networks. Social networks, of course. Particularly those involving members of social and political élites and activists involved in urban political protests. The city, with its cafés, clubs and bars, provides the perfect space within which such networks can form. And close networks of

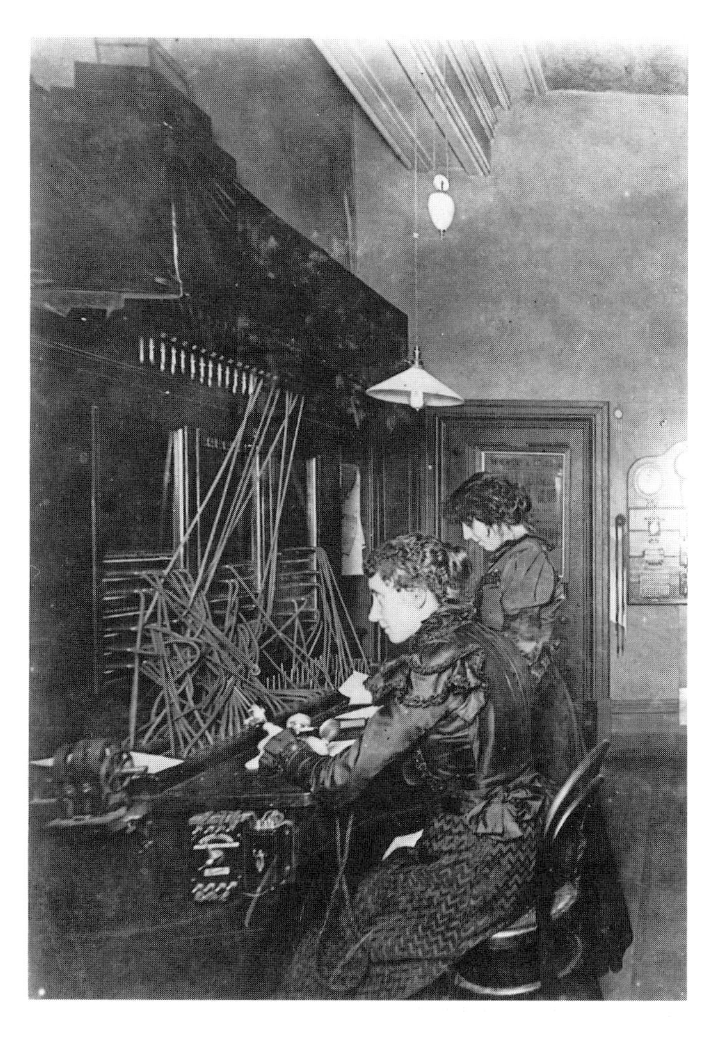

Figure 7 GPO telephone exchange, London. Reprinted with kind permission from the Museum of London

relations between professionals who, no longer managed in large organisations, depend on close personal relations with potential employers and customers. But also the extraordinary dense networks of transport, communication and information which make urban life possible. The activity of the city today is, according to many commentators, characterised by the density of its networks. We live, according to the urbanist Manual Castells, in a network society (1996). Social inclusion is a matter of being on the network. And the difference between the city and the countryside is a matter of degree. There is not the qualitative difference that existed in earlier times. In the city, the networks are simply denser, more intense and more tangled than elsewhere.

But how do networks work? There is the image of a network as something like a smoothly running system, in which connections are perfect. An electrical circuit with no breaks. This is part of the seductiveness of the image, and the sense of exclusion it generates for those who are outside. It's thought to be easy once you're on the network. Once you have gained access, whether on the basis of birth, status, skill, money or possession of the password. 'Where do you want to be today?' asks the advertisement for Microsoft. But networks aren't quite like that. The lines crackle. They break down. They snarl up. One accident or failure on the subway, or in the electricity supply, or on the roads, and whole areas of the city can be put out of action. Frustrated commuters sit in cars and underground trains or mope around at home unable to cook or watch television. *Static.*

People complain about such incidents, naturally enough. They blame them on the government, or the companies, or the national culture. No doubt the situation is better elsewhere. In smaller towns, where networks are thinner and less complex, and in places where the public services are efficient and well run, or where people are friendlier. Perhaps in other places, networks do work. The fantasy of leaving the city is no longer to return to 'nature', but just to simplify network connections.

But rather than seeing chaos and congestion as abnormalities perhaps we should see them as a normal feature of networks. To believe that networks run smoothly is to mistake the nature of networks. Networks do require constant repair work and attention. Constant reconfiguration in response to new circumstances and growth. Frequent checking to see that all the component parts are in place. Their multiple intersections can have complex and unexpected results.

In these circumstances, one urban experience is common. This is the need to be alert: to be aware of what may break down, and what is needed to avoid the consequences, or to repair them. Living in the city is a matter of developing a whole series of rather minor skills, which are infrequently acknowledged as such. What the relationship is between the published timetable and the real timetable. What the likely effects will be of the visit of a president or a public demonstration. Where to be able to meet people without fear of missing them in crowded parts of the city. How long to hold on the line. And living in the city is also a matter of developing certain psychological capacities. To not take it personally when somebody is late or cancels an appointment. To be alert to what might happen; and to be indifferent to what does. To have a sense of realism about what may be possible in practice, while possessing only a partial knowledge of what goes on elsewhere. To develop quite a sophisticated idea of what it might be possible to know, and what it might be possible for others to know.

The knowledge of the individual in the city is going to be limited. It demands a professional supplement. Reporters on local radio map the ensuing chaos, in real time. In reporting traffic congestion, they need regular slots and a permanent apparatus of monitors and surveillance cameras. The failure and congestion of one network establishes the need for others. Thus, the urban population is kept continuously informed and constantly updated, waiting to be reconnected and for the congestion to clear. For the city dweller there is always the psychological demand of network failure.

References

Castells, M. (1996) *The Rise of the Network Society. Volume 1: The Information Age: Economy, Society and Culture*, Oxford: Blackwell.

Nye, D. (1990) *Electrifying America: Social Meanings of a New Technology*, Cambridge, MA: MIT Press.

•NETWORKS (2)•

by Stephen Graham

I come across a common enough sign on the streets of urban Britain: '*This Area is Under 24 Hour CCTV Surveillance!*' Sure enough, I look up to see a CCTV camera scrutinising the urban street. But exactly who, or what, is at the other end of the camera's gaze? What would we find if we were able to enter the lens and follow the wires or wireless transmitters through to the final site of surveillance?

Nothing, perhaps? After all, many mock cameras have been installed – cheap simulations to deter burglars or criminals without the expense and hassle of having to watch the usually mind-numbingly dull panorama of, say, a backyard, all day and all night. Or perhaps the camera is connected to some local control centre in the basement of some adjacent building? Even then, while the videotapes might be recording its signals, the human operators may not actually be watching you; they can usually only actually physically watch a few of their screens. The camera may actually have been switched off for a break, time for a sandwich, say, to break up the endless hours of drudgery for the surveillor(s).

But there's really no way of telling where the gaze of the camera goes, where the eyes (possibly) watching you are watching from. Most people would probably assume that the camera is *locally* connected, tying the image in with that of the bored security guard half asleep in a nearby basement somewhere.

Such assumptions of local connectedness of CCTV emerge strongly in a (possibly apocryphal) story circulated by word of mouth in Newcastle in 1995, just after the installation of the UK's first wide area public CCTV system in a residential location (The City's West End). A set of very obvious cameras, mounted high up on spiked poles in the middle of the main streets of the area, had just been installed by the police to deter drug and property crime and to generally undermine the activities of criminal groups.

One group of young people, though, milling around in their neighbourhood, were angry at the prospect of constantly living under the gaze of cameras in their home areas. So they physically attacked the site where they thought the gazes of the cameras came together – a small building near the local community centre. Some damage was done, but to no avail. The cameras were actually being monitored in the City centre two miles away. Their gaze was 'local', just not as 'local' as they had thought.

But why, in an age of the rapidly growing power of telecommunications, stop at two miles? Why should the gaze not now be switched to a different country, a far-off time-zone, a networked space in some other city on some long-distant continent? Might our gaze into the CCTV cameras of the near future open up the prospect that the unknown human operator looking at us is on the other side of the world, a curious but anonymous meeting of eyes across global divisions of labour, an instant nexus between (northern) consumption and (southern) production?

Such logics, indeed, are already being explored, part of the relentless efforts by large corporations in the global North to switch routinised, low-wage occupations to cheap labour locations in the Caribbean, South East Asia and other 'developing' nations. Here, local tax inducements and land and property packages meet with very low labour costs and military-style work discipline to make up a tempting prospect for rationalising corporations and the companies that serve them.

But the possibility of globally networked CCTV is also being encouraged for the 'development' benefits it may bring the global South. For example, the World Bank suggested in 1994 that the millions of CCTV cameras monitoring shopping malls in North America should be switched across transoceanic optic fibre 'pipes' to be monitored in Africa – a major boost to economic development there. This (as far as I know) has yet to happen, but what a complex and ironic spin on the role of networks in 'globalisation'! Picture it. Black Africans monitoring the giant, postmodern consumer spaces of US cities, at the behest of the North American security companies that police such spaces. No doubt the brief by most such companies would be to scan especially closely the activities of young Black American men (those under most suspicion), whose ancestors were invariably slave emigrants from West Africa. And then, once 'suspicious' behaviour is spotted, instant communications back to private mall police forces to take disciplinary action. . . .

A more individualised global solution to street crime in the prosperous gentrified urban suburbs of the West has been suggested by the science fiction novelist Neal Stephenson (1995). Realising the boredom and expense of mounting some all-night neighbourhood watch programme in his own neighbourhood on the West Coast of the USA, he has developed the idea of 'Global Neighbourhood Watch'. This is based, essentially, on the idea of 'like-minded' computer-literate élites across the planet looking over each others' backs using digital CCTV cameras and computers linked to the Internet. In his plan, a series of neighbourhoods in, say, Seattle, London and Sydney – each conveniently eight to ten hours apart from the next in time-zones – would set up digital CCTV cameras in their neighbourhoods where crime is most frequent. When Seattle's Bohemians are asleep, London's will be working normally on their home or work PCs. The difference is that these PCs would also be linked over the Internet to the cameras in Seattle which would only be activated when something is seen moving at night. Once the camera is activated, a 'window' would pop instantly on to the computer screen in London, showing what's happening in Seattle, and allowing the London participant to either raise the alarm in the Seattle household or police station (when an attempted break-in is spotted) or ignore the window when it is triggered by a false alarm (say, by stray dogs and blowing rubbish).

'Windows are the best crime-prevention devices of all,' writes Stephenson; 'the virtual windows on a computer screen could serve the same function, linking distant places into a distributed neighbourhood. In a modern society, we (tellingly, the "we" here being readers of the bible of the so-called 'Digerati', *Wired* magazine) have more in common with fellow homeowners than we do with burglars living a few blocks way.' He speculates that pattern recognition software might soon develop to 'filter out dogs and tree branches', focusing only on 'crime, crime and crime again'. He anticipates (in line with the World Bank), 'that soon companies will establish warehouses in Malaysia where workers sit in front of screens keeping an eye on untold anonymous minivans.' Indeed, so entertaining might the endless, filtered, real crime scenes be, he believes, that 'they may even watch it for free'. Finally, to get things rolling, he puts up his e-mail address asking for readers in the UK or Australia to get in touch if they want to experiment by setting up their own Global Neighbourhood Watch.

It would be easy to dismiss this idea as the somewhat paranoid ramblings of a technologically enthusiastic science fiction writer, someone desperately trying to construct a global 'community' for like-minded 'Digerati' across the planet, while trying to sever his emotional, social and psychological connections with his surrounding city. Maybe. But next time you see a small camera, red light flashing, surveilling a gentrified neighbourhood on some dark evening, remember: it may not be false. There may *not* be a bored security guard in a local basement ignoring you while eating a sandwich. You might actually be under close scrutiny. And from much further away than you thought.

Reference

Stephenson, N. (1995) 'Global neighbourhood watch', *Wired UK*, December, p. 56.

•NOIR•

by Jane M. Jacobs

There are parts of the city that always feel as if they should be in black and white: those suspiciously quiet downtown areas cross-cut with impenetrable shadows and startling shards of bright light, those crumpled laneways, the left-over lobbies and darkened stairwells, those unhomely bungalows. These zones waver between the past, the present and the filmic. For film noir the city was a natural setting. Its visage has lingered in the contemporary urban imaginary, spawning various offspring: noir revival (e.g. Roman Polanski's *Chinatown*, 1974), future noir (e.g. Ridley Scott's *Blade Runner*, 1982/91), even a style of commentary aptly described as urban noir and best exemplified by the writings of Mike Davis (e.g. *City of Quartz*).

The incessant concern of the original film noir of the 1940s was the struggle between good and evil. The city was far more than an incidental setting for these moral dramas. Modern urban life – its disaffections, its unpredictable cosmopolitanism, its unknowable worlds within worlds, its opportunities for profiteering – brought these dramas into being.

But it is more than city space that is implicated in the moral architecture of film noir. These explorations of the shifting admixtures of good and evil are also worked through specific urban bodies. In the noir narrative it is almost always a 'white' body which suffers moral discoloration. The 'coloured' body – that of the Black American, the southern European migrant, the Mexican border crosser, the Chinatown Asiatic – is rarely part of the main plot. These racialized others appear as seemingly incidental additions. Yet they hover within the noir narrative as necessary amplifications of the embedded colour coding of virtue and vice in the modern city (see also Lott, 1997).

Noir may appear to have as its defining palette the intensities of black and white, but the 'dark soul' has very often been attached to an Asiatic body or place. Through this association the noir narrative embellishes a complex history of racialization within the western city – one that has depended upon coding the Asiatic body and Asiatic zones, such as Chinatowns, as the natural homes of all that is depraved, criminal, corrupt and vice ridden (see Anderson, 1991).

In Orson Welles's *The Lady from Shanghai* (1948) a grisly drama unfolds which ultimately sees the cannibalistic destruction of three disaffected, rich urbanites. Irrevocably entangled in this web of mutual destruction is a naive young sailor, Michael O'Hara (Orson Welles), also known somewhat appropriately as 'Black Irish'. The climax of this film comes when O'Hara, about to be wrongfully charged for murder, flees from the courtroom and hides out in Chinatown. He is followed there by his seductress, Elsa, one of the dysfunctional trio. It is during Elsa's pursuit of O'Hara through Chinatown that her true nature is revealed. Having once lived in Shanghai, Elsa speaks the language and calls on her various Chinese associates to help find the confused sailor. The ease with which Elsa moves through this space and harnesses its resources to her own ends reminds us of the dark moral interior of this seemingly whiter than white *femme fatale*. It is through the moral and ideological work done by Elsa being so at home in this 'other' place that we come to accept that she is indeed sinister. This moral imbrication is given specific expression not only through the palette of light and dark, but also through a racialized association which gives this white (innocent) body an Asianized (evil) interior.

A somewhat different struggle occurs between good and evil in Roman Polanski's *Chinatown* (1974). In this noir revival film the moral strife arises from corruption having infiltrated every corner of the city. That the ownership and supply of water is the conduit for this corruption only intensifies our sense of evil indiscriminately flowing through every institution, every pipe, every sewer, every body.

The moral crisis of *Chinatown* builds explicitly around a 'white' city father, his development aspirations and his family secret of incest. In the climax of the film, family drama and planning drama are drawn together in a tragic scene in Chinatown – that one part of the city where it is assumed such transgressive evil can flourish. Polansky's vision is of a city so sinister that its dark forces can no longer be pinned to identifiably Asiatic bodies nor located only in a few city blocks. Here is a white family whose intensity of evil outstrips that normally associated with and confined to a place called Chinatown.

In the two examples of film noir referred to thus far, the idea of Chinatown is central to the moral infrastructure of the narrative. Yet in both cases little of the action of the film is set in this 'other' world. The idea of Chinatown has an ambiguous presence in these dramas about whiteness and white city space – somehow central yet remaining marginal. This ambiguity of presence is replicated in a relatively recent set of urban commentaries which take inspiration from the film *Chinatown*, and which might best be described as the Chinatown essays. These urban noir commentaries revisit the theme of urban corruption and reactivate the figure of the good guy detective. Yet despite their titles, each of these essays has the idea of Chinatown and all it has stood for in the modern city well out of view.

Mike Davis's inaugural essay '*Chinatown*, Part Two?' (1989), which will form the focus of what follows, uncovers the real story behind the 'nightmare stage' of contemporary Los Angeles. Derek Gregory's '*Chinatown*, Part Three?' (1990) is an extended review of Ed Soja's sleuth work in *Postmodern Geographies* (1989). Finally, Rosalyn Deutsche's '*Chinatown*, Part Four?' (1993) queries the apparent temptation of associating contemporary urban theory with film noir, adding a specific complaint about the 'forgetting' of gender and sexuality.

Why is the idea of Chinatown virtually absent from the Chinatown essays? It might be that the processes of racialization that it stands for, pinned as they are to a particular place and people, no longer contribute something relevant to our understanding of the city or urban theory. Not even, perhaps, to the way we might account for the nature and effects of the 'Asiatic presence' in many contemporary western cities. Certainly Mike Davis's view of 1980s Los Angeles would suggest this to be the case.

For Davis, the evil which possesses contemporary LA involves a seemingly disembodied and deterritorialized process known as 'internationalization'. But just how disembodied (or deracialized) is this 'transnational' process of internationalization? The dark soul of this noir narrative resides in a new cast made up of 'Japanese mega-developers: Mitsui Fudosan, Sumitomo, Dai-Ichi Life, Mitsubishi and a dozen other major Japanese players' (pp. 61 and 71). Davis sees LA as being 'swamped by a tsunami of East Asian capital' (p. 72). Equally troubling are the new Korean entrepreneurs who, on the one hand, enliven the city with their 'Koreatown' ethnicity but, on the other, 'blight' it with their 'piratical real estate speculation' (p. 73).

Davis's 'Chinatown' speaks to a vision of Los Angeles already anticipated by the future noir classic, Ridley Scott's *Blade Runner* (1982/91). The famous Noodle Bar sequence at the opening of the film introduces us to the demographics of Los Angeles, 2019. Deckard (Harrison Ford) appears as an idiosyncratic white body in this seething mass of Asianized hybrids – their bodies made 'yellow' through the application of a specially devised makeup called 'Asian Blade Runner Blue'. It is the 'white' Deckard who is out of place in this new world, it is he who cannot order sushi properly nor understand the hybrid language 'Cityspeak'. But it is also this still-human, still-white body which is called upon to define and defend the contours of goodness in this future city.

In Davis's ambiguous vision of an Asianized LA it is not the moral fibre of 'white' LA that is at stake. Davis's worry is with the impacts that 'internationalization' is having on the working class, and specifically on racialized components of that class, the 'Black Los Angeles' and 'Latino urban culture', as Davis refers to them. This is a familiar Marxist narrative, albeit one which properly recognizes the complex processes of racialization and migration which constitute urban underclasses. But this is also an unfamiliar narrative in that it has a far more complex colour coding of good and evil than that given to us through the traditional palette of noir.

Urban planners now routinely transform once maligned Chinatowns into desirable components in the well-managed multicultural city. A sanitized and bounded Asian presence is what the western city requires. It would seem that an 'Asian presence' which is marauding, capitalistic, speculative and unpredictable defies such domestication. These 'Asianized transnationals' inhabit an unstable position that is both alien and omnipresent. And those who commentate upon their effects on the contemporary city often produce narratives that hover between paternalism and paranoia. Noir has always concerned itself with the boundary between good and evil, but it would appear that these latest urban transformations are opening out new zones of uncertainty for the politics of race and racism in the contemporary city.

References

Anderson, K. (1991) *Vancouver's Chinatown*, Montreal: McGill-Queens.

Davis, M. (1989) '*Chinatown*, Part Two?', *New Left Review*, July/August, pp. 61–86.

Davis, M. (1990) *City of Quartz*, London: Verso.
Deutsche, R. (1993) '*Chinatown*, Part Four?', *Assemblage*, 20, pp. 32–33.
Gregory, D. (1990) '*Chinatown*, Part Three?', *Strategies*, 3, pp. 40–104.
Lott, E. (1997) 'The whiteness of film noir', in M. Hill (ed.), *Whiteness: A Critical Reader*, New York: New York University Press, pp. 81–101.
Soja, E. (1989) *Postmodern Geographies: The Reassertion of Space in Critical Social Theory*, London: Verso.

•NOISE•

by Andrew Barry

For scientists, noise is not the same as sound. A volume of sound can be expressed in terms of pressure. Noise, by contrast, is primarily of interest in that it is heard. It involves a relationship between sound and *listener*. It is measured in decibels; an index which is intended, in quantitative terms, to give some indication of subjective experience and physiological effect. Zero decibels roughly corresponds to the threshold at which a noise can be heard by a normal person at a frequency of 1000 Hertz. Depending on the person, between fifty and eighty decibels can be annoying. Prolonged exposure to higher levels may result in hearing loss. Noise is not simply out there; external to the individual. It is a measure of the place of the body in an environment of sound.

In his manifesto 'The art of noises', published in Milan in 1913, the futurist composer Luigi Rossolo made clear the connection between noise and urban life. With the exception of a few natural phenomena which occurred in the countryside (such as thunder and waterfalls) noise was produced, according to Rossolo, by the city.

> Let us cross a large modern capital with our ears more sensitive than our eyes. We will delight in distinguishing the eddying of water, or air or gas in metal pipes, the muttering of motors that breathe and pulse with an indisputable animality, the throbbing of valves, the bustle of pistons, the shrieks of mechanical saws, the starting of trams on the tracks, the cracking of whips, the flapping of awnings and flags. We will amuse ourselves by orchestrating together in our imagination the din of rolling shop shutters, the varied hubbub of train stations, iron works, thread mills, printing presses, electrical plants and subways. (Russolo, 1913, p. 26)

For Russolo, far from seeking to block out this noise, modern composers should listen to it and learn from it. In doing so they should not attempt to produce pure sounds, which were, in his view, 'estranged from life'. Rather they should be inspired by the irregularity of noise, expanding on and developing the effects produced by the 'motors and machines of our industrial cities' (p. 29). The art of noise was an urban and technological aesthetic.

One aspiration of some of those associated with the twentieth-century musical avant-garde has been to develop not just an art of noise but a science. Many reckoned that a physical description of sound alone was insufficient to account for what was musically meaningful to the listener. What was required was a new science of musical perception – psychoacoustics – which would be able to determine what kinds of noises produced

Figure 8 Advertisement for Sony, London Underground, Winter 1997–98. Reprinted with kind permission from BMP DDB Ltd. Photograph taken by Steen Stundland

by new machines were of any musical value. To this end laboratories were established, in which the act of listening was studied in the pure space of a sound chamber. This listener herself was part of the experimental apparatus.

No doubt the science of psycho-acoustics is doomed to failure. At least insofar as its ambition is to provide an objective basis for future musical composition. For as Russolo himself implied, noise is much more than simply a relationship between an isolated listener and a sound. It is more than a psychological or psycho-acoustic phenomenon. It is also part of the experience of the urban. In isolating the experience of listening in the uncontaminated space of the laboratory, psycho-acoustics makes the movement back to the complex setting of the city street, or even the concert hall, almost impossible.

Measuring the experience and the volume of noise in the city has required a different approach. Rather than simply isolating the body in the laboratory, those concerned with the measurement of noise have to simulate the real experience of listening, however imperfectly. Regulations governing the noise of cars and motor cycles, for example, attempt to make the prescribed method of measurement in some way representative of their actual use. In this way, the technology of noise measurement perfectly exemplifies an argument made by the sociologist of science Bruno Latour, namely, that there are two processes at work in the production of scientific knowledge. On the one hand, scientific and technical knowledge becomes purified of any apparent social origins. There is no obvious visible trace of personality or place, emotion or economic interest in the design of a sound-level meter. They conform to international standards which are expected to have an effect everywhere. On the other hand, in being made more systematic and precise, scientific knowledge expresses its social and political content in a technical and non-human form. The listening experience of the pedestrian on the pavement is simulated in the form of a technical instrument.

If the modern city was once, as Russolo argued, above all a noisy place, then today we are witnessing something of the decline of noise. Today, noise exists, but it is increasingly blocked out, dampened down, or simply displaced by other noises. Its irregularity is reduced, whether in the supermarket, shopping mall or department store (through the use of carpets or piped muzak), in the home and office (through the use of double glazing or headphones), or on the streets (through the walkman and the car stereo). In short there has been a remarkable series of investments, both on the part of private individuals and the public authorities, in the regularisation of listening. Today, the philosopher Theodor Adorno's pronouncements on the banality of popular listening are regarded as élitist. But in this instance at least, Adorno may have been right. Today, perhaps, there is less a problem of a loss of hearing due to too much loud noise – measurement and regulation has reduced that – than a decline in the variety and quality of noises that are listened to. Form a society for noise appreciation.

Reference

Russolo, L. (1913) 'The art of noises: futurist manifesto', in *The Art of Noises*, monographs in musicology, no. 6 (1986), New York: Pendragon Press.

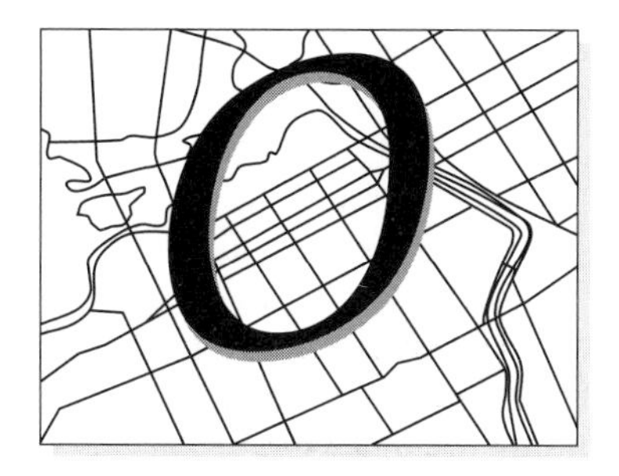

•Olympic Games, 2004•

by John Law
and
Ivan da Costa Marques

Looked at in one way the large condominiums in Barra da Tijuca are a good place to live. First, the houses and the apartments are modern. They have all the mod cons. This is First World living. Second, all the facilities that you might need are close at hand. There are shopping malls, cinemas, sports facilities, swimming pools. Again, this is the story of First World living. Third, it is safe: you live in walled compounds. There are guards on the gates. And in any case the poor are physically removed, living somewhere else. All of which corresponds to real enough needs, or at least desires, of many wealthy people. To live in a way that is insulated from the press of the absolute poverty of so many.

Until ten years ago, however, except for the large condominiums in Barra da Tijuca, gates and fences in Rio's streets were rare. Then suddenly, in the late 1980s and early 1990s, they started to appear. Today there are high metal fences, grills, gates in front of most of the middle income apartment blocks. For instance, to get into an apartment complex you drive up to the gate. If it's time to come home after work then the gate is open. Otherwise, the porter is sitting in a little booth. He either sees you, or you hoot. He presses a button. The gate starts to move. It slides on rails to the side. At a leisurely pace. You drive in, waving to the porter. And then it closes behind you.

Why have these ugly fences and gates sprouted everywhere? Barriers that are in sharp contrast with the natural beauty of the city. Indeed, make it difficult to see that beauty, masked as it is by what look like rows of gaols. Why are they there? Is it to stop burglary? Yes, in part. But also, often, indeed probably more importantly, it is to stop people sleeping in the porches of the buildings, to stop them sleeping in a space where they would be protected from the rain.

There are huge numbers of working people in Rio who prefer to sleep in doorways rather than going home. They end up living in the streets for most of the week. Probably they have a room with the family back somewhere in a *favela*, in one of the slums. But *favelas* in South Rio, or close to the centre, or close to any wealthy part of the city where there is employment for the poor, are also very expensive. Those that are cheaper are usually a long way from the rich places – for instance, along the road to the airport. And beyond. Yes, there are buses. But they are expensive. And often you have to change buses once, or even twice. Work it out. It can cost between R$3.00 and R$6.00 a day to travel to and from work. But the minimum wage in Brazil is R$120 a month. You could spend more than this simply travelling from home to work each day. It is simpler, cheaper, better, to find somewhere, a bit of greenery, a spot in one of the road tunnels, or a doorway where you can lie down for the night. Which is what thousands of people do.

The result? Take Leme, the beautiful place at the end of Copacabana Beach where a rocky mountain meets the ocean. In the early morning it's where all the cyclists, joggers and walkers turn round. There are exercise bars there too. Bronzed people of different ages are doing pull-ups and push-ups. Or lying on their backs with their legs in the air, mimicking the actions of cycling. Quite distracting it is. Bodies that are nearly naked in all sorts of poses. However, within twenty yards of these glossy and sexualised bodies there is a black boy. Boy? Well, perhaps he's 15. Lying on the grass and fast asleep, he's covered himself with a thin sheet. It is somehow wrong to say that he is 'huddled up'. He's certainly not protecting himself against the cold. It's seven in the morning, but the temperature is already 23°C. It says so on the digital sign thoughtfully provided by the city. So he's not huddled up. He's lying there, more or less curled up, relaxed, at rest, his possessions round about him. His flip-flop sandals. His bottle of water. A small bag. And, yes, his shoeshine kit. Sometimes people like him will try to wash clothes. Where does the water come from? And there is human excrement on the sidewalk. . . .

In 1996 Rio's population got really involved in the city's bid to host the Olympic Games in 2004. Politicians and the media mobilised thousands of people in a series of enthusiastic demonstrations. Many events were staged for the visiting international Olympic committee. Rio's middle classes took to the streets to support the bid. One Sunday a huge crowd got together to release hundreds of thousands of balloons at Copacabana Beach. On another Sunday they 'abraçaram a lagoa', 'hugged the lake' hand-in-hand, forming a human chain 7.2 km. round the Rodrigo de Freitas lake in South Rio. The natural beauty of the city combined with the ground-swell of goodwill were taken to be invincible reasons for holding the games in Rio. While the contrast between natural beauty and the horrors of social ugliness was denied by the middle classes.

So the middle classes were sure that Rio would be kept in the list of the five finalists by the Olympic committee. The city was at least as good as Buenos Aires, Johannesburg, Stockholm, Rome, Mexico City and Athens. And disappointment, when it came, came fast, and was equally quickly forgotten. It was like the result of a soccer match. Rio lost. Vague comments about poor infrastructure were offered as reasons. There was scarcely any reference to the way in which urban social ugliness nullified the city's natural beauty.

p
p
p

•PANORAMAS•

Imperial panorama (Point 12)

by Walter Benjamin

Just as all things, in a perpetual process of mingling and contamination, are losing their intrinsic character while ambiguity displaces authenticity, so is the city. Great cities – whose incomparably sustaining and reassuring power encloses those at work within them in the peace of a fortress and lifts from them, with the view of the horizon, awareness of the ever-vigilant elemental forces – are seen to be breached at all points by the invading countryside. Not by the landscape, but by what in untrammelled nature is most bitter: ploughed land, highways, night sky that the veil of vibrant redness no longer conceals. The insecurity of even the busy areas puts the city dweller in the opaque and truly dreadful situation in which he must assimilate, along with isolated monstrosities from the open country, the abortions of urban architectonics.

•PARKING•

by Anthony King

Parking has not just transformed the spaces of the city; through language, it's transformed the spaces in our head. In the thirteenth century, park, or parc, meant an enclosed tract of land, a space, used as a preserve for beasts of the chase. Not much has changed. In the eighteenth century, parks surrounding a country house in England were used for keeping cattle, sheep or deer, essential elements of the picturesque. Contemporary car-park designers could use these ideas: imagine herds of grazing Hondas, flocks of Fords, or a few rutting Datsuns, protected from marauding Jaguars by a posse of Vauxhall Cavaliers.

Initially compared with horses, cars were invested with horsepower and kept in motor-stables and motor-barns. Even today, American books refer – in a sexist, masculine way – to parking spaces as 'stalls' (as if they were urinals). It was the French who, instead of using *motable* or *motorium* (words around at the time), provided the horseless carriage (they renamed it 'automobile') with a 'garage' (*garer*, to shelter); in the United States, they also gave us 'ramp' (*ramper*, to climb). As the garage was clearly not a stable, the word allowed the rich to domesticate their cars, inviting them into their homes, so to speak; in Britain, the earliest 'attached' ones were in well-heeled suburbs, Hesketh Park in Southport, and Hampstead (1901/2). Since then, accommodating the car(s) during years of shrinking household size has become the single biggest influence on bourgeois house design.

It has been suggested that the space of the modern city is seen as being flat, hard and abstract. This is certainly true of parking space, except that it's usually concrete. And as it goes through life being trodden on, run over, spat upon (and worse) it probably feels neglected – which is why I'm addressing it here.

In the first hundred years of motoring it was the space of the city that adapted, in a myriad ways, and without much oppositional thought, to the space of the car. Whether above the ground, on the surface or below it, parking was to develop haphazardly, piecemeal and unplanned. In 1871, before the car was invented, (natural) park planner F. L. Olmsted had advocated the idea of a 'parking system' in the middle of Washington's broader streets, an early reference to both the term and the idea. But with the increase of vehicles, by 1900, the state (in the form of Congress) had begun to say where, and when, car-parking should be allowed.

The first public storage facility in the USA, a 'stable for . . . motor vehicles', was set up in Boston in 1899. In Britain, three years later, the Automobile Club established an eighty-car 'garage' in London's Queen Victoria Street. Both kept their vehicles firmly on the ground.

Before the 1920s, most workers or shoppers travelled into town by public transport – in trains, trams and buses; the few motorists, buggies or brewers' drays competed for public parking spaces at the kerb. But as Henry Ford's production-lines began to throttle the space of the CBD, legislation was passed in Britain (1925) to permit car-parking on selected highways. In the United States, meanwhile, a more revolutionary idea had developed: why not store them up in the air? The first multi-storey car-parks and 'mechanical garages' extruded upwards in the mid-1920s, providing drivers with new, rooftop views of the city. Vehicles were sometimes valet parked, the valet sliding to base via a fireman's pole. Patrons were wealthy plutocrats, anxious to preserve the paintwork on costly cars. In some regions, garages even contained rooms where chauffeurs played cards till their bosses decided to go home. But as the depression brought a collapse in land values, parking 'lots' (1924) opened up, bringing parking space down to earth again, as it were. From here, the final solution was to bury it, a ruse adopted in London with the Royal Auto Club's subterranean caverns (1925).

From time immemorial, cities had been built around the space of the human body, and then adapted to quadrupeds and carts. After the massive impact of the railways, the most disastrous influence on urban space was to occur between the 1940s and 1970s when cities began to reconfigure themselves around metal boxes on wheels. Into the language crept parking *lots* (UK 1958), *areas* (1961), *bays* (1962), *decks* (1970), *aprons* (1972), even parking *wardens* (1974), words and devices which both responded, but also succumbed, to the apparently unstoppable, space-pervading power of the car. In 1935, an insidious invention, intimidating as the time clock, had been introduced into the rapidly motorising American city. The first appeared in Oklahoma City, marking 20' of commodified space and time. Later to become a worldwide strategy for generating municipal income, parking *meters* first sprouted through London's pavements in the hot summer of 1958; fifteen years later, Toronto was minting $2 million a year from parking tags; by the 1990s, Los Angeles was being called 'the parking ticket capital of the world'.

Commitment to individualised car ownership as a form of mass transit is a matter of vulgar politics – private affluence and public squalor: the future as Los Angeles or bicycling Copenhagen. The advent of space-consuming, self-parking garages with separate ramps for 'in' and 'out' and hugely wider turning circles was a response, in the USA of the 1950s, to massively expanded car ownership, related suburbanisation, and women in two-car suburban households, following their men to work, driving to shop downtown. Inflated car sizes inflated parking space. In 1897, small cars (like the Lanchester) had measured 8'6" by 4'6". By the 1950s, when most British cars were 14' long, in the USA, they were (with fins and rock'n'roll) up to 19' long and 5'6" to 6'6" wide. These not only used extra gas but, with their accommodating size and power, encouraged people to eat more: by the 1990s, according to federal government reports, over half of the American population was seriously overweight (in Autumn 1999, Boeing decided to widen their coach class seats by four to five inches). In 1980, MIT city planners, accepting, without question, that the expansion of individual car ownership was inevitable (and that gas supplies were infinite), assuming one car for every two people, estimated that one-third of the space in a car-dependent city was needed for roads and parking; as each car needed 250 square feet, a population of 10,000 required 50 acres to park their personal chariots.

Suburbanisation, however, encouraged by the car, also meant city centre decline. This, ironically, spelt triumph for the car lot – asphalt patches, a shack, and pools of water after the rain. Territories for people to guard and make a living from. Car jockeys taking keys, parking in tight rows, squeezing the asphalt lot to yield a higher profit. Shopping trolleys on the perimeter, plastic bags blowing in the evening breeze.

Yet at the end of the twentieth century, the automobile in the city was clearly coming under attack. Where previously the city had adapted to the car, environmental protests demanded that the car adapt to the city. First came the introduction of zero-polluting electric and compressed air engines. Then, in the second decade of the new millennium, the real revolution: the plastic inflatable vehicle with clip-on solar-powered engine; deflated, the whole could be packed into a rucksack. Metal cars were banned and parking spaces converted into housing.

Till then, however, we might remember that parking also means romance. In Tokyo's love hotels, the management discreetly covers the number plates of cars parked in its garage. And how better to conclude than with poet Mary Clark's lines from 'Parking':

> I got to know what was soft
> and where the hard parts were
> in that upholstered bedroom.
> Every headlight was a worry.

•Pavements•

by Miles Ogborn

Pavements are a signifier of urban modernity. Their history is bound up with two histories of Enlightenment. One makes them part of the process whereby the city became cleaner, more orderly and, with the introduction of lamp-posts, 'enlightened'. The other, entwined with the first, is the reshaping of the city with the constitution of new versions of the public sphere which – in coffee houses, clubs, assembly rooms and newly paved streets – were critical to Enlightenment versions of rationality and democracy. These contexts can be read through one of the founding documents of modern streetscapes, John Spranger's (1754) *A Proposal or Plan for An Act of Parliament for the better Paving, Cleansing and Lighting of the Streets, Lanes, Courts and Alleys, and other Open Passages, and for the removal of Nuisances, as well within the several Parishes of the City and Liberty of Westminster*. In his proposals for Westminster, Spranger sets out a plan to achieve a particular geography. He envisaged convex city streets which would be paved with broad Purbeck stones and fitted with side gutters. These streets would be lined with raised 'Foot Ways' separated from the street by posts, paved with Purbeck squares and cleared of the 'Encroachments of Stalls, Sheds, Bulks, and other irregular Buildings'. (It should be noted that the term 'pavement' was not yet reserved in England for pedestrians, instead it referred to all paved surfaces. It was only when the space existed that it monopolised the name.) Complete with regularly spaced street lights and officers to clean the streets and pavements and to watch over them, a fundamental and recognisable fragment of the modern city was put in place.

This production of urban space was a political process. It signalled London's modernity and progress against other claims. As John Pugh put it in 1787, it was 'an undertaking which has introduced a degree of elegance and symmetry into the streets of the metropolis, that is the admiration of all Europe and far exceeds anything of the kind in the modern world.' The merchant, traveller and philanthropist Jonas Hanway went further. In an enthusiastic endorsement of Spranger's plan he interpreted the use of posts to separate 'footwalks' from the road as a sign of the comparative liberty of the English in direct contrast to their rivals for global hegemony:

> It is true they occupy a considerable space, but if we compare the streets of London with those of Paris, this distinction seems, upon the comparison, to carry with it a *kind* of proof, that we are a *free* people, and that the French are *not so*. The *Gentleman*, as well as the *Mechanic*, who walks the streets of *Paris*, is continually in danger of being run over, by every careless or imperious coachman, of whom there are many; and in fact these accidents frequently happen in that city, in so much that few people of distinction ever walk in the streets. (Jonas Hanway (1754) *A Letter to Mr John Spranger, On His Excellent Proposal for Paving, Cleansing and Lighting the Streets of Westminster*, London, pp. 21–22).

Pavements were read by Hanway as a sign of the superiority of the British in terms of their implicitly democratic ways and deep-rooted sense of liberty. Pavements and posts made sense in a polity which liked to think of itself as opposed to the hierarchical relations of absolutism. The gentlemen and mechanics within the posts on a Westminster pavement were both safe and free.

Indeed, Spranger's plan was itself a political scheme which had to find new ways of relating to the individual and the public while arranging both duties and liberties. Previously the streets had been the responsibility of householders who were required to pave the areas outside their houses and to fix lights to their walls to guide citizens out at night. For Spranger and many others, these uncoordinated individual duties – undertaken with very different levels of commitment, resources and skill – produced a chaotic, irregular and disorderly city. For the new streetscape to exist, pavements and all, householders were to be required by commissioners – composed of propertied inhabitants with the power to raise rates, and backed by the force of law – to pave in the way set out above. A new and important tier of urban governance was created which offered a regular system of administration rather than the sporadic punishment of householders by the courts, and a new relationship between individual inhabitants and 'the public' which was no longer defined in terms of the parish. For Spranger, this relationship was paramount to pavement politics:

> Every sensible Man, who lives under our most happy System of Government, must naturally be interested in all Things, that concern the Public; because he, though an Individual, sooner or later, more nearly or remotely, must be affected by every Degree of prosperity or Adversity, that attends the Community, of which he is a Member. This it is, that has made our political Constitution the chief Object of the Admiration, and Envy of the best Politicians on the Continent: That it is, that gives us a certain Connection, a Regard to the public Good, which is unknown in most other Countries; and whilst we are duly touch'd with a Sense of our Duty to the Public, every man in his Sphere will be ready and solicitous to promote the common Emolument, of which he is always sure to share.

Pavements and paving in eighteenth-century Westminster were therefore set within a new urban politics defined in terms of individuals and the 'publics' of which they were a part. For Spranger and Hanway, pavements must be a matter of public participation and public concern, they cannot be left to private individuals or rulers. They were, of course, constructing notions of individuals and publics which were structured by exclusions on the basis of gender and class. However, recalling this sense of the pavement as a space of public participation in the life of the city, while recognising and countering the limitations on these eighteenth-century versions of the public and on different people's use of the 'footways', means attending to how these city spaces can be made and occupied democratically today when the definitions and boundaries of public and private spaces and spheres are shifting once again and the cracks are appearing beneath our feet.

•PETROL STATIONS•

by Marc Augé

The world without language

On the street in Paris where I live there is still a small garage with three petrol pumps. If you stop there the owner and his employee will serve you while talking about the weather, the car traffic, or, more exceptionally, about the last results of the French football championships. This institution is archaic for three reasons – because the people in charge of it refuse any idea of 'self-service'; they give consideration to the fact that the spoken word is still the best means of communication, and refuse to admit that the whole matter of selling petrol is just about selling petrol.

Petrol stations adjusted to the taste of our times are radically different from the small garage on my street in all three points.

The ideal model of a present-time petrol station is a petrol station by the motorway. All petrol stations situated on the edges of villages by the local, national or regional road make a big effort to look just like the motorway stations, trying to find in such a reference a surplus of dignity, respectability and modernity. The first sign of respectability for a petrol station is the absence of service. Some billboards, panels, however, sometimes inform the tired traveller, old or disabled, that he or she can press the button and trigger a bell if he really wants to be served – which, after all, might seem in line with the 'Road Security' ('Securité Routière' = national public institute for better safety on the roads in France). ('On long distance trips, have a break every two hours.') But the usual usage is a help-yourself service, then move on towards the cash desk where a young man or woman awaits with patience your credit card, while checking everything – at a distance – the invoice amount and the number of your gold card. If you really want you can also wash yourself, your windscreen or check the level of oil in your car.

The drivers who stop to fill their tanks bustle about, each on their respective side (in front of those who arrived later and wait behind their steering wheels just to watch them leave); they ignore each other, just as they were doing shortly before when they would overtake each other silently, undisturbable profiles whose sole attention was on the cars in front, the speed on the meter and the rev count. A gesture suggests that they type in the number of their gold card into a small machine which gives the proof of their good faith and good memory ('valid code'), the buyers of petrol exchange just a vague greeting – sometimes just a mumble – with the cashier, who remains safely protected in his or her small tower. Each year they are more and more numerous, these entirely automatic machines which allow the petrol stations to operate all night without staff, which welcome the traveller like any other automatic machine of that sort, enquire about the desired kind of fuel, the credit card number, invite to insert the card, then take it back, and finally to help oneself before issuing a receipt and wishing a safe journey. It is true that perhaps in order to entertain the traveller and to avoid the feeling of solitude on his or her side which can be inspired by the silent messages on the screen of the teller and the muteness of the fellow passengers, there are also loudspeakers placed under the roof of the station that flood the area with the latest news and songs.

Some particularly cunning managers of certain petrol stations place the cash desks more often inside the vast shops located at some distance from the filling area. There, besides the hot and cold drinks and bathroom fittings, the passers-by are offered varied goods and especially the products of the region whose abundance and prominence could easily lead some naïves to mistake the boutiques of Shell, Esso or Elf for annexes of the Direction du Patrimoine au Ministère de la Culture (Department of the National Heritage of the Ministry of Culture). At least in France these petrol stations enjoy carrying a name borrowed from the local toponymy which is clearly inscribed in the territory ('Area of An (*male*) Owl', 'Area of The Three Stones' or other mysterious allusions to the peculiarities of the site and the depth of its history). In the meantime we can also watch television, send a fax, make a phone call to another corner of the world or linger over the spectacle of adverts and commercials that remind us, for example, that Disneyland is never too far away.

In order to allow a region the place it deserves, and apart from the peculiarities of the given place and the technologies of the global village, the designers of these petrol stations also give them a particular regional style: slate in Brittany, tiles in Midi, *faux colombages* in Normandy and Gascony, etc. In this way the scenery presents itself (*or* is properly set), the picture holds (*or* stands out). And we are in the very places themselves, prisoners of stereotype; we move along the roads as planned but we do not travel, at least as long as we only stick to the motorway or the service stations. It remains to ask ourselves whether, once we leave the network that guides us through the scenery of illusion all the way to our final destination, we shall find a different world, more genuine, more real; or whether by the infinity of cables and screens it is already infiltrated and penetrated by the seducing and empty message which does not cease to invite us to fill ourselves up with information, images, noises and illusions.

•PIGEONS•

by Steve Hinchliffe

Pigeons, cities and unnatural selection

There are pigeons in most city centres. In some ways they tell us where the centre of a city can be found. They are part of the place. But feral pigeons also upset a few too many urban sensibilities. They crowd into city squares and parks. They demand food from strangers, and have yet to learn the sacred nature of personal space. They show no respect for urban landscapes and memories, defacing monuments and buildings with their excrement. Their own appearance is shabby, their plumage is often soiled, they display gnarled limbs and their common foot atrophy suggests that all is not right in their world, or for that matter in our world. Like many non-human inhabitants of many cities, street pigeons are frequently treated as a homeless underclass, squatting illegally on the streets of an otherwise modern space. And pigeons have become expert squatters. Their superior spatial memory means that there aren't many places that pigeons won't go. Their ability to navigate in extreme conditions allows them to loiter in subway stations, home in on railway platforms, as well as frequenting the popular eating spots like parks and pavements. Cities are not simply a source of extra food, they also supply the feckless bird with shelter. Pigeons, it seems, are terrible at nest construction. They find the stability of urban buildings much more secure than trees or hedgerows. They also enjoy the warmth that some of those buildings seem to offer.

> *Pigeons*
> On shallow slates the pigeons shift together,
> Backing against a thin rain from the west
> Blown across each sunk head and settled
> feather.
> Huddling round the warm stack suits them
> best,
> Till winter daylight weakens, and they grow
> Hardly defined against the brickwork. Soon,
> Light from a small intense lopsided moon
> Shows them, black as their shadows, sleeping
> so.
>
> (Philip Larkin, 27 December 1955)

There are, of course, urban sympathies. Passers-by toss their excess bread to the needy. And the pleasure of sharing a meal with pigeons isn't limited to the charges of Mary Poppins and the urban wealthy. Some even go short themselves to contribute towards pigeon welfare. But urban pigeon benefit, like other welfare systems, has become a subject for critical review. Feeding feral pigeons ensures only that non-fatal diseases continue to be transmitted from generation to generation. And in helping some of the most needy, urban people are encouraging 'unnatural' selection. In a bid to reform the pigeon poor laws, we are told in true Malthusian, anti-welfarist tones that kindness can only prolong suffering.

Pigeons, cities and natural selection

Some of this sympathy stems from the long-standing relationship between urban inhabitants and pigeons. Pigeon-keeping became hugely popular in European cities in the eighteenth and nineteenth centuries. The birds' ability to 'home', or return to a favoured roost over long distances, and the ease with which they could be bred, made them ideal for racing and showing. Their popularity was also partly a result of economics: they were cheap to feed and straightforward to keep (although many escaped or failed to return, so boosting the growing feral population in cities). It was also a matter of space; lofts were easily accommodated in the smallest of backyards or allotment plots. The pleasure of keeping pigeons may well have served to foster a tenuous link to a rural, Arcadian past. But pigeons were also subject to such a vast experiment in selective breeding that it would be wrong to label this past as fixed or stable. Minute variations were selected by skilled fanciers to accentuate or minimise features of pigeon appearance and skeletal structure. As commentators of the time were inclined to point out, this display and transmutation of the species mirrored the more grotesque shapes and laced-up displays that were characteristic of Victorian costume fashion (see Figure 9).

The selection of pigeon traits by pigeon fanciers, rather paradoxically it might seem, gave pigeons a hand in the development of a theory of natural selection. We tend to downplay the importance of cities and their social institutions when we account for the success or otherwise of natural scientific theory. But Charles Darwin, who is more often associated with the finches of the Galapagos than he is with the pigeons of London, was a frequent visitor to the capital's pigeon clubs and societies. From his evenings at pigeon fancier meetings, some of which were held in gin palaces south of the river, Darwin managed to construct evidence for his ideas concerning evolution. The clubs, and the working-class clubs in particular, offered a less restricted forum at which Darwin could openly discuss ideas about the transmutation of species – a matter that was still largely taboo in the conservative halls of mid-nineteenth-century establishment science. In this sense, the pigeon clubs contributed to the city's heterogeneity, allowing Darwin the space to develop his ideas. Just as importantly, Darwin learned a great deal about pigeon breeding and its history from the fanciers. He observed their methods, listened to their discussions and watched how their competitive sport resulted in subtle changes of pigeon form, which, over time, had produced intentional and unintentional changes to the pigeon species. In short, Darwin had found an urban, social analogy for his theory of natural selection.

Darwin used the pigeon analogy to add weight to his notion that competition resulted in species change. The hand and eye of the pigeon fancier was essentially seen as a useful approximation to the invisible hand of natural law. Darwin's largely uncritical acceptance of the Malthusian notion of the ubiquity of competition in social and natural worlds only added to his suspicion that the fanciers provided him

Figure 9 Top: 'Five varieties of pigeon' from the *Illustrated London News*, 1851; bottom: 'Pigeons' from *Punch*, 1851

with a fitting analogue. The artificial world of pigeon breeding informed Darwin's view of the natural world of species survival and change.

The fate of urban natures and feral pigeons

Pigeons take part in various orderings of city worlds. Some are labelled unnatural, some natural. For all their differences, the two kinds of urban–pigeon relations discussed here share a number of similarities. While the differences can be exaggerated to sanction the anti-welfarism of the stop-the-pigeon campaign, the similarities can be drawn upon to destabilise these natural orders. The theory of natural selection, which was in part derived from the observations of the competitive fancy, is often used to devalue the unnatural selection going on in urban parks and on city streets where the feral pigeons make their home. But the city's role in the formation of the theory of natural selection can be highlighted as a means to threaten the legitimacy of pronouncements upon the unnaturalness of urban behaviour and species. Indeed, such a move may be necessary in order to radically rethink city-nature practices in the future.

Further reading

Palameta, B. (1996) 'Pigeons: the smart bird', in M. Dion and A. Rockman (eds), *Concrete Jungle*, New York: Juno Books.

Secord, J. (1981) 'Nature's fancy: Charles Darwin and the breeding of pigeons', *Isis* 72, pp. 163–186.

p
p
p

•PLANNING (1)•

by Arturo Escobar

> Your gaze scans the streets as if they were written pages: the city says everything you must think, makes you repeat her discourse, and while you believe you are visiting Tamara you are only recording the names with which she defines herself and all her parts. (Italo Calvino, *Invisible Cities*, 1972, p. 14)

Like most cities in the world, the Colombian city of Cali (population two million) is many cities in one. Not the divided city of the colonizer and of the colonized – as in certain Asian and African cities, such as that immortalized by Pontecorvo in *The Battle of Algiers*; not even the segregated city of the rich and the city of the poor, since both are inextricably interconnected, and in Cali (as in Caracas and unlike Bogotá) most 'good' neighborhoods have a 'bad' one close by, hill versus valley most times – creating a paradoxical dialectic of sameness and diversity, hatred and tolerance, fear and engagement between the poor and the rich, the unmarked and the marginal. For the middle classes, this dialectic of class and culture is most often felt as imposed on them by the hordes of the poor (as with the kids who sell you all kinds of little and often useless things at the traffic-lights; or the men who 'guard' your car in every block of the city, veritable 'owners' of each street block; or, more inconspicuously, the black women who pace the streets selling aguacates and chontaduros, two delicacies of the Cali middle class, announcing them with their enchanting voices, after having migrated to the city from their cherished homeland by the rivers of the Pacific Coast region); or experienced as a necessity (as with the maids and gardeners and handymen and, especially, the guards at the entrance of each office or dwelling building, all of whom come from 'the other side,' which can be a stone's throw away but certainly worlds apart, to serve the affluent for meager wages and lots of disrespect).

As with many Third World cities, Cali can be said to be fragmented into deeply interconnected 'formal' and 'informal' patches and parts. This division, invented by planners and urbanists, has come to make sense in unintended ways. In the formal city dwell the state, the banks, the office buildings, the schools and hospitals, the fancy restaurants, the police; even cars and scooters drive by with a certain confidence, despite the not too infrequent hold-ups staged by car robbers, occasionally with legendary skill. At night, private cars speed up at red lights, afraid of being cornered and robbed. But there, at the multiple 'other sides,' informality reigns: there is little or no police, little or no state, little or no recognizable sense of order . . . even if most times most things in these places, too, seem to be in place. Taxi cabs, for instance, are reluctant to go into many poor neighborhoods at night, or flatly refuse. Neither capital nor order have quite managed to structure these city sectors, even if the urban landscape reveals a dramatic social history in the making to the attentive eye, less openly political perhaps than lived in each practice, each dream, each struggle and each thought.

Many Third World cities derive their imaginaries from a special place that seems to subvert their very existence, and Cali is no exception to this rule. Mention 'Aguablanca' or 'El Distrito' (the district *par excellence*, out of many city districts) and most in the formal city will shudder, despite the fact that the district has been in existence for almost twenty years and its size has grown from a few thousand in 1980 to almost 400,000 today distributed across many neighborhoods in 1,200 hectares. Located close to the Cauca River – that gives its name to this valley of fertile lands, long ago appropriated by sugar cane barons – and under the river level, the area was prone to recurrent flooding, and thus difficult to urbanize. In the mid- to late 1980s, a program to extend public services to what soon came to be known as the Distrito de Aguablanca (DAB) was considered by many as 'unpayable.' Whatever improvements have taken place since, including land fills and self-construction, have been due much more to the individual and collective efforts of the *aguablanqueños* than to any concerted effort on the part of the state. About 50% of the DAB's households today have adequate services, although the costs of these improvements have been born disproportionately by the local people, given that the benefits of public infrastructure investment have been captured by private capital in another instance of the public sector's inability to cash in the surplus value it, itself, created. 'It's too expensive to be poor', say the *aguablanqueños*, commenting on the paradox that their settlements have taken much greater amounts of effort and money to regularize than any other part of the city.

Cali has witnessed periodic waves of squatting (*invasiones*) since the early 1950s. Urban studies suggest that 30% of the city was made up of 'informal' or 'subnormal' developments in 1960. The DAB covered an area larger than all its predecessors. It began as a tremendous wave of people and materials that doubled its size every year during the early years of its formation after 1980. The absence of an integral urban development for the sector is reflected, for the planners, in the lack of green areas, children's parks, cultural spaces, and adequate services. The steady increase in delinquency and drug use is also taken as a sign of the failure and inability on the state's part to satisfy the needs of the thousands of *destechados* (roofless people) who roam the city, many of whom have converged here. While 'social housing' programs do exist, these are still expensive for the very poor. The DAB itself was built by poor people from other neighborhoods of Cali who moved there in search of their 'own roof,' and by migrants from other parts of the country, especially rural areas. About 30% of the DAB inhabitants migrated there from the rivers of the Pacific Coast, pushed by economic forces and natural disasters, including sea quakes and earthquakes in 1979 and 1983. These mostly black migrants are among the poorest of the poor in the DAB. An enclave of the Pacific Coast black cultures, the DAB's black neighborhoods add to the city's racial dynamic, tainted by subtle and not-so-subtle kinds of racism and prejudice as it deals with its ineluctable and longstanding interculturality.

The district's oral history is already legendary. It is said that many households maintained a permanent flame in their stoves fueled by the natural gas that flowed from under the land fills on which the houses were built; it is said that during the early years of the invasion children and grown-ups would swim on weekends at a beautiful lagoon – Charco Azul – located in its midst (today no more than a reservoir of polluted waters); it is said that the clouds of mosquitoes were so thick that it was almost impossible to see during the day or to sleep at night. Floods were common, and everybody had to carry their good shoes and clothes in a bag to the outskirts of the district, where they would leave their soiled and muddy shoes and pants at a makeshift business for a hundred pesos a day, before going on to their places of work in the 'formal' city. New roads, during those days and even today, were swallowed up in a few years by the pace of activity, rains and floods. Despite all this, today a great part of the district has successfully joined the formal city, or so it seems, and more than half of its inhabitants work in the formal economy. But others come, and as some city leaders unreflexively fear, other aguablancas are being born. . . .

Planners in the 1980s were intent on controlling the growth of the district; many measures, besides the regularization of the settlements, were considered, including erecting physical barriers to its expansion, planting enclaves of city life to contain its growth, and supporting self-building projects with a standard building plan. Even the latter were unable to contain people's activity, as the low-cost homes given to people by the city housing agency were soon transformed by their dwellers vertically and horizontally in manifold ways. The planners' urbanistic and containment efforts were geared towards preventing 'other aguablancas' from emerging, to containing the growth of what urbanists began to call 'the other Cali,' with its 'explosive characteristics' and 'chaotic development.'

More recently, more enlightened planners have sought to abandon this policy. For them, at stake is the creation of territories – understood as a complex spatial, political, administrative, and cultural process – that is best left to those directly involved. Moreover, they think that the state's self-serving interventions – intended to governmentalize the poor and their space – is not only self-defeating in the face of the magnitude of the task at hand, but counterproductive in terms of people's needs and goals. Some sort of autonomous development – supported in some fashion by the state – is seen as more viable. The planner says:

> While it is true that any social process of creation of territory involves some level of violence, we see no other way out of this dilemma; people will have to develop their own form of understanding and management of conflict over needs and resources, including those from the state, in order to build the new existential territories in which they wish to live. But we know that the politicians will, in all certainty, impede these plans from coming to fruition, since what is at stake for them is maintaining a policy of occasional handouts and public works for their constituencies so they can keep their juicy slots in the state machine.

Navigating perilously between neo-liberal neglect and the support of self-directed development, and going against the grain of long-standing political clientelism and corruption, this handful of planners appears to intuit an alternative practice with little scope for success. In the meantime, the *aguablanqueños* go on with their lives against all odds, crafting a discourse in daily life that reverberates throughout the cityscape, making others think other thoughts and say other names, that also define what the city of Cali is and imagines itself to be.

Sources

Field days in Aguablanca (summer 1997); conversations with Arturo Samper, head of Cali's Municipal Planning Office; conversation with several *aguablanqueños*, past and present; DAB Development Plan (1984) and 1980s documents on the DAB, Municipal Planning Office.

•Planning (2)•

by Adrian Passmore

He pulled back a dusty sheet. His hand shook as it caressed the unfinished model. *Deliriums tremens.* He steadied for an imagined audience. This had begun a long time ago. He had taken the commission from raw lumber to where it was today. He had worked at little else although the vagaries of the world had regularly schemed to interrupt his progress. His thumb and forefinger pinched a corner, and traced the outline of a scaled wall, as was his job; that same thumb jerked with the direction change as it slid across the adjacent pavement. He was not architect of the commission, but sense said that it was his work and his city which was being hewn and planed in his father's old workshop. It was he who had cut the oak to form the base of the main civic square; predictable wood, but he had spent many weeks searching for the right piece. It was he who had vacillated between white beech and jungle mahogany for the waterfront – as if it mattered. It was he who had finally opted for the walnut veneer for that main square. He loved it for the subtle repetition of its grain when cut four-square. Burred echoes.

Poised there, with his eyes glazed into what he erringly considered his visionary look, he tried again to imagine how a new life could unfold when this model was released from the workshop and handed over to the agents. No longer would his slowed craft-life be punctuated with requests to tamper with the design: no more negligible widening of streets, no more edicts to relocate things, and no more demands for material changes. His look cracked, he was, in short, fed up with being asked to alter the project: in public he would claim that his integrity was being called into question (in private others would agree – it was becoming dinner-table talk to explore the blindness of his ways). His work, in truth, dissatisfied him – it was no longer pure since other people kept offering opinions and advice. He thought back, and looked to his father's tools and wondered if his unwillingness to deal with both the agents' and clients' requests lay with him or in the cold form of those implements. He tried to cast himself back into the distant past to recall if his father had done differently with them, but the memories were too worn to stand the scrutiny. He could only speculate, which was not today a part of his remit; his estate agent friend did that.

As he thought of a future without the model, and let his mind wander back to the present, he turned the palm of his hand upwards and examined the skin patterns of work. The sight of this still hand in front of his face bent his slow back as he searched for the light; as he did so, he levelled an eye to his favourite street. Since he had first pencilled it in, he had thought of this as his major achievement. It travelled uninterrupted in a thousand-yard stare. He had worked the polish of surfaces and the various grains to ensure that it looked straighter and more purposeful than even the master plan would have it. The micro architecture of this vista seemed his, which was strange considering he was himself so slow to venture through the door – even to his own backyard. His hand retraced the fine detail of the court-house, his back straightened, he stalled momentarily to blow the fine sawdust from the open spaces. With the glancing touch of a fingertip he bit back a crass joke about church domes, but sniggered to himself all the same. The blown dust collected against a step and rolled in an eddy behind an unfinished tower. Stepping back, he climbed carefully on to the fragile scaffold erected over the model.

From here it was different, he no longer thought about the separate elements with which he was completely bound up when his feet were on the workshop floor. Up here, only a reach from the prototype, he could survey and calculate the whole task. He could gauge how much more buffing the commercial district would take before being populated with the tiny wire and clay figures he loved to make. In a single scan he could see the glossing that would be necessary to allow the eye to travel the full range of streets without impedance by impertinent grain. But contrary to this finishing urge, he knew that he would never declare his city complete if he could not get over his antipathy to certain types of building: he'd always had a blockage over some of the outlying districts and other more central fading pockets where his hours of work rarely led to progress. He kidded himself that they were unfinished because it was all growing from the middle outwards and their turn would come eventually; but in reality he suspected, even if he would not acknowledge it, that their decay was part of the plan – strange in a model. Their potential coarseness had almost dissuaded him from continuing the commission: what sort of dumb-ass designed unfinished things? How could he build this beautiful city if some parts were perpetually tatty?

The worry made his brow furrow as looming tasks crept into his present world and chilled the still-warm thought of a model-completed future. If he had looked in a mirror he might have been consoled by the fact that this brow furrowing was more uneven than anything he had ever been asked to make, but instead he rolled over on to his back and dreamt his favourite dream. It was the dream where everything condensed to a single gesture, more intricate and intimate, yet monolithic, than any other: he called it his city dream. It was as sophisticated and accomplished a thing as the calendar of girls hanging on the workshop wall. He felt the flatness of the wood and it was the same as the flowing path of his father's plane; simultaneously, he shifted free of gravity across the creaking scaffold and it was as easy a motion as his eye made down that favourite cut street. This dream smelled good. It smelled whole. And since he dreamed it only when he was awake it could always serve the same chosen tranquillising purposes. It put him at an ease that only the ancient feel before sleep: tomorrow, he hoped, would begin like this. He looked down and stroked his city some more – it hardened to his touch. He sparked a cigarette and blew a clever smoke ring that wisped quickly away. Something nagged that made him suspect he was out of step with his world, but on reflection he didn't care – the work was contained enough to be doable, even if the workshop was draughty and the scaffolding appeared to give a little more each year. They stirred and stiffened together with an unoiled arthritic creak – a noise from the street disturbed him.

•POLICE STATIONS•

by Eugene McLaughlin

In November 1997 two London police stations briefly captured the headlines, and for me the accompanying stories exemplified how police stations acquire local reputations and carry their own histories. If you look at the contents page of a *London*

A to Z you will see that ▲ represents a police station. It is one of the few public institutions to be granted its own identification marker. This particular urban institution can conjure up similar and yet quite divergent meanings and realities for different people. A police station can be a place of refuge and a visible representation of the order of the city. However, an individual's presence in a police station, unless s/he is a police officer, inevitably signifies trouble or potential trouble because s/he is there as a suspect, victim, witness to wrongdoing or informer. Anyone 'helping the police with their inquiries' who has moved from the cramped, chaotic front lobby space of the police station to the more private parts for questioning, detention, etc. immediately recognises that one is moving between two very different worlds and that one's sense of self is radically altered. The sense of disorientation is heightened if one is taken to an interrogation room or a cell. The fact that making contact with the police can be a daunting and traumatic task is, I suspect, why the citizens of most cities prefer to think of police stations as 'off limits' and have as few dealings as is legally possible with these particular buildings. However, for certain social groups the local 'nick' plays a very prominent role in their lives.

On 14 November 1997 one of the *London A to Z* markers ceased to have any meaning because the Metropolitan Police closed it. The last full operational day of Vine Street station received considerable media coverage and the commissioner of the force, at a gathering of officers, past and present, memorialised its passing. An article in a Metropolitan Police newspaper informed readers that nostalgia pervaded the proceedings with officers trading fond stories of crime and beat policing in 'the good old days'.

So why was Vine Street station so important to the collective memory of the police? It was the oldest police station in London, symbolising the police presence in the West End. Indeed, it pre-dated the modern police, because in the seventeenth century the site was a 'lock-up' facility known as the Parish Watch House of St James. In September 1829, with the controversial formation of the Metropolitan Police, it was turned into a station of C district. The district was mapped into carefully delineated divisions, subdivisions, sections and numbered beats. Constables walked their beats in a fixed time according to an appointed route night and day, and were engaged in the disciplinary ordering of time and space. If he deemed an individual to be 'out of place' s/he would be stopped, questioned and searched. If people did not 'move on' when they were told to, they ran the risk of being arrested for being suspicious persons; loitering with intent; causing an obstruction; committing a nuisance; using abusive or insulting language; importuning or soliciting; being drunk and disorderly; or engaging in behaviour likely to cause a breech of the peace. The article reminded readers that the new policing met with considerable hostility and that in May 1833 a colleague of Vine Street's officers, P.C. Culley, was one of the first metropolitan police officers to be murdered in the line of duty. What really shocked the government and police authorities of the day was the fact that a jury returned a verdict of 'justifiable homicide', providing one of the lasting examples of just how unpopular the new police were with many sections of London society. And the opening of new police stations in certain parts of the metropolis was the equivalent of planting forts in hostile territory.

However, the article failed to mention the lasting unpopularity and controversial reputation of Vine Street station among certain Londoners. For example, in the 1980s, Shane McGowan of *The Pogues* (a London Irish punk-folk group) also memorialised Vine Street station, this time from the perspective of a young Irish down-and-out:

> One evening as I was lying down by Leicester Square
> I was picked up by the coppers and kicked in the balls
> Between the metal doors at Vine Street I was beaten and mauled
> And they ruined my good looks for the old main drag.
>
> (*The Old Main Drag*)

The other London police station, which attracted media attention in 1997, was Stoke Newington in the north-east of the city. One of its officers made front page news when, at the end of one of the longest and most expensive cases involving the Metropolitan Police, he was convicted of assaulting a student. The outcome of the case confirmed the station's reputation as a 'bent nick' which was home to rough justice and rogue cops. Stoke Newington, perhaps more than any other in the capital, has cast a long shadow over relations between the police and local black communities and two very different versions of reality swirl around it. Although both sides can agree that police–community relations are marked by considerable hostility and suspicion, they disagree on why this is the case. As far as Metropolitan Police officers are concerned, the strains and tensions resultant from high levels of social deprivation and the racial and social mix of the inhabitants make for an incredibly difficult policing environment where the potential for serious disorder is ever present.

The area routinely occupies 'top spot' in the force's league table for reported crimes of violence, burglaries and street robberies. It's litter-strewn and graffitied back streets are home to more than their fair share of notorious night-clubs and illegal shebeens and rival armed gangs who are jockeying for control of the area's lucrative drug and prostitution trade. The hostile, dangerous and thankless nature of the policing task calls for a robust 'law and order' response and it is inevitable that on occasions younger officers, under pressure, will make a 'wrong call'. The Metropolitan Police also believes that local criminals, especially the 'Yardies', and political trouble-makers have made concerted efforts to undermine the authority and effectiveness of local officers by tarnishing the station's reputation. Given these circumstances, it is not surprising, according to senior officers, that the rank and file construct a cocoon of defensive loyalties and *ad hoc* support systems that allow them to cope with the atmosphere of impending violence that is associated with certain beats.

The local black community advances a very different interpretation of why Stoke Newington police station has such a notorious reputation. It argues that this station has fostered a highly racialised 'canteen culture' of corruption and violence that each generation of officers seems to be socialised into. Throughout the 1970s there was a conspiracy of silence to cover up or explain away: the deaths of black people being held in custody at the station; harassment and racist abuse; framing suspects to boost clear-up rates; and the activities of openly corrupt officers. There were also complaints that officers on the beat behaved as if they were a law unto themselves rather than a community-based police force.

In 1983 there was fresh controversy when a police investigation failed to clarify the circumstances in which Colin Roach, a young black man, died in the foyer of the station of a single gunshot wound. For the rest of the decade the front of the station became a symbolic site for protests about local police actions. Local people's views received some confirmation when Paul Harrison, a respected journalist, spent some time researching life in the neighbourhood. He described the station in terms that conjured up images of *Fort Apache – The Bronx*, with its 'isolated and beleaguered' officers going forth 'like commandos, equipped with all the latest technology, into enemy held territory'. And in an attempt to calm community fears about police stations such as Stoke Newington the government passed legislation which allowed for the appointment of panels of lay visitors who could offer an independent assessment of conditions within these closed worlds.

In the early 1990s rumours started that the station had tipped over into outright criminality with a cohort of officers who were running and protecting their own drug dealers on the local 'frontline'. Court cases began to collapse because the testimony of Stoke Newington officers was deemed to be unreliable. A number of local people successfully appealed against their convictions because certain officers could not be regarded as credible witnesses. Major internal investigations were launched and extensive efforts made to clear up the station, with senior officers promising to learn about the nature and gravity of poor police–community relations. Given the very different stories attached to Stoke Newington police station, it will be very interesting to see how it is memorialised if the Metropolitan Police ever decide that it no longer meets its needs. However, in the meantime its reputation as a dangerous place for black people remains, and the ▲ signifies somewhere that should be avoided.

Further reading

Harrison, P. (1983) *Inside the Inner City*, London: Penguin.

Independent Committee of Inquiry (1988) *Policing in Hackney 1945–1984*, London: Karia Press.

Metropolitan Police (1997) 'Vine Street dates back to the beadles', *The Job*, 14 November.

Poguetry: The Lyrics of Shane MacGowan, London: Faber and Faber.

•POOLS (SWIMMING)•

by Harvey Molotch

Once upon a time, before air-conditioning, swimming pools were a way to beat the heat; a cool dip was the only cool there was. But now swimming pools are mostly things to be next to as in 'lying out by the pool'. That's why the biggest problem in accessing pool water, for those who might actually want to go in, is wading through the sea of carcinogenic bodies that block the way.

Part of the intrigue of the pool is the chance to see other people's bodies – sometimes, as in the case of convention-hotel pools, people you have just been with in a room without windows and a malfunctioning slide projector. They had no thighs, body hair or stretch marks and few, if any, protuberances. The other side of the poolside coin, of course, is that you also have to show yours, a prospect that is welcomed or feared depending on where one is on the exhibitionist–voyeur continuum.

Journey-to-pool raises acute self-presentation issues. Hotels seldom make clear the protocol for going from room to water, and in some cases there is no choice but to enter the elevator *en speedo*, sharing that always closed and threatening space with people wearing wing-tips and talking about market swings. Sometimes the pool is not on the ground floor but the 'terrace level' which means displaying the *derrière*, including the waddle of the way to those left behind.

Proto-athletes with real ambitions for their bodies need pools, whatever floor they may be on, that have a long straightaway (say 25 meters) and cool temperatures

(80° F or under). You need lifeguards who keep children out of lap lanes – completely and totally – and other pool users who know the etiquette (swim counter-clockwise in the lane, except among the British who, despite having at one time held dominion over most of the world's pools, swim alone in the other direction). When it all comes together, which is rare, the lap swimmer has a royal experience. For the best hotel pool in the world I nominate The Palace, not the one owned by Leona Helmsley in New York, but the one created by Mme Chiang Kaishek in Taipei. She knew how to do things right. There, by the largest pagoda-style roof in the world – it's gotta be big to cover that many hotel rooms and meeting halls – sits a 50-meter Olympic pool, chilled against the humid air of the South China Sea. With monumental red lacquer architecture all around, it's like being the martini olive served in a USA-style Chinese Restaurant, like my local 'Jimmy's New China Inn'. Not necessarily a delight for everyone, but a refreshing personal best for me.

For the classicist swimmer, a highlight would be the pool at Pamukkale in Turkey, where they have flooded an ancient agora with waters from the adjacent natural carbonated hot spring (that's a lot of features). A scatter of variously submerged genuine Grecian bric-à-brac simulate a home aquarium waterscape that makes for fun human diversions like diving under toppled fluted columns or perching on a stranded entablature or Corinthian capital (eat your heart out, Lord Byron). Because of the carbonation, you can see little bubbles all over your skin, another nice touch from the gods.

Public swimming pools in the USA tend to be pretty businesslike: rectangular and more likely to be cool than in other places. They can also have social policy functions. Chicago's Mayor Daley, fearing ghetto uprisings in a late 1960s summer, ordered all city swimming pools to open extra hours (and fire hydrants spraying into the streets) to simmer things down. Black comedian Dick Gregory, appreciating the gesture, said, 'there isn't going to be a dry nigger in all of Chicago.'

In Vancouver, BC, social democrats and athleticism nicely combine in excellent public sports facilities, including the giant (150 yards long) Kitsilano outdoor pool with its gently curving sides, giving the pool the shape of a USA-Canadian football with the ends cut off. This configuration encourages a single oval for all the swimmers to travel counter-clockwise in common, rather than in parallel lanes. The circuit thus produced is wide enough for faster swimmers to maneuver to the outside or inside of the route traveled by the more 'relaxed' swimmers (as the lifeguard so gently termed the slow people when explaining to me how it all worked). Everyone is moving together, like a school of human fish who can have only vague knowledge of how many meters they have done, their 'times,' or who came in first: relaxed West Coast Canada.

The Soviets built the biggest pools with least effect. A pool in Budapest claimed to be the world's largest, and it was, when I saw it *circa* 1960, filled with a vast crowd moving through the chaos. Water, water everywhere but not a drop for laps. There was particular pandemonium when a child misplaced her parent, and no wonder: Red Square but with the added intrigue of going under.

The British were late in figuring out how to run a proper facility – 'sport' in Britain was historically a way to be well bred, not a better way to breathe, 'fitness' a basis for empire, not a state of the body. When Labour was last in Before Blair, it created the national 'Sport for All' program that dotted the whole nation with modern facilities. But it looked more like 'Snack for All,' to judge by the relative density of the café compared to any other part of the building. There were crisps and sweets of enormous variety, and even beer. Nobody got out of there healthier than when they went in. At least there wasn't a lot of smoke, as with a basement public pool café in downtown Paris where *après-swim* is in a room with zero circulation and dense second-hand Gauloise.

There are now master swim 'teams' throughout the world – people of all ages and many levels of ability who want to swim regularly. Websites now can direct people, including visitors, to the local team's time and place of practice. Most welcome all comers; swimmers tend to be a friendly lot. As with other special interest city activities, it's a way to meet people when traveling or moving to a new town.

Team names are an index of which fish are most feared or admired at any given time: sharks are everywhere, but there seems to be a taboo against being what you ate as a child: tuna, salmon, and filet of sole go untouched. Gay and lesbian teams, also global, make the most of the naming opportunity: 'Different Strokes' in San Diego, 'Out to Swim' in London, 'A Contre Courant' in Montreal, 'Wet Ones' in Sydney; the 'Aquatics Club' in the USA capital wears the logo 'AC/DC', and a Hawaiian male-female swim group calls itself 'Mixed Fruit.'

Gay or straight, in or out, rectangle or round, pear-shaped or kidney, hotel or public, secular or godly, pools are means for camaraderie, presentation of self, adding spirit and changing the pace and face of city life. Air-conditioning could never do them in because they do too much; as with their predecessors at Caracalla or today at my Mom's condo in Miami, they hold more than water and expose more than skin.

•POVERTY•

Consumer and employment data

City	*Unemployment (1990)*	*Per capita income* ($US)	*Calorific intake* (CAL per day)
Abidjan	35%	1,208	2,038
Bangkok	145,400 people (1989)	3,235	2,500–2,800
Beijing	0.5% (48-hour week)	337	n/a
Bucharest	n/a	1,793	2,462
Buenos Aires	8.8%–5.7% (1988)	7,851	2,450
Cairo	2.3% / 7.5% (potentially unemployed)	n/a	2,381
Delhi	794,000 (1989, applications at employment exchanges)	391	1,900
Istanbul	n/a	n/a	2,500
Jakarta	121,325 (seeking employment, 1988)	816.7	1,709

continued . . .

Consumer and employment data (continued)

City	*Unemployment (1990)*	*Per capita income* ($US)	*Calorific intake* (CAL per day)
Kuala Lumpur	n/a	n/a	n/a
Lima	7.9% (1989; only 18.3% adequately employed)	n/a	n/a
London	4.5%	16,839.9 (1987)	2,190
Madrid	10.3%	13,000	2,800
Mexico City	3.7% (1989)	n/a	n/a
Montreal (Quebec)	9.3% (1989)	14,823	2,163
Nairobi	19.9% total 27.3% of women need jobs	n/a	n/a
New York	5.8% (1989)	11,188	n/a
Paris	10.4%	23,597	2,053.8
Rome	7.3% (1989)	17,309	n/a
Sào Paulo	8.2%/8.7% city/metro region (ratio of unemployed to economically active)	3,860	n/a
Seoul	5.5% (1989) 8.6% (1985) (59-hour week)	n/a	n/a
Stockholm	0.6% (1989)	17,085	n/a
Tokyo	n/a	27,211	1,999
Vienna	5.4% (1989)	24,231	2,500–3,600

Source: Governor of Tokyo and Summit Conference of Major Cities of the World, March 1991, *Major Cities of the World* (Tokyo: Tokyo Metropolitan Government). Cited in Sassen, S. (1994) *Cities in a World Economy* (Thousand Oaks: Pine Forge Press), pp. 128–129.

•PRESERVATION•

Taxidermy

by Dolores Hayden

From two Greek roots, arrangement/of skin.

The art of preparing and preserving the skins of animals, and stuffing and mounting them so as to present the appearance, attitude, etc., of the living animal – OED

This young man can't get a second date.
His fingers reek of formaldehyde and ammonia,
he's been stuffing birds he snared in Harvard Yard.
Young women find him eccentric, obsessed
with feathers, beaks, beady eyes.

A few years later, he's hunting grizzly bears
on his ranch, he prefers the Dakota Territory
to the old family brownstone on 20th Street,
the summer place at Oyster Bay,
the new house on 57th Street.
I think he's out west to strategize.
He needs to rebuild his political career,
he's made too many enemies in New York.

But I digress, forget self-preservation,
this is about preservation, taxidermy if you will,
and his boyhood home. The family brownstone,
sold, becomes a shop. Neighborhoods are changing.
Then the brownstone-with-shop is demolished,
another store is built on the spot,
although his uncle's brownstone still stands next door.

Our main character doesn't notice – he's busy,
he's an assistant secretary of the Navy,
he's governor, he's nominated for vice-president,
he wins, he succeeds an assassinated president.
He busts trusts, wins the Nobel Prize for peace,
writes half a dozen books, explores Brazil,
gets the River of Doubt renamed for himself,
Rio Teodoro, river of Theodore Roosevelt.
'The only person who makes no mistakes,' he says,
'is the person who never does anything.'

Now I'm getting to the real story, about the mistake.
After he dies in 1919, the women of his family
raze the shop to rebuild his boyhood home –
boyhood being about motherhood, preservation
being women's work. Rich women's work, mostly.

And for these rich women, it's a religion.
With wood and stone they recreate the Nursery, the gym,
his mother's bedroom with its heavy satinwood bed,
the parlor with its horsehair chairs and sofas.

They stuff the house with old taxidermy projects,
hunting trophies – lions, zebra, antelope, deer.
They frame two smiling wives and six children,
mount old campaign ribbons, dangle pendants
shaped like elephants, new brooms sweeping clean,
and – I didn't see it but it is probably there – a Bull Moose.

p
p
p

'The women did it all,' my guide said proudly.
'But,' I said to her,' they broke the rules.
An old building has to be preserved. This is a fake.'
'The architect not only copied a demolished building,
she gutted his uncle's house which was standing
to design a museum for Teddy's stuff.'

I was complaining to an employee
of the Department of the Interior.
If the taxidermist and big game hunter
hadn't been such a supporter of the Interior,
all this might not have happened.

Pay them a visit at 28 East 20th Street.
You'll imagine young Teddy, upstairs
in his bedroom, stuffing a sparrow,
and cook, down in the basement kitchen,
making a meat-and-potatoes dinner
to run up the dumbwaiter to the family
assembled in the dark dining room
around the mahogany table. They sit uneasily
on straight chairs with scratchy horsehair seats.
Pretty soon you, too, will feel impatient
to chase bright-feathered birds in Harvard Yard
or ride out to the Dakota Territory, or paddle
down the River of Doubt, which will one day bear your name.

'Taxidermy' originally appeared in *Michigan Quarterly Review* (2000).

•PUBLIC TOILETS•

by Gargi Bhattacharyya

Of course, we are all too tired already of the idea of the flâneur. After so much talk, it is impossible to see any longer the especial cleverness of the analogy. Now we expect strolling to be an insistently urban, built-up, population heavy kind of experience – pastures have gone out of the window altogether.

The flâneur is a strange creature, all resilient feet and saucer eyes, able to move and look at leisure. The urban experience which this implies denies the call of more mundane physicalities altogether. The flâneur feels no discomfort at heat or cold (thanks to the modified atmosphere of the arcade cum shopping mall), and although we assume enthusiastic participation in the voyeuristic pleasures of café living, there is no indication that the diuretic outcomes of this activity have a space in this mythic city.

Instead, the uncomfortable necessity of a toilet is passed over in most accounts of modern public life. The liberatory promises of public urban living take pleasure in the life of the crowd, but prefer to ignore the far-from-home bodily functioning which this interminable strolling must accommodate. But, as we all know, such calls cannot be ignored forever.

George Jennings installed public lavatories in the Crystal Palace for the Great Exhibition of 1851. He introduced his 'monkey closets' in the retiring rooms at Hyde Park and then later at Sydenham, demonstrating to the public the possible comforts offered by the public convenience. By the 1890s, Jennings and his followers had introduced the benefits of the public convenience to thirty-six towns and 'many others' across England. His customers included thirty railway companies in England and his urinals could be seen and used in the streets of Paris, Berlin, Florence, Madrid, Frankfurt, Hong Kong and Sydney. Despite initial resistance, the public convenience becomes a great success.

However, convenience alone did not ensure acceptance for Jennings' designs. Success lay in his negotiation with the particular coyness of urban living. Jennings found his invention difficult to sell at the Great Exhibition – although many used his conveniences as part of their day of exotic spectacle, this acknowledgement of the body and its needs did not integrate easily into the public culture of the nineteenth-century European city. However obvious and unpleasant the signs of this oversight became, building conveniences solely for this purpose would admit the needs of the body, rather than reworking, containing, veiling the body in more usual ways.

So, cleverly, Jennings' plans included ways around this embarrassment, ways of integrating bodily necessity into the urban landscape, as if this too could become part of the peekaboo pleasure of the city:

> many conveniences were built underground, with cast iron arches, railing or pergolas to mark their whereabouts. Those built above ground were distinctive little buildings in their own right with their finials, pillars, panels and enhancing lamps. (Lambton, 1978, p. 10)

As the twentieth century ends, many European cities retain these decorative little buildings as pretty artefacts, not actual conveniences. The properly public toilet, open to all and maintained as a communal facility, once again seems hard to find.

Once a sign of the socialised space of the city, now these amenities are shut down or dangerous. Instead of enabling a clean ordering of the body even when away from home, the still public toilet encapsulates the jumble of dirt and disease which threatens us in the city. Parents warn their children – you can't be too careful – who knows what lurks in those dark recesses? So obediently we don't speak to strangers, don't sit on the seat, wait until we get home. It seems that nothing can keep these little publicly maintained niches of private space in check, so, playing on our sense of danger, hard-pushed councils shut them down.

In 1996 the Welsh Consumer Council published a report damning the state of public toilets across Welsh cities – a situation certainly echoed across the British Isles, if not beyond. The survey showed that public toilets in the street were uncomfortable – too often without paper, soap or even running water – and ill-equipped – too little provision for people with disabilities and parents of small children, not enough drinking water, sanitary products or condoms. Worse still, there was often no toilet at all. Across Britain another result of the twenty-year spending squeeze on local government has been a gradual closing down of public amenities such as the toilet. Once again the need to excrete is pushed out of the public realm, a privilege for which you now pay twenty pence in the railway stations of London. The much hyped fears about the possibility of public sex act as a veil for something more fearful still.

> The splitting off of our consciousness from our own shit is the deepest training in order; it tells us what must happen privately and under wraps. The relation that is drummed into people with regard to their own excretions provides the model for their behaviour with all sorts of refuse in their lives. (Sloterdijk, 1988, p. 151)

Now, when both rich and poor worlds find it hard to deal with the assorted refuse of our lives, bodily functions are once again excised from orderly public life. Pissing and shitting become more private again – a privilege for customers only – and the body is told to wait. But, of course, we should know from the urine soaking the subways and alleyways across our cities that the body never waits.

References

Lambton, L. (1978) *Temples of Convenience*, London: Gordon Fraser.

Sloterdijk, P. (1988) *Critique of Cynical Reason*, London: Verso.

•PURSUITS•

by Jane Rendell

Rambling describes incoherent movement, 'to wander in discourse (spoken or written): to write or talk incoherently or without natural sequence of ideas'. As a mode of movement, rambling is unrestrained, random and distracted; 'a walk (formerly any excursion or journey) without any definite route or pleasure' (*Oxford English Dictionary*). The wandering form of the ramble is not entirely random, but defined in relation to purpose – the pursuit of pleasure. In the early nineteenth century this was specifically connected to the city, to urban space; only later was the term associated with rural excursions.

Rambling has its origins in 'spy' texts published in the sixteenth century onwards. These urban narratives told of various country gentlemen's initiations to the adventures of city life under the guidance of a street-wise urban relative, wise to the delights and entertainments, as well as tricks and frauds, of the city. These texts were supposedly aimed at educating and warning 'Johnny Raws' from the country of the corrupting dangers and sophisticated criminality of the town. But they also worked to titillate the reader, by stimulating desire for the denied pleasures of the urban. Spy texts favoured the pursuit of usually hidden aspects of urban life. These provided a source of excitement for the reader by carefully balancing the visible and the non-visible, and controlling exactly what was revealed. The discovery of new forms of knowledge was a vital aspect of the pursuit of pleasure for the 'spy'. This might involve uncovering aspects of the already known in different, unexpected ways, as well as the discovery of new sites and activities. The dangerous aspect of city life was important as both a warning and a temptation. The spy focus on exploration, detection and the exposure of criminal codes, continued into the first decades of the nineteenth century.

During the eighteenth century, rambles started to appear alongside spy tales. Rambling texts represent urban explorations, journeys of discovery to unknown parts of the city. 'This day has been wholly devoted to a ramble about London, to look at curiosities.' In early nineteenth-century rambles the emphasis shifted to highlight the pleasure gained from visual display and fashionable aspects of urban pursuits. The visual relation between the explorer and the city changed, from the voyeur, or secretive spy, seeing but not wanting to be seen, to the narcissistic rambler, looking but also desiring to be looked at. The thrills of undercover urban exploration and the joys of exposure were replaced by a new excitement in the obvious display of body and clothes through promenade, and the pleasures of looking at views of the urban designed to be seen as spectacle. Coinciding with the return of men to London from the Napoleonic Wars, French fashions and flamboyant military costume played an important role in defining urban masculinity. Rambles provided urban dwellers with a chance to admire themselves and the city. And in contrast to earlier spy stories, primarily scripted descriptions with black and white illustrations, visual pleasure for the reader was enhanced through the provision of coloured images interleaved with the written text.

In their intention to introduce the reader to the city, rambling texts could be described as guidebooks. The rambler visits traditional sites of leisure – assembly rooms, opera houses and theatres, parks, clubs, sporting and drinking venues – as well as the emerging spaces of commodity capitalism – shopping streets, arcades, exchanges, bazaars. Although the places described are real, sites are described not as formal or stylised objects but in terms of their discovery by pleasure seekers. But rambles are only semi-documentary in terms of the pursuit of urban knowledge. Rather than give accurate and objective information, the intention is to provide sensation, to produce a pleasurable response. There is no plot or story-line, no character development; instead the intentions and motivations are spatial and thematic. Rather than select and organise sites by location or type, rambles link places as part of a journey. Pleasure is invoked through representing aspects of the city, both real and imagined, in relation to one another. But most importantly pleasure is produced by introducing the new: 'pleasure and novelty were his constant pursuits by day or by night' (Egan, 1821).

The ramble also represented the emergence of a new kind of masculinity – an urban masculinity connected with knowledge of the city. The ramble was a form of initiation into urban manhood for young bachelors, usually members of the aristocracy, men of leisure in search of pleasure in the city. The male rambler articulated his gender through social and spatial practices; language and dress codes, modes of looking and moving, and various activities involved in the pursuit of pleasure – gambling, drinking, sporting, and the consumption, exchange and display of commodities. But the rambler's pursuit of pleasure was in many cases specifically the pursuit of sexual pleasure – 'to go about in search of sex' (see Partridge, 1984). Like the ranger – 'penis', 'rover, wanderer, rake', and ranging, an activity involving 'intriguing with a variety of women' (see Partridge, 1984, p. 959), rambling was connected with sexual pursuit. Derived from male sporting culture, contemporary magazines concerned with sex, and offering lists, locations and descriptions of prostitutes had the words rambling or ranging in their titles. Many of the publishers had links with pornography, some published prostitutes' memoirs.

> Our motto is be gay and free
> Make Love and Joy your choicest treasures
> Look on our book of glee
> And Ramble over scenes of Pleasure.
>
> (From 'The Rambler in London', *The Rambler's Magazine*, 1 April 1824, vol. 1, no. 1)

The rambler's identity is constructed in relation to other men and women. The rambler is contradicted and reinforced through other urban masculinities; the Corinthian, or upper-class sporting gentleman; the bruiser, or working-class boxer; and the dandy, or aspiring man of fashion; and urban femininity, represented by the figure of the 'cyprian'. The most popular term for the urban female was a cyprian. As such, she was seen as 'belonging to Cyprus, an island in the eastern Mediterranean, famous in ancient times for the worship of Aphrodite or Venus'. Thus she was defined as a goddess of love; as 'licentious, lewd' and, in the eighteenth and nineteenth centuries, unsurprisingly the term was also 'applied to prostitutes' (see *Oxford English Dictionary*). Clearly, rambling can be described as a gendered activity undertaken by male pleasure seekers in pursuit of cyprians.

In pursuit of pleasure, in constant motion, rambling represents the city as multiple and changing sites of desire. In traversing the city, looking in its open and its interior spaces for adventure and entertainment, the rambler re-maps the city, both conceptually and physically, as spaces of social interactions, rather than as a series of static objects. Rambling represents urban space as fluid and complex, varying according to time and specific urban location. Examining the relation between the figures and spaces of the ramble enables us to explore the production, reproduction and representation of space through pleasurable leisure activities – consumption, exchange and display, and through looking and moving.

> We have already taken a promiscuous ramble from the West towards the East, and it has afforded some amusement; but our stock is abundant, and many objects of curiosity are still in view. (Badcock, 1821–1822, pp. 198–199)

References

Badcock, J. (1821–22) *Real Life in London*, London: Jones and Co.

Egan, P. (1821) *Life in London or the Day and Night Scenes of Jerry Hawthorn Esq. and his elegant friend Corinthian Tom, accompanied by Bob Logic, the Oxonian, in their Rambles and Sprees through the Metropolis*, London: Sherwood, Neely and Jones.

Grose, F. (1788) *A Classical Dictionary*, London.

Partridge, E. (1984) *Dictionary of Slang and Unconventional English*, London: Routledge and Kegan Paul.

•QUALITY OF LIFE•

Quality-of-life information, 1987–91[a]

City	*Number of housing units*	*Criminal offences per year*	*Number of day nurseries*	*Number of hospital beds*	*Number of doctors*	*Annual traffic deaths*
Abidjan	n/a	49,000	n/a	2,091	n/a	1,019
Bangkok	1,084,583	23,021 (1988)	n/a	n/a	4,861 (1988)	917 (1988)
Bucharest	n/a	5,087 (Jan–June, 1990)	348	27,352	3,599	180 (Jan–June, 1990)
Buenos Aires	1,216,325	110,243 (1988)	n/a	24,807 (1988)	n/a	3,750 (1988)
Cairo	1,734,100	n/a	184	24,901	3,132	n/a
Delhi	n/a	n/a	50 (1984–85)	19,000	n/a	1,581
Istanbul	n/a	n/a	n/a	24,173	13,534	n/a
Jakarta	1,783,194	35,270 (1988)	1,785 (1988–89)	15,377 (1988–89)	n/a	451 (1987)
Kuala Lumpur	n/a	7,000	273	4,226	3,608	62
Lima	872,550	86,530	n/a	14,873	13,114	832
London	2,782,000 (1986)	620,000 (1988)	9,100	n/a	7,553	742

continued . . .

City	*Number of housing units*	*Criminal offences per year*	*Number of day nurseries*	*Number of hospital beds*	*Number of doctors*	*Annual traffic deaths*
Madrid	308,733	99,394	4,619	13,937	19,984	454
Mexico City	2,017,972	71,598	191 (public)	21,551	22,488	n/a
Montreal	443,560	209,193	358	39,421	6,136	67
Nairobi	n/a	n/a	23	5,696	n/a	n/a
New York	2,992,169 (1992)	626,182 (1992)	3,000	34,863	24,279	566 (1992)
Paris	1,094,000	198,287 (1990)	465	88,972	35,352	104
Rome	1,015,769	379,320	137	38,773	n/a	n/a
São Paulo	1,312,107	n/a	564	239,504	18,544	8,863
Seoul	1,506,167	324,529	1,857	26,697	15,618	1,365
Stockholm	382,000	n/a	1,610	21,580	4,200	24
Tokyo	4,818,000	221,431	1,603	136,759	26,670	488
Vienna	n/a	34,460	14,996	21,720	8,221	81

[a] Figures are for 1991 unless otherwise specified.

Source: Governor of Tokyo and Summit Conference of *Major Cities of the World*, March 1991, *Major Cities of the World* (Tokyo: Tokyo Metropolitan Government). Cited in Sassen, S. (1994) *Cities in a World Economy* (Thousand Oaks: Pine Forge Press), pp. 132–133.

•QUEEN VICTORIA•

by Jenny Robinson

Nearly every major town in the United Kingdom erected a monument or statue to Queen Victoria during and soon after her reign. The Victorian era was in general a time of enormous public interest in celebrating some dignitary or local hero through sculpture, usually through public subscription.

The Mall, in London, stretching from Buckingham Palace to Admiralty Arch and Trafalgar Square, lays out the entire (dead) Empire in her honour. South Africa: an Ostrich and a monkey; Canada: wheat, fruit and a seal (Read, 1982). Her statue stands at the head of the Mall outside Buckingham Palace. For a while it stood with a shiny white nose, replacing the old one which some young vandal-hero now stores among his London memorabilia. Darke (1991), recalling this, comments that the statue now makes a 'superior roundabout' where London traffic comes to a standstill, jostling with the tourists reading their guides, London/Londres/. . . .

Around her sun-always-rising empire, too, public servants, settlers, local leaders made plans to decorate their city with her slim/lithe/dour/maternal/imperial/old/domestic stone form. Not graced with a visit by this home-bound empress they had to be content with her stone form.

What are we to make of this Queen, littered across cities from India to Australia? What do statues like this do to us, what do we do with them?

The tourist guides no doubt remind us of Queen Victoria's greatness, of her empire at the height of British global dominance. The statue speaks of power, calls a nation together to remember her fame and to recall the glorious days when colonies, scattered across the world, paid her homage. They weren't content with simply erecting a statue. On her death, the nobles and politicians sought to make 'some great architectural and scenic change in the quarter of London selected' (Read, 1982, p. 40). The memory of the Queen would inspire a sense of grandeur. They produced a 'total environmental experience' (p. 379) in which tourists now throng, through which city-dwellers pass *en route* from shops to the office, along which ceremonial marches have attracted millions to the spectacles of a reduced royalty. So perhaps the statues are excitedly speaking, with uncontrollable outpourings of speech, littering the pavements with their scattered and glorious pasts, aggressively tripping us up on our travels around the city to remind us of their greatness.

The statues speak power, then, but is this all?

For power has to hide its methods and its effects to be glorious. None of the colonised are represented in the Queen Victoria memorial outside Buckingham Palace. Ostriches, monkeys, leopards and eagles, various crops and fruit. The human dead are forgotten, as are the wars and famines, diseases and dangers faced by the conquered and by settlers. The dirty hands she could never wash clean are pure and white on the statue. The pleas and cries she refused to hear as she silenced the empire's requests hang in the air, unanswered, haunting.

Her statues could also be thought of as a form of public hysteria, memorialising and paralysing the past to make sure we forget. Statues to silence the ravages and traumas of history. Paralysed in and paralysing the city, they cover over the past, burying our traumas (death, duty, horror) in their silence. The obscene, Lefebvre suggests, is banished; and this 'exclusion from the scene is pronounced silently by space itself' (1991, p. 226).

But perhaps the statues also incite us to another sort of speech, as we re-cast the memories, recover and replace their stone(y) silence, their repressed significance, with our own stories. . . .

> The Irish national monument to Queen Victoria was carefully dismantled before the IRA could get to it, and now surveys a courtyard in a somewhat disjointed manner. The Indians, too, found the commemorative relics of the Imperial Raj politically distasteful, but dismantled most of the monuments before any serious damage was done. Many of those in Calcutta were removed to the police barracks at Barrackpore, some miles out of the city, and are still stored in the courtyard there. (Read, 1982, p. 40)

Then again, perhaps the statues are simply ignored. In the former capital of one of her ex-colonies, she stands regally outside what was once the legislature building. It is the centre-piece of tours of Victorian Pietermaritzburg – although downtown Pietermaritzburg is no longer the kind of place nervous tourists are recommended to visit. Around her, people go about their daily life. Guards in the building behind stare past her all day; passers-by don't give her so much as a glance.

Children sit in the shade of the stone

sharing the remains of a Wimpy burger and chips from a broken polystyrene container. Noses dripping, sticky with glue, they jump up to ask a passer-by for 50 cents.

He looks away, shaking his head.

She watches?

The grand colonial redbrick town hall – where 'development' is being planned for the new Greater Pietermaritzburg Area, Umsundusi – looks over her, looks over the street children. It also overlooks her, forgetting her now the colonial past seems irrelevant and a new era is underway. And the street children, are they forgotten too?

She looked over her empire in stone; she overlooked her empire. Chiefs and communities often sent off petitions to Queen Victoria, even made pilgrimages to ask for assistance from the fair and noble Queen. They were hoping for some relief from the exploits of her subjects. In return? Silence. No reply.

So perhaps it's fair enough that we overlook her too; even alright if we forget to put her away out of sight or can't be bothered to blow her up now that we have no use for her, or for our memories of her. Or that we leave her statue, and that of her beloved, hiding in forgotten parks, sheltering in the last corners of Empire, crumbling in the rain, dwarfed and imprisoned by high-rise buildings. She has become neighbour to street children and beggars, to workers sleeping off the waking night, waiting for the first bus home. All ignoring her.

Thirty years living in my home town and I didn't know where to find her. Then I vaguely remembered the stone companions of the pigeons I used to feed as a child in the town square. They've been chased away now – she remains, silent, stony, forgotten. She hides away in what is now Freedom Square – renamed since the white founder of Durban was no longer someone to remember in our everyday directions. He languishes there too, with other white luminaries. Forgotten. Ignored.

All rulers must have to forget many things they've done, to freeze the horrific memories of their actions in inaccessible parts of their minds. Statues are good resting places for their memories, then: immobilised, forgetting, forgotten.

Queen Victoria overlooks the growing poverty of her former realm. Her heirs, today's neo-imperialists, continue these habits – only without statues. Silent, unobserved, impersonal, making policies on the run, tapping data into their universal economic equations, producing reports of statuesque proportions, passing by, not looking. Perhaps they don't manage to visit either. . . .

If we don't get around to destroying their arrogant, phallic presence in cathartic moments of revolt, dismantling statues and countering their claims to greatness, perhaps it is best if we just pass them by. Better that we mimic their forgetful silence than be immobilised in the street in their stead, angry, crying, our limbs (also) paralysed by the memories of what they did; of what they do. . . .

References

Darke, J. (1991) *The Monument Guide to England and Wales*, London: MacDonald and Co.

Lefebvre, H. (1991) *The Production of Space*, Oxford: Blackwell.

Read, B. (1982) *Victorian Sculpture*, New Haven, CT, and London: Yale University Press.

•Restaurants•

by Gail Satler

Tasting the global economy

Restaurants, as they are inscribed in advanced urban areas, reflect the intricate global forces that create and sustain the city in which they are found, and in turn respond to the unique economic, social and psychological infrastructure we view as the city. In many ways, restaurants are global cities in microcosm in that they offer sites within which we can observe the paradoxes that define globalization. This sense of globalization is often overlooked when the focus is primarily on dominant or most apparent economic or cultural features rather than the multiplicity of sub-economies and (work) cultures that constitute these places. Here, I seek to claim or reclaim the notion that the geography and composition of the global economy produces complex dualities: a spatially dispersed yet globally integrated organization of economic activity, tightly knit command centers and spatially dispersed production sites, sites offering the highest level of skill and product as well as places in which casualization dominates, sites where the dominant culture and formal economic processes permeate as well as sites where the informal economy and other cultures can be found to prevail. What we have is a terrain of vital and viable contestation.

In New York City, the command center in which dominant restaurants reside continues to be centralized in Manhattan. Here we find the largest number of highest grossing independent, often high-end restaurants of all styles. These often spawn 'clones' by way of their chefs, owners or chef/owners, which become situated in other parts of the city, often close by, in other cities within the USA, for example, Los Angeles, San Francisco, Miami, or to global cities around the world, most notably London and Hong Kong. What has also occurred over the past fifteen years or so is the tremendous growth or infiltration of more casual and moderately priced dining. *Casual dining* is defined by the industry as 'informal dining with style'. Cafés, bistros, and trattorias are examples of this trend. More deeply, both the continued growth of high end restaurants as well as the new emergence of less formal dining trends in

Manhattan are due in part to the influx of professionals both single and married as residents as well as visitors, and a vital immigrant population.

For patrons, there is perhaps a growing desire and need to create and seek out places in which we can feel comfort(able); spaces that balance or accommodate the desires of each within the socially ordered and defined sites of restaurants; places that find the balance of improvisation within a structure we *know*, and places which are affordable for frequent dining out. For owners, casual dining provides a way to keep down costs in what continues to be an extremely labor-intensive business. Casualization allows the menu to be eclectic and lends itself to seasonal as well as other types of change needed to keep down costs. Even in restaurants that are not specifically defined as casual, this type of dining has been incorporated by, for example, offering several lower-priced selections on the menu by utilizing different or less costly ingredients and preparations so that less expensive labor can be utilized, offering *prix fixé* meals as part of the menu, and/or creating a more informal atmosphere through design which allows for a more informal and less expensive service. Manhattan's restaurants reflect a very critical feature of globalization – increased polarity of classes.

In the outer boroughs and areas that straddle Manhattan, the impact of globalization takes a somewhat different form. The clientele in these areas consists mostly of locals and is perceived as such. However, since these places reflect the broader global changes in employment opportunities and immigration patterns, what has become more pronounced is the expansion and entrenchment of ethnic enclaves – *true neighborhoods as they were traditionally perceived.* In some cases these have replaced those once considered to be the mainstay and domain of Manhattan. As neighborhoods of Manhattan become more gentrified, ethnic communities are driven out. And while it may appear that what is occurring is eviction, economics seem to show that in actuality we are witnessing an example of *agglomeration and dispersal* patterns that reinforce the dominant culture in the production and command centers in Manhattan while creating strong and viable but different cultures which exist alongside the dominant one in these outer areas of the city. This may be a way of handling not only the increasing polarity in classes created by globalization, but also the ever-growing, intense competition among the twelve to fifteen thousand restaurants that exist in the city. Rather than view the global economy or globalization as increasing independence, or convergence, we may instead view it as a growing and healthy dependency between the high and low ends of the spectrum and between the dominant and other cultures that inhabit the city. The form is new and unique.

One might ask whether this will allow for continued diversity and variety, or will it channel our dining options into the more dominant types and forms of restaurants consistent with dominant tastes? By valuing or giving status to the less dominant cultures by way of restaurant types, cuisine and foods, the answer would appear to be yes to both. Intimate neighborhood or niche restaurants continue to hold their place in the city amid the mega-restaurants and chains, not just in the outer boroughs and regions, as discussed, but in the heart of Manhattan as well. They, in fact, are growing in number and in popularity as are trips to outlying areas to experience different foods and cuisine. And as these values and needs intersect with values and needs of the larger global landscape, we will continue to see changes that reflect 'other' forms of culture in the economic, social and political spheres of city life that we tend to read only through parameters set by the dominant culture. The landscape of the city will expand and deepen, shifting the place which the command center holds and providing truly different spaces that become viable options for residents. We will not only be what we eat, we will be where we eat as well.

•Roads•

by Ivan da Costa Marques
and
John Law

one: Cyclist

It is hot and the driver has probably been driving the bus for several hours. Indeed, he is probably working an illegal double shift – not that anyone would care, for the laws are for appearance, 'para inglés ver', 'for the sight of the English' only. It is summer and very hot. Buses in Rio have no power steering, no air-conditioning, and no automatic transmission. The engine is right beside the driver. More heat and noise. The conditions of work are alienating, difficult, trying. The driver's being driven mad. Ahead there's a boy riding his bicycle. It's in the way of the bus. What is going to happen?

At the last moment the driver uses the weight of his body to turn the heavy wheel. The bus nudges out to the left. Just enough. It misses the boy. It misses the boy by an inch, so it's fortunate that he's riding dead straight. Not weaving about. An accident narrowly avoided. Because the driver is exhausted and he's done just enough . . . no more than the minimum. And had there been an accident? Then he would probably have felt that he had done nothing wrong. He'd done what was needed.

Why? What is this politics of near-invisibility? Of denial? No doubt it is partly a lack of money for power steering, air-conditioning, or automatic transmission. At any rate, the 'Mafia' who own the buses say that they can't afford high technology when fares start at US$0.55. And in any case, they add, the roads are so bad that fancy equipment gets broken the moment the buses take to the streets. Perhaps they are right.

But there is more. For no one really obeys the traffic rules. And then again the young men who drive the buses work double shifts to make more money – and no doubt have an exaggerated view of their ability to drive when exhausted. In short, no one cares. And cyclists, not to mention pedestrians, are nearly invisible. They are denied, nearly denied, in a network of partial connections and disconnections.

two: Motorway in Gávea

When they expanded Rio west and south in the early 1980s they planned to build a new motorway across the campus of PUC, the Catholic University. This is an élite school run by the Jesuits. But PUC turned out to have good political connections. It pressured the authorities, and finally succeeded in getting the course of the proposed road changed. So now it skirts the edge of the university, hidden from the campus by a concrete wall, and indeed a tunnel – forms of insulation from noise, vibration, dust and smell not afforded to those without such good connections. But if you travel south, just before passing the campus you'll see that the motorway goes through some apartments right in the middle of a long, low block of flats. To be precise, it passes through the place where the apartments used to be. A hole in the middle of this building. So today there is this most surprising situation: a motorway on which several thousand vehicles drive each day through a residential building. With all the pollution from the motorway. Not to mention the collapse in property values.

Are people angry about this? The answer is: not really. Or if they are, then not in large numbers. What one might think of as the 'level of civility' is so low that this engineering violence is denied. More or less. True, the people in the block of flats aren't really poor. They don't belong to the slums. But neither are they upper class – or as rich as the students of the PUC or other more select neighbourhoods in Gávea. Instead they are denied, nearly denied, in a network of partial connections and disconnections.

three: Older drivers

'Sure there are speed limits, but who's checking? When in Rio drive as fast as you like! You will not be stopped for speeding in the city, but always keep your eyes open since you are not the only one going over one hundred kilometres per hour!' 'Lanes don't mean anything. Pass wherever there is space and on either side.' Squeezing is another way of saying 'Let me in.' Do not squeeze a bus. It will gladly squeeze you back. Remember! No one pays any attention to blinkers – theirs or anyone else's.' 'It is wise to always keep a eye on your rear-view mirror. In the blink of an eye you will find a car, one that had previously been nowhere in sight, just centimetres from your rear bumper flashing his brights. This is the Carioca way of letting you know that you are expected to move over and let him pass. If you don't, he will just squeeze in next to you anyway and pass into the seemingly non-existent space in front of you' (quotations from Priscilla Ann Goslin (1991) *How to be a Carioca*, Rio de Janeiro: Netuno Ltda, p. 48ff.)

The irony may be exaggerated. But it suggests that the privilege of being 'behind the wheel' belongs to the young in Rio. And the risk of insanity by mid-life is serious unless you quickly develop a proper attitude towards driving.

A Brazilian friend is driving. 'I see you practising Zen meditation as you drive', I tell him, as the cars weave in and out between the lanes of fast-moving traffic in ways that appear incredibly dangerous but also seem to lead to remarkably few accidents. Or as the buses hurtle from lane to lane, seeking the least advantage. When some particularly dramatic manoeuvre takes place, my friend, speaking English for my benefit, says, 'Exciting!' and laughs a little. I like the way he says it. Quietly, with a smile, placing his emphasis on the middle syllable: 'Ex-*ci*-ting!' But he doesn't drive in an exciting way. Not at all. Gently. Courteously. Watching the near misses around him with good humour. Often enough with amusement in his eyes. Which is why I tease him about meditation. 'Well, in this country if you really *cared* about getting there quickly then you would die of frustration. Or high blood pressure. The only way to drive,' he says, 'is, so to speak, to take it as it comes. And then to allow yourself twice as long as you think you might need for your journey.' I think he is treating this as a metaphor which extends far beyond driving.

But in fact it is almost impossible for older people to cope with such aggressive driving. Their bodies, their reflexes, are not quick enough. Old people are denied their right to be behind the wheel in Rio. They are denied, nearly denied, in a network of partial connections and disconnections.

ROUNDABOUTS

by Marc Augé

The revenge of the local

For the British, the roundabout is an old institution, almost historic. It is a national tradition, like driving on the left. But – unlike driving on the left – it is conquering the world, most particularly France. In France for the last ten years we have realised that the best way to prevent cars from going too fast is to stop them, and that the best way to stop them is to make then turn in an anticlockwise circle. Or, more exactly, to invite them to join the circle, only after the others have passed – well, running to its full term, a true cultural revolution, to respect priority to the left.

Cultural revolution indeed, in a country where traffic priority has always been to the right, unless a mandatory sign, usually one that's familiar to us all, indicates that one is on a main road (and does not have to worry about the right or the left) while informing others that they are on a secondary road (and *do* have to take notice of the right and left simultaneously). The French crossroads were the metaphor of a centralised and hierarchic country (all the main roads coming from the capital being, by definition, priority). This rigid rule grates against the right of free judgement from which the French driver should, in principle, benefit. He still enjoys this freedom in Paris, on squares such as L'Etoile or La Bastille. Here the roundabouts have no signs, and only the driver's skill and vigilance allow him to establish and impose his priority – plunging as close as possible towards the outside of the roundabout's circumference to keep his pursuers and adversaries to his left. The delights of 'the regatta' that are part of the charm of driving in France and in Paris – although surprising to the foreigners' eyes.

In such a context, we can conceive that the sudden appearance of roundabouts has been a shock. Indeed it was, literally, a few years ago when incredulous drivers who refused to acknowledge roundabouts, or who couldn't believe their eyes, planted their vehicles in the central floral arrangements, or crashed them against statues or other artistic constructions supposedly decorating them. The roundabout created a semantic shock in driving manuals and official instructions, which spoke in terms of 'gyratory crossroads' to avoid any Anglicism and to protect the French language. This notion seems to consider as solved the famous problem of 'squaring the circle'.

Finally, the roundabout created a psychological shock to all those, of whom I am one, who came to see it as a sign of creeping modernism and therefore proof that Great Britain was once again ahead of the game.

Today we know that communication and traffic are moving faster, we move quicker from one place to another. Information moves at the speed of light. In the world of motoring we're not there yet, although on the motorways in Europe we can drive quickly from one point to another. Motorway interchanges prevent any risk of coming together, and cars are circulating in the large European arterial roads like blood in a healthy body, without threat of cardiac arrest. The roundabout corresponds to the necessity of organising the circulation, and keeping it flowing, but at a lower level – to the driver who comes off a motorway and, impelled by his accumulated momentum, thinks he is going to reach the next town at the same speed. The roundabout imposes a check, by giving priority to the quieter stream of local traffic. It is the revenge of the local on the global.

Today in Europe each small town, each hamlet, can make drivers travel in circles, thereby obliging them to glance at the signs which indicate and promote local places of interest. The roundabout imposes the same democracy of the road on the tractor coming from a country lane as it does on the limousine exiting the motorway.

In these days of 'The Global Village', roundabouts remind us of the reality of the traditional village, and the necessity to think of global and local systems together. The roundabout's too-fast, repetitive merry-go-round momentum can tire drivers out. But it can also suggest the idea of pulling over and stopping by the side of the road.

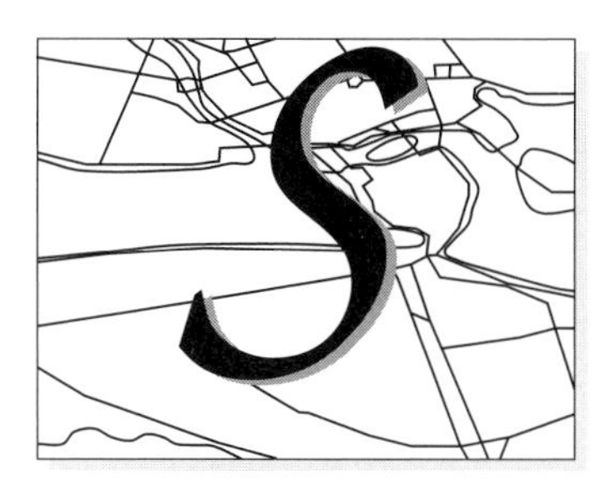

•Sacredness•

by Jane M. Jacobs

Where is the sacred in the modern city? There are those who might imagine that the capitalist city would become a largely secular thing, yet we know that sacredness survives and even flourishes. Perhaps it is more appropriate to ask: what form(s) does the sacred take in contemporary cities? For the French sociologist, Michel Leiris, the modern sacred resided in the spaces of everyday life. He found sacredness in the 'humblest' of spots: his parent's bedroom, a brothel, a racecourse, even the bathroom of his childhood where he and his brothers would secretly conjure mythical and magical tales about animals, detectives, soldiers, pilots and kings (Leiris, 1938).

For most of us, sacredness in the city is attached to a far more precise urban geography of mosques, temples or churches. These places of worship manifest as sacred spaces not simply through the magic of faith, but also the logics of property and the edifice. These sacred spaces are what the College of Sociology would call the 'officially sacred' (Leiris, 1938, p. 24). They have much in common with, and are often still part of, the state apparatus. They claim precise, firm and constant centres from which a powerful zone of influence reaches out across the rest of the city, and sometimes even the rest of the world.

In many modern cities the state and religion are technically separated. Yet often enough the sanctity of places of worship is supported by the state apparatus. For example, the state may guarantee certain allocations of land for sanctified use, and it may protect those spaces of spiritual development that already exist from other, more secular development forces. Places of worship that are sufficiently old or architecturally unique may be listed on official registers as heritage monuments. In a time when even religion is touched by economic rationalism, this 'added value' is often all that ensures the survival of these buildings after decommissioning from service to the sacred.

Transnational migration flows and postcolonial formations mean that the sacred geographies of the contemporary city are often characterised by their religious heterogeneity. As migrants settle in their new homes they transform urban space to meet their religious needs. A good example of this process exists in the East End of

London. On narrow Brick Lane an elegant Georgian building still stands, protected, of course, by heritage listing. It began its life as a French Calvinist church, servicing the needs of Huguenot settlers. In the late 1800s it was transformed into a synagogue for Jewish settlers who had fled from persecution in Russia. Now it is the London Jamme Masjud, a mosque that daily calls to prayer Bangladeshis who have come to London as postcolonial migrant settlers.

Often, settler communities are only able to claim precarious sacred spaces from the secular zones of the city. In the Australian city of Melbourne, for example, Sri Lankan Buddhists have transformed an ordinary suburban house into a temple. The interior of the house has been adjusted to accommodate the various requirements of Buddhist worship; an illuminated sign adorned with the Buddhist colours has been placed on the fence, and in the garden a shade house has been carefully constructed to ensure the appropriate micro-climate for the all-important sacred Bodhi-tree (the symbolic tree of knowledge and enlightenment).

There is little of the seemingly secular city which is untouched by sacredness. Take as an example the residential high-rise, that building form which for many contemporary urbanists stands as the paradigmatic expression of rationalist efficiency and urban alienation. Even here the sacred can flourish. Urban sociologist Chua Beng-Huat has documented the various ways high-rise residents in Singapore transform the interiors of these standardised apartments to make space for the sacred. Chinese householders, for example, will adjust interior arrangements in order to create the requisite internal symmetry to sustain the symbolic order of the home/universe. Hindu Indian householders may appropriate store-rooms for prayer rooms, while the Islamic Malay household may introduce beaded curtains to signal the necessary gender partitioning of the home (Chua, 1997).

The state acquires a different relationship to the sacred in the multicultural and multi-religious city. Here the state often functions as an agent which facilitates and manages religious diversity. For example, the geographer Lily Kong has shown how the Singaporean state self-consciously positions itself as a secular apparatus in relation to the religious diversity of its citizens. In a nation where there are five key religious groups (Buddhist, Taoist, Christian, Muslim and Hindu), careful and even-handed planning has been required to create a balanced geography of religious diversity (Kong, 1993).

Not all sacred space is articulated by way of the logic of property and edifice. In the case of the 'traditional' beliefs and practices of indigenous Australians, this apparent lack of property and architecture positioned them as primitives in the colonial framework of the world. The cities of the Australian colonies were founded upon the blank spaces created by the assumption that indigenous Australians were uncivilised, of the land but not mastering the land. In contemporary Australia, as in other former colonies, indigenous people are making postcolonial claims to have their rights to land recognised, including the protection of their sacred sites. Until recently it was assumed that urban areas were immune to such claims. Here the erasure of an indigenous interest had been quick and thorough, and the cities that grew fully occupied the appropriated spaces. But urbanisation has not kept the indigenous sacred at bay.

In the 1980s an unexpected claim to protect the resting place of the Waugul (a dreamtime creation snake) sent city planners in Perth into a flurry of activity. Fearing that this might be the first of many such claims they attempted to compile a complete record of all Aboriginal sites of significance in the metropolitan region. But the indigenous sacred defied the logic of planners. It was not clear what the precise boundaries of the Waugal site were, or even the extent and number of Waugal sites.

Some secret sacred sites were simply not available for charting on to multiple land-use maps. How can the state plan for a sacredness that does not have property or edifice and has the unpredictable geography of these postcolonial returns? The attempts to map the complete number of indigenous sacred sites might well have been done in a postcolonial spirit of recognition. But this mapping frenzy also marked an anxious colonial reterritorialisation which sought to maintain far more manageable boundaries between the modern city and the disturbing effects of the Aboriginal sacred.

References

Chua, B-H. (1997) *Political Legitimacy and Housing: Stakeholding in Singapore*, London and New York: Routledge.

Kong, L. (1993) 'Ideological hegemony and the political symbolism of religious buildings in Singapore', *Environment and Planning D: Society and Space*, 11, pp. 23–45.

Leiris, M. (1938) 'The sacred in everyday life', reprinted in D. Hollier (ed.) (1988) *The College of Sociology 1937–39*, Minneapolis: University of Minnesota Press, pp. 24–31.

•Satellite dishes•

by Stephen Graham

The Internet, camcorders, multimedia and virtual reality games are now filling our urban spaces with a continuous choreography of electronic visions, an exploding new electronic visual culture. Before the early 1990s, however, it was very difficult to visualise the many IT-related developments going on in cities, especially those surrounding economic development.

To anyone carefully watching cities, small, incremental changes were certainly afoot. Roads were dug up and optic fibres were secretly strung and woven beneath the urban fabric. Telecommunications towers mushroomed atop tall buildings. Banks of personal computers could be glimpsed, green screens aglow, behind many windows and façades. Bank cashiers were slowly displaced by 'holes in the wall' (ATMs). And yet, despite all the hype in the 1980s about the transformation of urban areas into 'information cities', 'telecities', and 'technopolises', the fabric of cities basically looked suspiciously similar to that of the previous twenty years. In most cities it wasn't clear exactly where the gleaming new edifices, the futuristic spaces, the icons of the city in the 'information age' *were*. Beyond the mushrooming 'global financial capitals' of London, New York and Tokyo, it all seemed rather too familiar for the rhetoric to have much credibility.

A wide range of local projects, developed to try and use IT for economic development, 'electronic democracy', and new community networks, similarly proved to be too *invisible* – hidden within subterranean flows of data linking small metal boxes with flashing lights. Were we supposed to shift our society, economy, and culture into some disembodied, on-line realm, without any parallel shift in the material geography of cities? This problem was especially difficult because municipal politics had always heavily emphasised the symbolic importance of 'cranes on the sky line' and the visual power of major physical developments for urban boosterism.

For planners and development agencies across the world, this presented a problem as they strived desperately to use the glamour and hype of technologies as vehicles to re-image their city spaces. Advertising campaigns and new slogans, such as '*Manchester – the Information City!*' and '*Barcelona – The On-Line City*', created potent images of modernity and technological transformation. But physical symbols of technological advancement were urgently required to back up such rhetoric.

Enter the so-called 'teleport' – a mixture of advanced office campus, high-quality telecommunications links and, key to it all, banks of satellite dishes. For the thirty or more cities that developed teleports, these banks of angled satellite dishes at last provided tangible and visible symbols of the new urban modernity of the 'telecommunications revolution'. At last it seemed possible to build a physical development to concretely anchor the wild rhetoric of technology-based urban marketing in the 1980s.

In the 1980s every city worth its salt wanted a teleport. There were yearly teleport conferences. An organisation called the World Teleport Association toured the cities of the world encouraging what the French called teleport 'fever', selling the message to mayors and planners that if their city was not part of the growing network of teleports, then it was to be a technological backwater with no future. It was argued that the teleports were an 'information-age airport', a new installation linking the local into global webs of flow and power in the media, finance, and telecommunications industries. Only teleports would support dynamic sites for competition and innovation in advanced services linking computing, telecommunications and media industries. Teleports were, it was argued, to become entry points into the global 'space of flows' while acting as nexus points between global property firms and consultants and local political coalitions and development agencies in particular urban districts.

The more Messianic of teleport consultants promised that the projects would ensure anything from international solidarity through democracy, social cohesion, civic pride, property markets and so on. Anything, in fact, that would whip up the civic paranoia to the extent that public subsidies, cheap land and lucrative finance packages would fall the way of specialist teleport consultants and developers. So, with varying amounts of public subsidy, cities as diverse as Cologne, Poitiers, Amsterdam, San Francisco, and Sunderland built their gleaming teleports, proclaiming themselves, like Roubaix in France, to be on the 'networks of the future'.

What the teleport boosterists never quite managed to explain, however, was this: if the 'information age' was to make distance and geography redundant, why would large-scale developments, rooted to particular spots and hooked into specific IT infrastructures, have such a hold over the globalising dynamics of the world economy? Surely all major cities would soon be wired up with something close to the latest in telecommunications infrastructure anyway, given an increasingly competitive industry eager to seek out their concentrated demand? Gradually the realisation dawned that teleports were as much symbol as substance, as much the result of local effort to project a global technological and modern identity, as some functional requisite of international capital.

Major cities were clearly at risk of fighting hard to become 'competitive' by positioning themselves on the glitzy new global teleport network, only to find the advantages to be little more than a charade, even a short-lived, embarrassing and expensive anachronism. In the more peripheral and disadvantaged cities, the teleport, in a wave of Messianic and deterministic rhetoric, became the latest in a long line of serially reproduced quick fixes to restructuring crises and uneven development. As with the convention centres, malls, heritage zones and mega-airports that preceded them, they were often of questionable use in supporting urban revitalisation. Many were built but hardly used. 'What if somebody built a teleport,' asked the American geographer, Ron Abler, 'and nobody came?' Thankfully, this widening realisation has since promoted cities to concentrate on IT applications, rather than making huge investments into hard telecommunications infrastructure.

The best example of the teleport as a triumph of technological symbolism over the aterritorial logic of advanced telecommunications came from Edinburgh, capital of Scotland. Keen to secure its future as one of Europe's pre-eminent legal and financial centres, and eager to overcome its geographic peripherality, Edinburgh's planners and policy makers started to consider a teleport in the mid-1980s. They looked in particular to the futuristic new Maybury Business Park on the edge of the city. They then engaged consultants who talked to all manner of telecommunications providers about the feasibility of the project. Eventually, however, it became clear that, in actual fact, there was absolutely no need for a brand new satellite ground station in Edinburgh. British Telecommunications would upgrade their networks, certainly, but any use of satellite services could easily be accommodated by distant, unseen ground stations switched in instantly from other parts of the UK (and, horror of horrors, using the one in London, the global finance capital that was 'centre' to Edinburgh's 'periphery'). 'The location of a dish on the site' reported an EC study in 1991, 'would be symbolic rather than of practical use.'

The Edinburgh dish would thus have been an urban icon of local–global connectivity that might as well have been made from cornflakes packets, as it would not have been connected to anything more than a concrete slab. However, as long as people *thought* it was some super-modern springboard to 'the Global Village'; some crucial node in arcane, unimaginable flows of corporate data; some instantaneous transit point for billions of electronic dollars; some real-time nexus for switching Gigabytes of media product around the planet, then it would have succeeded.

In a way it's a shame, then, that the Edinburgh teleport was never 'built' – the project, in fact, became a much more vague and invisible 'telezone' based on upgrading the hidden lines and switches under the city to world standard. For had it been built it would, perhaps, have been the ultimate symbol of the tensions between attempts to politically enclose cities and the space-transcending effects of telecommunications networks; from between the urban boosterist's need to contain benefits and build futuristic symbols, and the internationalising and space-transcending logics of new technology. Jean Baudrillard would have loved it. The perfect simulation: A hyper-real node in a network that never existed.

•SCHEMA•

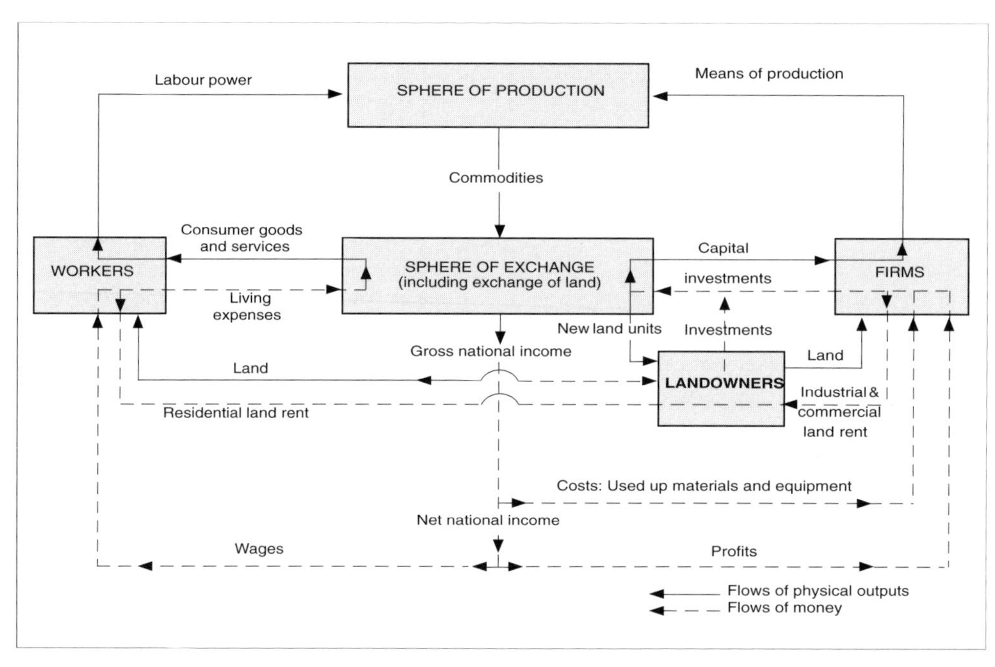

Figure 10 Commodity production: a simple schema of interrelations of capital, labour and land

Source: S. Roweis and A.J. Scott (1978) 'The urban land question', in M. Dear and A.J. Scott (eds) (1981) *Urbanization and Urban Planning in Capitalist Society*, London: Methuen.

•SCOOTERS•

by Arturo Escobar

> One only knows a spot once one has experienced it in as many dimensions as possible. You have to have approached a place from all four cardinal points if you want to take it in, and what's more, you also have to have left it from all these points. Otherwise, it will quite unexpectedly cross your path three or four times before you are prepared to discover it. (Walter Benjamin, *Moscow Diary*)

There are over 100,000 scooters in the city of Cali (population: two million), probably more than in any other Latin American city of its size. It is said that the city's relatively mild tropical climate – never cold, rarely unbearably hot and without lasting rains – is behind the high use of this mode of transportation, but surely there are many other factors as well. Drive a car in any of the main city streets and you cannot fail to notice the ubiquitous presence of these small machines, their sudden turns and twists, their annoying moves, rarely sticking to the rules, or so you think. In this already car-centred city – not to speak of the large buses, now mostly for the working class – you would wish that the swarming scooters would just go away.

But who are these *motociclistas* (drivers of scooters and motor cycles of all kinds) who, defying any reasonable sense of safety, venture everyday into the streets of this city with a number of cars that by now exceeds the city's capacity and that seems to ceaselessly expand? How do they experience space, time and speed? Which places do they connect in their meandering through the semi-regular city streets? What is their sense of body, territory and time? Which evasions and subversions do they succeed in practising when confronted with the fixed city map? What types of social relations do they establish with fellow riders and with drivers of cars? What does it mean to be a woman or man on one of these machines? What sorts of relations do they establish with their machines? In sum, which 'tactics of the habitat' and practices of body and space shape their days on the streets?

It is August 1996, and, based on a new study that found *motociclistas* to be a high-risk group ('accounting for 35% of all deaths in traffic accidents, amounting to a loss of 15,382 years of potential productive life', etc.), the mayor's office issues a decree making the use of helmets mandatory. According to the same study, only 1.7% of all *motociclistas* wear helmets. Confronted with this 'terribly unfair' and one-sided decision, hundreds of *motociclistas* demonstrated in the city streets for several days, erecting blockades and burning tires, all to no avail. That helmets were uncomfortable and hot; that wearing them you could not hear the surrounding traffic, making driving unsafe; that they were too expensive for most (more than US$40 for a decent one); that in this city where one can die easily of so many different causes, why worry about wearing a helmet or not . . . all these reasons, adduced by the protesters, were rejected by the mayor, armed with the alarming statistics of the many deaths and injuries that could be prevented by the simple fact of protecting your head. As the campaign of fear continued and fines started to be enforced in full – and since perhaps most people knew deep down that helmets are a good thing – people began to conform to the new law, and a year later it had been accepted by most.

'I was once in an accident,' my friend, a motor cycle rider, said,

> 'and when I awakened at the hospital I saw a bruised face in the mirror I was at pains to recognize as my own. I had recently got my second-hand Kawasaki 100 and going down a hill I stumbled upon a big hole on the street. The fear of another accident lasted for a long time, especially after I started riding with my daughter; but I've never had an accident with her on the back seat. I am a careful rider, and most women are. It is true that men are more careless; they don't like to give the right of way, not even to cars. For them, I guess, it is a question of competition. It is also true that us, women, are not as good drivers as men, but we are more careful; we follow the rules and always make sure that it is safe to ride, since in case of an accident it is we who suffer the most. We even look constantly in the mirror to see what's happening at our backs; you have to! And you cannot ride with anger or anguish, because you need all the concentration you can have. There is also a lot of solidarity among all *motociclistas*, especially in relation to the larger vehicles on the street. Car drivers, well, they are like in a bubble, you know? For us they are totally impersonal, faceless, and we don't relate to them at all. It's a different world. As for buses, well, they are the kings of the roads; they are the evil to be avoided at all costs.'

Most *motociclistas* are between the ages of 20 and 40. There is a clear class basis to the division between them and the owners of private cars. That's why many riders believe that the *motocicletas* are 'noble' machines. *Motocicletas* (or just *motos*: scooters, motor cycles, lambrettas, etc.) are the 'salvation of the poor' in a growingly complex city life where, as my friend says, you often have to juggle home, grocery shopping, several workplaces, entertainment, friends, and the like. You can get a Honda 100 (or the equivalent Kawasaki or Suzuki, which are the most popular), for instance, for two million pesos (about US$2,000), and it's easy to get a bank loan to buy it, be it new or used. Some of the smaller ones, especially Italian and German, are a bit cheaper; they are almost disposable, and a lot of women use them (women account for about 30% of all riders).

Most people decide to buy a *motocicleta* in order to get to their places of work, which more often than not today entails holding two or three jobs in different parts of the city. For women, this creates new situations and needs.

> 'You usually have to carry a good pair of shoes for work, since the left shoe gets worn out easily because of the gears. Most women also have to wear pants, unless your *moto* is a lambretta type, in which case you can wear a skirt; your hair is always a mess, because of the wind and the smog; some women put on makeup when they get to work; I don't. And of course you have to wash your face carefully at night to remove the dust and the soot. What does it mean to be a woman rider in relation to men? Of course, it is important to know that it is not something that only men can do; it also gives you a sense of independence, a vindication of your rights. Whether you are a man or a woman matters in a positive way, I would say; we have good relations in the street. Men, for instance, are always helpful in case of a breakdown; they are *caballeros* [gentlemen]. If you happen to stop at a light next to a man, there is always the possibility of a certain flirtation, the sweet game of being two, even if only for the few seconds until you have to be again on the move. I would say that relationships between the sexes are most times friendly when you are on the street. I wonder to what extent this is related also to a certain eroticism women feel when you have the gas tank (of the machine) between your legs. It gives you a good body feeling, plus it looks attractive, I think. It makes you think! It might also be related to a feeling of freedom, of being exposed to the elements, the wind and the sun, even if of course there is also the pollution. There is a sense of adventure, even of sacrifice (when it is too hot or too cool) that adds a certain sensuality to the whole experience. Finally, you are keyed to your machine, you form a bond with it.'

Riders also have greater freedom than cars in inhabiting the city on the move:

> 'If you are in a rush, you have to use the main avenues, which are always congested and with lots of smog; otherwise, you can take side roads, and then it can become a leisurely ride. At night, of course, you have to keep to the main roads, and try not to stop at all! On the side roads you don't have so many encounters with the police, who really enjoy giving you tickets for not respecting the rules, like when you go a bit on a sidewalk to avoid a traffic jam. Did I tell you that there is a law that prohibits men to ride with another man in the back seat? I think this law came about to stop the *sicarios* [paid killers, who usually rode in *motos* two at a time], really, but it has also helped a lot to reduce the number of robbed *motocicletas*.'

And as she prepared to ride away, I asked her if she would exchange her Kawasaki for a car; she said:

> 'no way, unless I were to move to the countryside, in which case I would buy me a jeep'

And she left, swallowing the city in yet a new way while contemplating her dream.

Sources

Interview with Gabriela Morana, Kawasaki 100 owner and music teacher; author's experience as a car driver in Cali; newspaper coverage of 1996 demonstrations against the helmet law.

•SEEDS•

by Arturo Escobar

It is not the voice that commands the story: it is the ear . . . Falsehood is never in words; it is in things. (Italo Calvino, *Invisible Cities*, pp. 135 and 62)

You may arrive in many parts of downtown Cali and think that trees have ceased to exist. Along the river that crosses the city – greatly diminished by deforestation upstream and effluents of all types – you might still intuit a glorious past of big and beautiful trees out of the few that remain, a past of leisurely swimming in pristine ponds that continue to exist only in the memory of the nostalgic cadre of elders who still sit on the benches of the downtown parks to enjoy the most sensuous breeze in the late afternoons as they lament the passing of their river and, in fact, their entire city with the advent of the 'culture of cement and the fast buck', particularly after the fateful date of 1971, when the city hosted the Pan American games and there began the construction frenzy that has not stopped since, fueled in the late 1980s and early 1990s by the bonanza propitiated by the Cali Cartel. And yet you might stumble upon a growing number of young or not-so-young dreamers of a different city, where trees will be plentiful and spaces more welcoming, and where the extreme differences between the poor and the rich would have been greatly attenuated so that everybody could live with a new sense of peace and a trusting smile on their face. Like the one I met on my last trip to the city – a friend from old times, indeed – who took me around Cali in search of the oddest places in which to plant trees, and who prefaced our joint field-day with the following words:

'The first thing I have to say is that when I was little my mother told me the story of the founding of the world; to describe the Eden she told me of a paradise full of trees of all kinds, some of which even had little bags with capuccinos, fountain pens, cromhorns and other wonderful things, besides all the fruits of the world. Then I understood that from a seed grows a tree, and failed to see why if we were expelled from Paradise we had not managed to make it again, and this time in the right way, without avenging angels, deceitful serpents and the fearful tree of the science of the good and the bad. So that since I was a kid I would go around my neighborhood throwing seeds on lawns and plant pots, thinking that seeds are like little pills that could make the earth well. One day, once grown and still trying to stick to my dream, I took a cab and found to my dismay at the end of the ride that I had left my money at home. Then I told the driver thus: "I don't have any money, my friend; isn't it true that this is like a declaration of war?" Anybody else might have died on the spot, but I happened to have my seeds with me, including one of an avocado tree which, as I told the driver, has an entire tree closeted in it, so that he could plant it and soon he would have a tall and generous tree and he could eat avocados for free for the rest of his life, and he would be rich because seeds are the currency of God, as each fruit could lead to another tree. . . . And I proceeded to give him other seeds, each with its proper scientific name and conditions for optimal growth, and he dropped me off happy at my destination and occasionally when I meet him in the austere Cali streets he still takes me to places for free, content with having my tales of the seeds of the world in exchange.'

'So it was that I developed this crazy idea of having cab drivers become the harbingers of my strategy to replant the entire city with trees; because I am intent on making of the cab drivers of this city the most ecological group in the planet. Every time I take a cab, knowing that I have a captive audience for some time, I give him (since it's mostly male drivers in Cali, in contrast with Bogotá, where female drivers are more common) the latest version of the ecoterrorist narrative I've developed throughout the years. I tell him first about global warming: "look at all the exhaust pipes everywhere putting out smoke, and not only here but in all cities of the world, day and night; because the planet is like an aquarium, and if all the fish shit and shit and the water is not renewed we will all be soon breathing shit." And I tell him about the ozone layer, "and if we lose this sensitive protection, what will it be like for our children", and I look at the expression in his eyes, and if he shows signs of terror I am certain to assume he has children and my task becomes easier. And when I estimate that two-thirds of the trip have gone by, I start the panoply of solutions. I start by telling him about a leaf as big as a soccer stadium to teach him about cells and chloroplast that become activated with light and how by mixing up sunlight with the smog of our car they manufacture sugar and expel oxygen, free of charge. By now he understands very well and even contributes that "so, even all the weeds that grow against all odds in the sidewalks and that people will pull out so that the cement can shine are also producing this oxygen, oh my god!" And finally, before getting off, I offered him seeds, and not infrequently they accept them eagerly as part of the meter fee.

'Other times I asked the driver a question they never expect, such as: "What would happen if after a wink of your eye all of a sudden you are 65 years old? You cab drivers are facing a grave threat" – and at this point he looks around, thinking it might be somebody who is about to steal his hard-earned money, as often times occurs. No, I am quick to say, the gravest danger is when you stop at the light and the bus to your right points its exhaust pipe at your window and spits its harmful smoke, and soon I have him at 65 and dying of cancer and having to sell his car and his family home to pay a doctor who will tell him that he is about to die anyway. And every day I make up a new story – you could too, we all could – so that with the passage of time their vision gets greener and greener so that at least they will not try to cheat when the next smog control for the car comes around.

'I have come to develop what I call the theory of gnomization, an absurd idea that grew out of imagining how the city would look like if it were inhabited by gnomes instead of humans. Gnomes would not step on other species as we do to open the way for our parking lots, avenues, shopping centers and so forth. The gnomization theory consists of populating the visible space with new species, especially those that can help with pollution control and that are easy to grow. In dark nights I go out like Count Dracula – the flaps of my coat up to cover my face, and with a bag of seeds hidden underneath – and deposit seeds in the cracks of the pavement, since each crack is an ecological niche, and a few days later, after a rain has fallen, little green things start to appear; and since people no longer have eyes for anything natural, well, there they remain. So that little by little the sidewalks of this or that block start to get filled with microgardens, and it's not unheard of to see a few days later a home owner bent on his knees nurturing his garden in the middle of the pavement and the trash.'

'To "gnomize" technically means to break the monotony of cement with patches of colors and greens, to fill in all possible spaces and occupy empty niches with a functional species, and then think about other species that could feed on the first one, that is, to create small trophic chains that can disrupt this other 'logic of the concrete' we have come to depend on. I even dreamt of secretly populating the city with carnivorous plants from the Amazon that would devour the city one night in its sleep, but we have plenty of these malevolent dreams. I also thought that we could gnomize the insect-ridden heads of many a poor kid in the slums of the city, so that there would be birds that would land on the children's heads to eat up the insects, and larger birds who would prey on the small ones. . . . Or parasite plants that would feed on the humidity that sometimes affects the walls of poor people's homes. . . . To say it differently, gnomization entails imagining trophic chains that could digest the pollution we create, and going from street to street making friends among cab drivers, old people, children and gardeners to entrust them all with the task of sowing seeds "treacherously" – as I say with an air of complicity – and preaching the word that one cannot kill what one does not know, and how come we kill weeds and plants without even knowing what they are? And if we have the need to kill some weeds and pests, then perhaps we should think of fumigating them with dry jalapeño seeds – you know, you can dry up the seeds, mash them finely and then put them in those fumigating machines one carries on one's back, don't you think?

'Good ecology is good imagination, that is my motto. I already have a large collection of seeds of fruit trees from wondrous places, from Madagascar to Burma, from Sumatra to Senegal and the many places in between, all gifts smuggled by friends from their travels on my insistent request. It is inexcusable that in these tropical countries people are dying of hunger, when we could fill in the streets with fruit trees of all sorts. Remember that Monet's most famous work was of a garden in a farm he bought not for business but to use the land itself as a canvas to be imprinted with the colors of nature. That's what I mean by reconstructing the Paradise. That's what I want to do with the city. The English – good gardeners that they are – might understand my dream. We must come to see it in our tropical lands, instead of feeling ashamed of our peasant past in our obsession with urban things. We must come to develop a pedagogy of nature and peace, to counteract the state-sponsored pedagogy of violence where violence becomes big business for those intent on destroying our lives. Perhaps this is what I've been talking about all along in my taxi cab rides in this ambiguous city caught between hope and despair, between the growing demand for efficiency and a certain attachment to the enjoyment of life. So I go on planting more seeds in the city's fissures and cracks, searching for a surreptitious smile and a nurturing hand.'

S
S
S

Source note

The story is real. The seedsman in question is Adolfo Montaño, Professor of Music at the Universidad del Valle, founder and director of children's choirs, author of unfinished novels and symphonies, *frenáptero* and daily inventor of worlds.

•SERVICES•

by John Allen

In the 'smart' new buildings of any big North American city, especially New York, Los Angeles or San Francisco, they know a clean floor when they see one. It has to shine. More than that, it has to shine in a way that leaves you in no doubt that the surface is clean. It should literally be 'in your face'. With toilets and restrooms, it's different. They should 'smell' clean, in an obvious, rather pungent sort of way. The contract cleaners who service these corporate buildings know this, as do those who manage and 'work' in these towers of steel and glass. Yet in Copenhagen or Stockholm, they know something different. A clean surface in these cities is not a shiny surface but a smooth, clear surface: a lavatory does not have to assault the nose to be judged clean, rather it should present itself to the eye as such. In São Paulo and Mexico City, the map of sensitivities is different again. Depending upon which city you are in, there are different routes to cleanliness.

None of this, of course, is lost on the big cleaning multinationals (most of whom are Scandinavian in origin, interestingly) who service these and many other cities through a global network of contracts. Alongside the more familiar cluster of commercial services which have come to dominate the skyline of the major cities – the big financial institutions, the law firms, the advertising agencies, the corporate information producers and the like – the less glamorous service firms are equally global in character and likewise attuned to local tastes. Whether catering for the diverse eating habits of the many international banking groups in the major nodes of global finance, or maintaining the security arrangements for a sprawling property-based conglomerate, or fulfilling contracts to clean the offices of a worldwide legal partnership in multiple locations, the low-profile service multinationals differentiate space every bit as much as their more glamorous global counterparts.

At a glance, it may look as if a global service city such as Tokyo, Hong Kong or Singapore represents the kind of single-minded space that many associate with financial dealing, property management, corporate business and, of course, pristine shopping complexes. A lingering look, however, would reveal the many routine servicing roles which not only coexist in such spaces, but positively enable the few to perform their well-paid business roles. The issue, quite straightforwardly, is one of power, representational power: where the ability of certain groups to code a space in their likeness makes it difficult for others to be seen. As Henri Lefebvre (1991) would say, a dominant coding of space provides a sense of 'membership' for those involved; it prescribes a certain use for a space and an appropriate style in which it may be used. Any other use or anyone else is simply 'out of place'. Look up at a corporate high-rise building in one of the busy, global city centres, whether it be in East Asia, North America or Europe, and the symbolism will exclude all bar the dominant service users.

It is not just within a city's buildings, however, that one finds the tangled worlds of service routinely suppressed in one way or another. Cities, in an economic sense, are essentially made up of services and service workers who cross-cut one another throughout the day and night. The home deliveries, the internet shoppers, the child-minders, the gardening helps, the petrol pump attendants, the supermarket check-out staff, the bar staff, the cinema attendants, the street cleaners, the all-night security guards, the motorbike couriers, the taxi drivers, and many more, each have their own time and rhythm of work in the city. They are part of what moves and part of what stays put in the city, part of what is acknowledged and part of what, to all intents and purposes, remains outside of the field of recognition.

Somewhat ironically, in the cities of the developing world, in Jakarta or Manila, or Nairobi or Kampala, the contribution of both formal and informal services to city life appears altogether more transparent. The market traders, the street vendors, the messengers, the porters and the like take their space alongside the ubiquitous global activities of business and finance. There is nothing particularly democratic about such visible arrangements, everybody knows who is servicing who and whose time is deemed to be of more importance in the run of the day's services. To paraphrase Andre Gorz (1989), the distinction between servile work and service work is evident for all to see.

In the service-rich cities of the western world, where Manuel Castells' (1989, 1996) image of the 'informational' city appears to hold out the vision of the future, any attempt to register the comings and goings of those who fall closer to the servile rather than the service end of a city's economy remains decidedly evasive. In one of London's prestigious Royal Parks, for instance, bordered to the east by an imposing nineteenth-century terrace complete with neo-classical façade (designed by Sir John Nash no less), the job of cleaning the park's toilets has been out-sourced to ISS, a Danish service multinational which operates in more than thirty countries across four continents. While both the Danish company and the predominantly ethnic workforce involved, in these quality conscious times, know a clean toilet floor when they see one, the global service itself remains steadfastly one of the many outside of the city's consciousness. After all, if it's London, it must be finance.

References

Castells, M. (1989) *The Informational City*, Oxford: Blackwell.
Castells, M. (1996) *The Rise of The Network Society*, Oxford: Blackwell.
Gorz, A. (1989) *Critique of Economic Reason*, London and New York: Verso.
Lefebvre, H. (1991) *The Production of Space*, Oxford: Blackwell.

•SEWAGE•

by Gargi Bhattacharyya

Think of sewage and you picture waste, mess, smell, all the extra things which no one needs or wants. This is a cause of pollution, a source of disease, not anything of interest to the snappy guidebooks in city living. This stuff is strictly extra, in excess to the urban rather than the urban itself.

We already know that lots of life generates lots of waste. The concentrated populations of the city make their own nasty by-products. The incomplete attempts to contain and dispose of this ever-replaced waste punctuate city life with eruptions of shit, smell, effluent.

But, of course, the survival of the urban rests on the ability to manage the waste of crowded city living. Without proper drains, so many people living in such close proximity to each other become a danger to themselves. This is another of those great lessons of the nineteenth century – that cities must have an efficient network of drains. Without this, everyday life became unpleasant and disease flooded city streets. And for the nineteenth century the disease of poor drains is cholera, scary mass killer of all and sundry. Yet for a while, the English, and perhaps white people generally, don't realise they are at risk. Lawrence Wright writes of English perceptions of their racial immunity to cholera:

> Though it had spread slowly westwards from India and had by 1830 reached European Russia, cholera was thought of as an Asiatic disease incapable of attacking a decent Englishman, until it struck London with sensational effect in 1832, and again at intervals until 1866. (Wright, 1980, p. 104)

It takes some time for the English to understand that the ideal conditions for epidemic cholera have developed in fast-growing London, that decency and Englishness are no protection against a disease which does not realise itself to be Asiatic. It takes the painful death of many Londoners for this to be learned and the benefits of good sewers to be appreciated. Even then, with the development of a system of drains, poorly engineered but connected drains could also carry disease, rather than prevent its spread.

The turning point comes with Joseph Bazalgette, Chief Engineer to the Board of Works, who conceived the plan of eighty-three miles of large intercepting sewers, draining over a hundred square miles of buildings across London. This system was opened in 1865, and with it cholera begins to calm. Although often forgotten, sanitary engineering has been a key technology of urbanisation. Without the development of this profession of waste management, our death-trap cities would be hard pushed to sustain any life at all. This is a high point of modernity, at least as impressive an example of the development of engineering as a suspension bridge. Once built, efficient sewers become another of those western boasts, a way in which the British in particular mark their difference and superiority from the rest of the world.

Of course, the sewers are the hidden underground beyond time. Unlike the carefully counted, quantified, rationalised into the time–space orders of capital and modernity terrains of the above-ground city, the sewers are the randomness of what is left over. These are the mythic bowels of the city, repository of unacknowledged histories. This is where all sorts of family dirt ends up.

Common law can take a drain to be any channel which is made to carry away water, sewage or faecal matter. A sewer is a bigger or public version of this. Public Health acts from 1848 until the Water Act 1989 take the basic principle that sewers should be designed to drain buildings and built objects such as roads, rather than land itself. Sewers exist as an adjunct to the built environment, part of the underground system which sustains the urban world. Here the strange distortions of the body demanded by the particular disciplines of city living can be cleaned up and piped away. If a concentration of bodies leads to a concentration of waste, properly designed and serviced sewers can ensure that we forget these nasty by-products of the crowd. In the hotter and poorer urban, this ongoing battle is often lost, and heat and poverty let us know how insistently the smell of death clings to us all. The richer West, however, still marks its progress by the ability to mask these uncomfortable everyday reminders of mortality. Until now.

The 1989 Water Act was designed to privatise the British water industry – to take this public utility out of the hands of the (local) state and into the profit-seeking hands of private enterprise.

Now when we hear that sewage may be leaking into our water supply, that effluent is backing up through drains and taps, that bodily waste is bubbling up all over, it is less clear who should take responsibility. The infrastructure designed to maintain and contain the modern body seems to be crumbling around us, leaving us sloshing about in our own shit with only the trite told-you-sos of postmodernism and assorted end-of-millennium angst as consolation. We could blame Thatcherism and the combined meanness and ebullience which believed that essential services could only be run efficiently through the market. But deep down we all already know that that time was just another marker of changes much bigger than ourselves. The long night of the new–old right showed us how much our world had already changed. After the endless ascendancy of the West, now we had to admit that there was no more expansion, no more new markets, no more easy money to be made anywhere. The bitter belt-tightening of that realisation culminated in that broad set of happenings we call the rise of the new right and the death of a certain set of dreams about progress.

As proof of this historic sea-change, the West once again finds it hard to provide clean water. The excessive waste of the rich world begins to creep back into spaces which should be clean. In Britain from 1993 to 1997, the Drinking Water Inspectorate recorded 350 serious drinking water pollution incidents. As the West declines, we all find that our own waste may be a risk to our health again and that family dirt always sticks eventually.

Reference

Wright, L. (1980) *Clean and Decent*, London: Routledge & Kegan Paul.

•SIMCITY•

by Stephen Graham

One of the most popular computer games of recent years has been *SimCity 2000*™. A game of strategy, the player must regulate and manage all aspects of 'simulated' city life and attempt to build up from scratch a giant metropolis facing all the tensions and contradictions of capitalist urbanisation: riots, poverty, economic collapse, natural disasters, corruption scandals, financial crises, infrastructural breakdown, etc. A whole suite of data about the city is displayed in 'real time', allowing decisions to be made which feed back, cybernetically, on to the next iteration of city form, city life and city politics. Superficially at least, the relationships are so 'lifelike' and convincing that *SimCity 2000*™ is widely used to familiarise new planning students with how cities 'work'.

But are 'real' cities starting to mimic *SimCity*™? Certainly, similar virtual simulations are quickly emerging in the business world to support new organisational and management strategies for retailing, utilities, financial services, media and telecom firms, information bureaux and transport companies. Here the aim is to capture as much data about the social, demographic and geographic makeup of cities through 'data mining' or 'data warehousing' – that is, bringing many, disparate data sources together in one integrated database. This information is then displayed and visualised in a variety of (geographical) ways, presenting, in effect, an urban simulation which can then be used for building scenarios, developing strategies, tracking competitive firms, and reorganising services so that profits are maximised and costs reduced.

Thus, changes to 'real' cities seem to echo, and co-evolve with, the engineering of virtual reality simulation games such as *SimCity 2000*™. In both, God-like operators can access more and more comprehensive information sets about the processes of urban life, allowing them to create more and more realistic simulations of the real-time dynamics of urban areas. The development of cities is then 'programmed' through computerised models to see the results visualise before the eyes through direct cybernetic feedback – the 'city as laboratory'. In both 'virtual reality' and 'urban reality', Julian Bleecker has argued that the result seems to be the emergence of a more and more dystopian urban 'reality', approximating increasingly closely to the many postmodern, *Blade Runner*-style scenarios of total corporate power, absolute surveillance and urban decay.

This is, of course, a parody of a complex situation. City life is about all the messy confusion of social practices and cannot simply be 'programmed' by computer. But examples are emerging where data about consumption, geography, transport flows, consumer preferences and human identities are automatically captured, visualised in computers, and used, cybernetically, to feed back on to shaping social practices of urban restructuring, the allocation of life chances, social control and location decision making. Through wide-area tracking systems, covering whole cities, regions, nations and international transport routes, the behaviour of human subjects is increasingly becoming aggregated into detailed, time-space simulations, which offer radically new possibilities for tracking and social control.

This growing cybernetic automation of city life is very significant because it tends to abstract economic and social power from its consequences. It completely privileges the level of the individual from any broader social definitions of identity. It threatens major injustice, dangerous mistakes and growing social polarisation. And it ushers in ever-more fine-grained systems which can be used for social control, social surveillance and disaggregation.

Take an example: so-called Road Transport Informatics (RTI), the application of computers and telecommunications to make highways 'smart'. The control capabilities that new surveillance and simulation technologies bring are of central importance here in supporting a shift from 'dead', public, electro-mechanical highways to 'smart', digitally controlled, and, increasingly, privatised highways. Virtual electronic networks of automated sensors, Closed Circuit TV, tracking and charging devices, computers and electronic maps are being laid over established road transport networks, allowing private firms to operate them profitably. Road networks, with all their complexity of flow and pattern, increasingly become *SimCity*™-like simulations, fed data on which vehicles are going where in real time by all the sensing devices.

An excellent case of the capabilities of RTI can be found in the construction of a new private highway network (number 407) around Toronto. Built to ease congestion on the world's busiest highway, to which it runs parallel, in-car transponders will automatically charge all users of the highway around $1 per 11 km. trip, without necessitating them to stop. Tariffs will vary automatically, to peak around peak commute periods, to ensure that use of the highway never exceeds pre-defined limits, so allowing moving traffic to be guaranteed. Cars without transponders will be automatically photographed and their owners tracked. From 2000 over $100 million per year is expected in tolls; and speed limits may even be higher on the highway than for other state highways to make the road even more of a 'premium' attraction. The consortium which built the road is now selling off all the key development sites along it to the highest bidder, for malls, affluent neighbourhoods, business parks and logistics, creating, in effect, a second-tier land-use transportation system for the élite interests in Toronto.

Electronic road systems like Highway 407 are starting to directly supply the data that can be processed into real-life simulations of the entire movement patterns of whole cities. It might be possible to maintain continuous *SimCity*™-like visualisations of the whole movement patterns of a city in real time. Were such toll systems to extend over whole urban highway networks it is not fanciful, as the transport critic Garfinkel argues, to imagine that a single digital tape could soon 'store the record of an entire day's commute in a major metropolitan area, including 10 million hour-long trips, recoding every car's position every second (a total of 576 million bytes) and a photograph of every car every five minutes (a total of 3.6 Gigabytes)' (Garfinkel, 1996).

Using such tapes to directly build up real-time simulations of city transport patterns might then, in turn, be used to support precise, profit-driven, infrastructural planning, the detailed analysis of people's driving habits, the tracking of criminals, or people considered 'suspicious' by the police, and the reselling of this valuable information as a marketing tool to a whole host of information bureaux and direct marketing firms. Already, many cash-strapped US municipalities are considering how to capture such data for resale in the lucrative information marketplace.

Further reading

Garfinkel, T. (1996) 'Why driver privacy must be part of ITS', in L. Branscomb and J. Keller (eds), *Converging Infrastructures: Intelligent Transport and the National Information Infrastructure*, Cambridge, MA: MIT Press, pp. 324–340.

Graham, S. (1999) 'Spaces of surveillant-simulation: new technologies, digital representations and material geographies', *Environment and Planning D: Society and Space*.

•SIZE•

The world's largest urban agglomerations in 1990

Urban agglomeration	*Population* (thousands)	*Annual average increment in population, 1980–1990* (000s)	*Average annual growth rate, 1980–1990* (%)	*Estimated population, 2015* (000s)
Tokyo	25,013	316	1.4	28,701
New York	16,056	46	0.3	17,636
Mexico City	15,085	120	0.8	18,786
São Paulo	14,847	275	2.1	20,783
Shanghai	13,452	171	1.4	23,382
Bombay	12,223	416	4.2	27,373
Los Angeles	11,456	193	1.9	14,274
Beijing	10,872	184	1.9	19,423
Calcutta	10,741	171	1.8	17,621
Buenos Aires	10,623	72	0.7	12,376
Seoul	10,558	228	2.5	13,139
Osaka	10,482	49	0.5	10,601
Rio de Janeiro	9,515	73	0.8	11,554
Paris	9,334	40	0.4	9,591
Tianjin	9,253	199	2.4	16,998

continued . . .

The world's largest urban agglomerations in 1990

Urban agglomeration	*Population* (thousands)	*Annual average increment in population, 1980–1990* (000s)	*Average annual growth rate, 1980–1990* (%)	*Estimated population, 2015* (000s)
Jakarta	9,250	327	4.4	21,170
Moscow	9,048	91	1.1	9,306
Cairo	8,633	178	2.3	14,494
Delhi	8,171	261	3.9	17,553
Manila	7,968	201	3.0	14,711
Karachi	7,965	294	4.7	20,616
Lagos	7,742	336	5.8	24,437
London	7,335	–41	–0.5	7,335
Chicago	6,792	1	0.0	7,528
Istanbul	6,507	211	4.0	12,345
Lima	6,475	204	3.9	10,526
Essen	6,353	2	0.0	6,526
Teheran	6,351	128	2.3	10,211
Bangkok	5,894	117	2.2	10,557
Dhaka	5,877	267	6.2	18,964

Source: United Nations, 1995, *World Urbanization Prospects: The 1994 revision* (New York: Population Division). Cited in United Nations Centre for Human Settlements (HABITAT), 1996, *An Urbanizing World: Global Report on Human Settlements 1996* (Oxford: Oxford University Press), pp. 16–17; 451–456.

Note: Population figures will vary according to the boundary chosen to define the metropolitan area. Figures have been estimated from various official and unofficial sources.

•SKATEBOARDING•

by Iain Borden

It has often been remarked that decades of urban technology have unwittingly created a concrete playground of immense potential, and that it sometimes takes the mind of a 12-year-old to realise its potential. This comment refers to the urban practice of skateboarding.

The skateboard passed into mass public consciousness in the summer of 1977; visible on every sidewalk, office plaza, parking lot and suburban road, the skateboard became an unavoidable part of urban life. After a decline in popularity, skateboarding underwent a renaissance in the mid-1980s and its practitioners are now more committed and more numerous; there are skateboarders in every city right across the globe. And skate subculture offers a complete alternative to conventional city life, replete with its own music, clothes and language. Skaters have a sense of self- and collective identity, and a way of engaging with the urban realm unique within the modern city.

Skateboarding first arose in the 1960s on the sweeping roads of calm suburban subdivisions. Surfers killed time by replicating ocean moves on smooth tarmac. But in the 1970s skaters quickly developed an urban character, first appropriating deserted swimming pools, drainage channels and schoolyard banks, and then reaching new physical and technical heights in purpose-designed skateparks in cities across South America and North America, Europe and the East. Skaters have used these specialist facilities to evolve a supremely body-centric set of spaces, creating spiralling forces of movement that act centrifugally, extending out from the body to the edge condition of the terrain beneath, and then centripetally, pulling body, board and terrain together into one dynamic flow.

But since the 1980s, skateboarding has taken on a more aggressive, and more political identity and space. After the closure of many of the skateparks, skaters were forced into the streets, and here they now create on a daily basis a radical subversion of the intended use of architecture. Around 1984, skaters began radically extending skateboarding on to the most quotidian and conventional elements of the urban landscape. Using as their basic move the 'ollie', in which the skater unweights the front of the skateboard to make it pop up into the air, they ride up on to walls of buildings, steps and street furniture. As Stacy Peralta, former professional skateboarder, described it, '[f]or urban skaters the city is the hardware on their trip'. Skaters across the world now skate not just on the sidewalk but over fire hydrants, on to bus benches and planters, across kerbs and down handrails.

This kind of street-style skateboarding is antagonistic towards the urban environment. But, beyond simple accusations that skaters cause physical damage to persons and to property, there is a more significant dimension to this seeming aggression: in redefining space for themselves, skateboarders threaten accepted definitions of space, taking over space conceptually as well as physically, and so striking at the very heart of what everyone else understands by the city. Skaters emphasise the micro-experience of the city – sounds, textures, time-space rhythms, smells – while simultaneously challenging the ideas of the city as architect-designed corporate and retail structures; make their own edit of discontinuous building elements and spaces, suggesting that use values are more important than exchange values, and that play is a more creative production than labourist work. Skaters produce an overtly political space, a pleasure ground carved out of the city as a kind of continuous reaffirmation of the notion that beneath the pavement lies the beach.

Nor is this by any means confined to America. Beside famous skateboarding cities like Los Angeles, Philadelphia and San Francisco, just about every developed city now has its own cohort of skaters. São Paulo, Frankfurt, Mexico City, Gothenburg, London, Madrid, Marseilles and Tokyo are just a few of the most energetic skate centres. And skaters also communicate globally – skateboarding on the Internet is growing exponentially, offering interviews, product reviews, video clips, still images, skateable locations and music.

Unsurprisingly, this kind of activity does not go unchallenged. Because skaters test the boundaries of the urban environment, using its elements in ways neither practised nor understood by others, they meet with repression and legislation. Some cities have placed curfews and outright bans on skateboarding. More usually, skaters encounter experiences similar to those of the homeless, frequently in areas of semi-controlled space hovering between private and public domains. Like the homeless, skateboarders occupy the space in front of mini-mall stores and office plazas without engaging in the economic activity of the building and, as a result, owners and building managers have either treated skaters as trespassers, or have cited the marks caused by skateboards as proof of criminal damage. Skaters face fines, bans and even imprisonment – they are subjected to spatial censorship.

Anti-skateboarding legislation is rarely systematic, but the nervousness of the status quo when faced with skaters highlights a confrontation between counter-culture and hegemonic social practices. Ultimately, being banned from the public domain becomes simply one more obstacle for skaters to overcome, causing skateboarders to campaign that 'Skateboarding Is Not A Crime'. Alternatively, such repression simply adds to the anarchist tendency within skateboarding, reinforcing the cry of 'Skate and Destroy'. Either way, skateboarders are part of a long process in the history of cities: a fight by the unempowered and disenfranchised for a distinctive social space of their own.

In doing so, skaters do not accept cities as they are – rather than trying to section off their own little piece of land, skaters want to use all of the city, and in a particularly unconventional and active manner. By using their youth, and in particular by using their own bodily pleasure, skateboarders create their own space, their own cities, their own architecture.

Further reading

Borden, I (1996) 'Beneath the pavement, the beach: skateboarding, architecture and the urban realm', in I. Borden, J. Kerr, J. Rendell and A. Pivaro (eds), *Strangely Familiar: Narratives of Architecture in the City*, London: Routledge.

Borden, I. (1998) 'An affirmation of urban life: skateboarding and socio-spatial censorship in the late twentieth century city', *Archis*, 5, pp. 46–51.

Borden, I. (1998) 'Body architecture: skateboarding and the creation of super-architectural space', in J. Hill (ed.), *Occupying Architecture: Between the Architect and the User*, London: Routledge, pp. 195–216.

Borden, I. (2000) 'Another pavement, another beach: skateboarding and the performative critique of architecture', in I. Borden, J. Kerr, J. Rendell with A. Pivaro (eds), *The Unknown City: Contesting Architecture and Social Space*, Cambridge, MA: MIT Press.

Friedman, G. E. (1994) *Fuck You Heroes: Glen E. Friedman Photographs, 1976–1991*, New York: Burning Flags.

Friedman, G. E. (1996) *Fuck You Too: The Extras + More Scrapbook*, Los Angeles, CA: 2.13.61 Publications and Burning Flags.

Current skateboard magazines include *Big Brother*, *Sidewalk Surfer*, *Thrasher* and *Trans-World Skateboarding*.

•SLUMS•

by Ivan da Costa Marques
and
John Law

Fear

How do you drive from the middle-class areas in South Rio – Zona Sul – to unfashionable North Rio – Zona Norte? For instance, to the Federal University of Rio de Janeiro? Answer. There is a motorway. And one of the short cuts runs through a corner of a *favela*. A slum.

> 'In the daytime I sometimes go that way. But not after dark.'
>
> 'Why not?'
>
> 'I'm afraid. It's unsafe. Even just for two blocks'.

We're going down a shabby street. Mess everywhere. Reinforced concrete frames, filled in with cheap brick. Gaps for windows. There's a store where they sell basic construction material like bricks and cement. There's a butcher advertising a kilo of chicken as a bargain. There are people, old people, middle-aged people, young people, children . . . young women dressed in rags, young men in torn shorts, all in flip-flops, standing, walking, sitting on stairs at the entrance of narrow alleys that go into the *favela*. It is a mixture of poverty, semi-nudity, sensuality, promiscuous combination and filth. People do not look friendly. (How could they not look unfriendly?) There are animals: many cats and dogs, but also a few goats, pigs, and two or three cows and horses (and certainly throngs of rats that we can't see). All, humans and animals, are sick, undernourished. All within a stone's throw of the motorway – and close to the Federal University.

> 'If the car breaks down right here, I don't know what would happen. They are the have-nots, and I don't know what their attitude would be. If they were to find us, intruders that we are, at their mercy.'

Smell

And then we turn a corner and we see a dump, there under a motorway overpass. Next to the *favela*. Not that there is any wall between the dump and the slum. Rubbish accumulates here for a few days, and then trucks from the City Garbage Service come to pick it up. But in the meanwhile it lies in the open, fermenting in the semi-tropical heat. All kinds of rubbish. Piles of it. Inorganic rubbish: the remnants of old furniture and home appliances, waste paper and cardboard, hundreds of plastic bottles, bags and wrappings, old cans. Organic rubbish: coconut shells, rotten fruit and vegetables, waste from the markets, and also the waste from butchers: pieces of meat and – judging by their size – bones from cows. And then, a further horrifying thought, the possibility of medical waste – for the law for disposing of this is full of loopholes, is hardly enforced, and there are so many small private and poorly equipped hospitals. So the dump is filled with rotting this and rotting that. While piglets and people are rooting. And it smells. It smells just terrible. It sickens you, makes you want to vomit, this repulsive smell of putrefaction.

And then, further ahead, very close to this short cut entrance to the motorway, mixed up with the people and the animals, a fire is burning with an acrid and unclean smoke. A smoke which makes the eyes water. A smoke with a smell that will linger in the nostrils when we are a dozen miles away. Another smell from the same rubbish. The smell of the putrid when it burns. A smell known to everyone on the motorway at certain times and on certain days.

And then this: the boys are middle class, and grown-up. They're students. And now, for the first time, they are visiting a *favela*, a slum. Their grandmother's maid is having a celebration. They are invited. And they are apprehensive. Because they have never been inside a *favela* in their lives before. Even though this is *Rocinha*, a very large *favela* and certainly the richest in Rio which climbs the hills in Gávea on one side and comes down to *São Conrado* on the other side. Even though it has post offices, all kinds of shops and the banks have branches there. Even though they are with friends. When they go to the party they enjoy it. Except for this. They come away with a memory of the smell. For yes, indeed, it smells, it smells in the *favela*. For while there are pipes for the water – overhead pipes on posts, like telephone lines – there are no drains. The sewage flows in open ditches. Which stink. A third kind of smell. Most unpleasant. Even in a luxurious *favela* like Rocinha!

Denial

A few years ago a building firm announced the construction of a set of expensive flats in São Conrado. The flats were intended for rich businessmen. The brochure showed the flats, close to the shore, in front of the beach. And, behind the flats the brochure showed a nice green area climbing the hill. But this was strange. Indeed, it was a denial. Because there was no green area. Instead there was a *favela*, Rocinha. This *favela* – the one visited by the boys – is one of the largest in Rio. More than a hundred thousand people live there.

Was the building firm unusual, particularly outrageous? The answer is no. For the *favelas* that cover a large part of the city do not show up on the official maps of Rio. Instead, like the brochure, these have nice green patches. Nice green patches which, given the complex topography of the city, its hills and its mountains, lie right alongside middle- or upper-class neighbourhoods.

What is happening? Here is a part of the answer. Until very recently the *favelas* were regarded as temporary dwellings. These temporary dwellings, it is true, house three million of the ten million inhabitants of Rio. But even so, they were treated as something that would shortly be removed. No official property tax was being levied on them. So obviously they did not exist.

•SOUVENIRS•

Travel Souvenirs

by Walter Benjamin

Atrani. – The gently rising, curved baroque staircase leading to the church. The railing behind the church. The litanies of the old women at the 'Ave Maria': preparing to die first-class. If you turn around, the church verges like God Himself on the sea. Each morning the Christian Era crumbles the rock, but between the walls below, the night falls always into the four old Roman quarters. Alleyways like air shafts. A well in the market-place. In the late afternoon women about it. Then, in solitude: archaic plashing.

Navy. – The beauty of the tall sailing ships is unique. Not only has their outline remained unchanged for centuries, but also they appear in the most immutable landscape: at sea, silhouetted against the horizon.

Versailles Façade. – It is as if this *château* had been forgotten where hundreds of years ago it was placed *Par Ordre du Roi* for only two hours as the movable scenery for a *féerie*. Of its splendour it keeps none for itself, giving it undivided to that royal condition which it concludes. Before this backdrop it becomes a stage on which the tragedy of absolute monarchy was performed as an allegorical ballet. Yet today it is only the wall in the shade of which one seeks to enjoy the prospect into blue distance created by Le Notre.

Heidelberg Castle. – Ruins jutting into the sky can appear doubly beautiful on clear days when, in their windows or above their contours, the gaze meets passing clouds. Through the transient spectacle it opens in the sky, destruction reaffirms the eternity of these fallen stones.

Seville, Alcazar. – An architecture that follows fantasy's first impulse. It is undeflected by practical considerations. These rooms provide only for dreams and festivities, their consummation. Here dance and silence become the *leitmotifs*, since all human movement is absorbed by the soundless tumult of the ornament.

Marseilles, Cathedral. – On the least frequented, sunniest square stands the cathedral. This place is deserted, despite the proximity at its feet of La Joliette, the harbour, to the south, and a proletarian district to the north. As a reloading point for intangible, unfathomable goods, the bleak building stands between quay and warehouse. Nearly forty years were spent on it. But when all was complete, in 1893, place and time had conspired victoriously in this monument against its architects and sponsors, and the wealth of the clergy had given rise to a gigantic railway station that could never be opened to traffic. The façade gives an indication of the waiting rooms within, where passengers of the first to fourth classes (though before God they are all equal), wedged among their spiritual possessions as between cases, sit reading hymnbooks that, with their concordances and cross references, look very much like international timetables. Extracts from the railway traffic regulations in the form of pastoral letters hang on the walls, tariffs for the discount on special trips in Satan's luxury train are consulted, and cabinets where the long-distance traveller can discreetly wash are kept in readiness as confessionals. This is the Marseilles religion station. Sleeping cars to eternity depart from here at Mass times.

Freiburg Minster. – The special sense of a town is formed in part for its inhabitants – and perhaps even in the memory of the traveller who has stayed there – by the timbre and intervals with which its tower-clocks begin to chime.

Moscow, St Basil's. – What the Byzantine Madonna carries on her arm is only a life-size wooden doll. Her expression of pain before a Christ whose childhood remains only suggested, represented, is more intense than any she could display with a realistic image of a boy.

Boscotrecase. – The distinction of the stone-pine forest: its roof is formed without interlacements.

Naples, Museo Nazionale. – Archaic statues offer in their smiles the consciousness of their bodies to the onlooker, as a child holds out to us freshly picked flowers untied and unarranged; later art laces its expressions more tightly, like the adult who binds the lasting bouquet with cutting grasses.

Florence, Baptistry. – On the portal the 'Spes' by Andrea de Pisano. Sitting, she helplessly stretches her arms for a fruit that remains beyond her reach. And yet she is winged. Nothing is more true.

Sky. – As I stepped from a house in a dream the night sky met my eyes. It shed intense radiance. For in this plenitude of stars the images of the constellations stood sensuously present. A Lion, a Maiden, a Scale and many others shone lividly down, dense clusters of stars, upon the earth. No moon was to be seen.

•SQUARES•

by Ash Amin

S
S
S

In Classical Greece, and later in Renaissance Italy, the city became synonymous with political life. The public square or forum was a place for free citizens to influence political opinion and governance decisions. The square came to be seen as a place of civic engagement, education and participation in public life – an important symbol of political democracy. It represented a politics based in the active participation of citizens, different from other political forms such as authoritarian rule, the rule of law, or representative democracy. Consequently, as an arena that nurtured public consciousness and civic responsibility, the square assumed a civilizing influence, set apart from the barbarism, ignorance, and tribalism or individualism of life beyond the city.

Such an understanding of the square or any other open space has long disappeared. We no longer associate it with civilization and citizenship; indeed, it seems almost ridiculous to do so. The quiet square at best is the place for a leisurely stroll or meeting friends, or a place of escape if too threatening in its emptiness. The busy square is a place of incessant noise and movement if given over to traffic, or a market-place, shopping area, and place of consumption if secure and amenable for pedestrians. This is not to say that the square has ceased to be a place of politics. In many cities of the world, the main square is where mass demonstrations congregate, where the public protest, where political regimes respond, repress, or renew. And, less visibly, the square is also a zone of tolerance for public campaigns of one sort or another, as well as minority interests likely to be persecuted elsewhere. But the contemporary 'political' square is a place of protest and confrontation, not a school for cultivating civic virtue and good citizenship.

Interestingly, though, at the end of the second millennium – echoing developments at its start and its middle – there is a strong revival, at least in the West, of the classical idea of good citizenship as an urban phenomenon. One set of claimants includes civic communitarians and urban utopians. The former, who lament the loss of the responsible citizen in an age dominated by market individualism or the all-powerful state, propose the restoration of 'social capital' through projects designed to strengthen civic associations (e.g. voluntary organizations), community spirit (e.g. ethics and civics on the school curriculum) and public responsibility (e.g. socially useful work). The latter, who lament the costs associated with rising urban crime and other forms of social breakdown such as unemployment, alienation, loneliness, and violence, suggest that a contributing factor is the destruction or degradation of public spaces and propose, for example, the restoration of squares as places of social and civic interaction.

Urban planners have not been slow to respond to these ideas. The restoration of public spaces is very much part of the contemporary rhetoric and practice of urban regeneration in virtually every city in the western world. City centres, shopping areas, squares, heritage areas, and other open spaces are becoming pedestrianized, restored, and animated with cafés, restaurants, markets, shops, festivals, spectacles, and so on. So much so, that the city centres of most cities look increasingly uniform. Much of this, without doubt, has to do with the hard-nosed economics of urban rejuvenation based in the city as a centre of retail, leisure and cultural consumerism (see also *Economic Assets*, pp. 68–70, this volume). However, the rejuvenation is also about encouraging face-to-face contact and congenial social interaction as a means of fostering a local sense of place, common aspirations, and civic pride – in short, good citizenship.

But, it has to be asked whether the regeneration projects are encouraging the return of the good citizen or, rather, the rise of the new consumer and bon viveur, who is as socially irresponsible as the alienated urban misfit. There is no doubt that the pedestrianized square bustling with shoppers, amblers, and pleasure seekers is a safe and pleasant environment, vastly preferable to the semi-derelict square that nobody crosses for fear of their personal safety. Viewed from the outside, it is fairly obvious where most people would prefer to be. But, what does being in the revamped square mean? What does public encounter mean in this context?

I remain sceptical that the prime, or indeed even secondary, experience is that of good and responsible citizenship. After all, the main attraction of the consumer-oriented public place is that it allows individuals, or the tribes they come with, to satisfy their personal and sensory desires through pleasurable encounters and market transactions. At worst, the outcome is to reinforce self-centredness (individual or tribal) rather than a regard for others, if the perception of the 'commons' is that it simply enhances personal utility. At best, the experience is likely to improve appreciation of the value of public goods, leading to a more positive civic appreciation of the 'commons'. But this culture too is far removed from the culture of classical political citizenship, defined as being part of, and taking part in, collective decision making of one sort or another. The revamped square is a long way from being the cradle of participatory democracy.

Perhaps the more fundamental question is whether cities, and their public places in particular, need to reinvent classical ideals of citizenship. Is this not a misplaced romantic anachronism? In classical times, the square was a political arena partly because there were few other civic institutions, and partly because political life was less institutionalized. Today, participatory democracy occurs through and beyond the town hall and other civic institutions, as citizens scrutinize and confront the state, formulate their independent actions through a variety of civic organizations, and comment in general through intermediaries such as newspapers, leaflets, and the televisual media. It may well be that these contemporary institutions have ceased to operate in an open and genuinely consultative way, in preference for more top-down and elitist practices, especially in the context of public retreat from political life. But this is a criticism that should lead to reforms to make the institutions more open to grass-roots influence and more democratic in their decision making, and a discussion of ways in which civic involvement in politics might be enhanced.

If the urban problem is seen in these terms, as a deficit of democratic participation and collective orientation, then neither the contemporary pizza-piazza, nor the classical citizens' piazza will do. The former is all about consumption, and the latter is all about a small number of wise men able to practise a civic politics because everybody else is excluded as a non-citizen! Faced with these options, I think I prefer the noise, disorganization, and unpredictability of the ordinary contemporary square of popular protest or tolerance for diverse usages and groups, even as a basis for building citizenship. If urban renewal is about nurturing alternatives and creative potential, then part of the task is to find ways of giving a voice to subaltern, excluded and marginalized citizens, so that they can offer possibilities in areas where the mainstream has let down or failed to reach the entire city. The protest square is a small but important symbol of the open city.

•STAIRCASES•

by Joe Kerr

From Lenin to Bevin, a flight from memory

Bevin Court, an ostensibly modest post-war social housing scheme, can be located after a careful search in a discrete corner of the inner London area of Finsbury. Little on the exterior of this austere block would suggest that such an effort is worth making, but buried in the midst of the building is an extraordinary space: a complex circular staircase rising the full eight stories of the block, whose dynamic geometry powerfully evokes the heroic architectural language of early Soviet Constructivism. While the evident incongruity of such an extravagant gesture in the expected context of welfare utility is sufficient to arouse curiosity, there is nothing tangible to be seen here that could prepare the imagination for the traces of epic events and memories that this staircase carries.

For a historical excavation of the site leads to the 'uncovering' of the Lenin Memorial, which had briefly occupied the same space, and whose remains are buried nearby. This tiny monument, which represented a brief but poignant episode in the public memorialisation of London, also served to link what is an unprepossessing and wholly local corner of the city into ideological and physical conflicts of global significance. The Memorial was built in wartime London, close to Lenin's bombed former residence, as part of that outpouring of public sympathy and support for the Soviet Union in its heroic resistance to the armies of the Third Reich. Unveiled by the Russian ambassador Maisky and his wife on 22 April 1942 – Lenin's birthday – it was equally plausibly a monument to the tactical pragmatism of its most prominent sponsor, the British Government. For the ceremony, which was filmed for *Pathé News*, the surrounding bombed-out houses were draped in red flags, and the Memorial was unveiled from behind curtains made of the Union Jack and the Soviet flag, while an army band played the *Internationale*. Maisky spoke of 'the mutual understanding between our two peoples', and was cheered to the echo, in the words of the London *Times* newspaper.

The monument was the work of the eminent émigré Russian architect Berthold

Lubetkin, who before the war had spoken of the desire of Soviet architects 'not simply to build architecturally, but to build *socialistically* as well'. Ironically, this modest seventeen-ton concrete and stone memorial was the closest any architect in Britain came to fulfilling such an ambition. The monument featured a plaster bust of Lenin in a niche, dramatically lit from above through red glass. Otherwise, no overt ideological references were apparent, save a symbolic length of broken chain anchored to the base. The visible and conscious challenge to dominant ideology is not a possibility often offered within architecture. In this case the extraordinary freedom of circumstances thrown up by the short-term expediency of the State, coupled with the iconographic certitudes and freedoms offered by a monument in contrast to the normal complexities of architectural typologies, provided Lubetkin with the unique opportunity to effect a subversive political purpose. It was this of course which also ensured its inevitable destruction.

A popular place of pilgrimage for the British Left, the monument quickly gained unexpected notoriety through a series of covert fascist attacks on it, in the course of which it was damaged, and daubed with such slogans as 'Communism is Jewish' – a disturbing challenge to popular conceptions of London at war. Steps taken to guard against further incidents included casting twenty-four replica concrete busts in readiness to replace any damage, and placing a twenty-four-hour police guard on the site. But as the tide of war gradually turned, it no longer became necessary or desirable to acknowledge such a public debt to communist Russia, and the monument fell into official disfavour, eventually being ignominiously buried by the architect close to the housing block he was erecting on the site.

A blurred photograph is the only remaining record of the strange death of the Lenin Memorial, an image whose imperfect detail seems to replicate the indistinct quality of memory itself, as if this Box Brownie snap had somehow inscribed the visual recollections of an actual witness. The burial of the unwanted memorial has left no other official record, and this uncertain knowledge – even the known date of destruction is approximate – adds considerably to the feeling that this was a furtive act, the consequence of a sordid betrayal. It is common to argue, as does Iain Sinclair, that 'memorials are a way of forgetting, reducing generational guilt to a grid of albino chess pieces, bloodless stalagmites' (*Lights Out For The Territory* (1997), London: Granta, p. 9). But in the case of the Lenin Memorial it would seem that the opposite holds true, that the destruction of the monument and its erasure from official narratives of wartime London represents a significant loss of memory – both of a particular event in the history of class struggle, and of the general death of post-war idealism that accompanied the onset of cold war conflict. We no longer care to remember London's spaces of revolution – albeit of other people's revolution – and we have expunged from popular memory the evidence of our once close links with the peoples of the Soviet Union.

In one last subtle but telling act of censorship, the name of the block which now stands on the site was changed by just two letters from *Lenin Court* to *Bevin Court*, in an eloquent expression of the retreat from an internationalist future into the parochialism of the post-war settlement. But other physical traces of these events and conflicts have remained, in the way things do linger beyond the span of human recollection, as palimpsests within the complex layers of the urban fabric. The staircase itself stands largely unrecognised as the architect's own private memorial, a personal marker for the place that had become so intimately connected to his former compatriot. Also, a short distance away, from an upper window of the Marx Memorial Library, peers the bust of Lenin formerly owned by Lubetkin, which was his model for the monument. But with no apparent intended sense of irony, the bust is now celebrated as a piece of local colour on the official Clerkenwell Heritage Trail!

Thus within a very small span of years, this otherwise unremarkable place in the city was subjected to violent disruption both of its physical fabric and its perceived ideological importance. The space and its meanings were successively made and re-made in a localised, abstracted and parodic enactment of global struggle. After half a century the grievous wounds inflicted on the city's fabric by that conflict have largely been healed, but the suppression of significant memory that is the legacy of the war and its aftermath is a loss that can never be redressed.

•Standing Around•

by Harvey Molotch

S
S
S

The rapid pace of urban life gives people who 'just stand around' some special significance, but the meaning depends on context and historic moment. In Tokyo labor costs are high, except for young women who are dolled up to stand next to elevator doors offering a big smile to each approaching person – butterflies pinned for commercial display. Inside the cars, as they await the marriage that will enclose them elsewhere, the women press the floor buttons that riders in other national contexts push for themselves. Tokyo also has men paid to stand around, as in the subway stations. The precise timing of the train system tolerates no unanticipated events: a fainted body blocking a train entry could affect a schedule. So there are people to take care of whatever problem arises: this requires a sense of duty to the organization's larger mission of movement rather than following the details of a job description. This orientation to *general* duty is the opposite of work-to-rule as the British unionists do when they want to frustrate their employers completely. Tokyo subway expediters can be trusted with just-in-time human delivery.

In Soviet Moscow just before the end, there were men standing around in front of the hotels and concert halls. Their job was not to hail cabs or open doors but to make a big deal out of who came and went. Pathetic in that glasnost meant that nothing terrible could happen to people who disobeyed, they were left to act out the script of the old regime, hoping for some Marlboros as tribute to a lapsed authority. The betweenness of the moment was evident in the blustery ambivalence of their attitude and stance.

In Cairo and Delhi, perhaps as colonial residue, extra 'guards' at museums and tourist sites stand to exact tolls here and there from those who wish to pass. In cities all over the world there are entrepreneurs of the more casual encounter who take you to a perfume factory or some other facility not on your route but routinely on theirs. They work *ad hoc*, like much of the surrounding society's ways of getting things done, adjusting the merchandise and the texture of the pitch to the situation at hand.

In New York, doormen perform needless tasks for the privileged to show them their privilege is intact; there are few doormen (or bellhops – think about that word) in more egalitarian Stockholm where ordinarily people will open doors or lift luggage for someone who needs help. Other people standing around New York (and many other cities) excel at standing in just the right place at just the right time, as when they wait outside the ATM enclosures and request the person who has just filled up on cash if they have anything to spare. They are also good at asking other questions, especially of women ('you married, honey?') or offering complements ('nice legs'). These moves may not get them money, but it's a way for men to pass time, often at the expense of the (usually) higher-class women they don't mind disturbing. US democracy gives those relegated to homeless impoverishment the right to patter and scrounge in public places, just as the uniformed doormen have license to deploy their own mix of free-enterprise charm and coercion.

In the poorest world cities, of course, the useless waste of people's energies is everywhere evident even among the able-bodied, never mind the obviously sick and those with the invisible ailments (bad backs, arthritis, weak hearts) that keep them from the only kind of work available. Where there is contact with the privileged (as in the newly 'opened' Havana), the great wealth disjuncture means that vast numbers stand around on the off-chance that their bodies might be of some use to the few who can pay – to dig a ditch, move something from one place to another, or have sex.

By looking at who is standing around and what they are or are not up to, one can read something about one's surroundings. Next time you find yourself in a different city and standing around – something tourists practice in their particular way – contrast your stance with the others about you who also appear to be doing nothing.

S
S
S

•STATIONERS•

by Walter Benjamin

Street-plan. – I know one who is absent-minded. Whereas the names of my suppliers, the location of my documents, the addresses of my friends and acquaintances, the hour of a rendez-vous are at my finger-tips, in her political concepts, party slogans, declarations and commands are firmly lodged. She lives in a city of watchwords and inhabits a quarter of conspiratorial and fraternal terms, where every alleyway shows its colour and every word has a password for its echo.

List of wishes. – 'Does not the reed the world – With sweetness fill – May no less gracious word – Flow from my quill!' This follows 'Blessed Yearning' like a pearl that has rolled from a freshly-opened oyster-shell.

Pocket diary. – Few things are more characteristic of the Nordic man than that, when in love, he must above all and at all costs be alone with himself, must first contemplate, enjoy his feeling in solitude, before going to the woman to declare it.

Paper-weight. – Place de la Concorde: The Obelisk. What was carved in it four thousand years ago today stands at the centre in the greatest of city squares. Had that been foretold to him – what a triumph for the Pharaoh! The foremost Western cultural empire will one day bear at its centre the memorial of his rule. How does this apotheosis appear in reality? Not one among the tens of thousands who pass by pauses; not one among the tens of thousands who pause can read the inscription.

In such manner does all fame redeem its pledges, and no oracle can match its guile. For the immortal stands like this obelisk: regulating the spiritual traffic that surges thunderously about him, and the inscription he bears helps no one.

•Streets•

by Adrian Passmore

I'll scream the melody, you hum along.

Plain old me, plain old street. I feel boring and straight. The same things parked down my side, guarded by sentries of similar lights thrown by similar lamps. I can be anywhere and everywhere, I conform enough to be any one of a million streets: I know this cannot really be the case, but I want to be a model. I want to be the very example of the slim, sophisticated and urbane. I want to stretch out and languish and be wide in all the right places. But all of my ambitions are small fry compared to the dull power of sameness which grips me: but I *will* still be a model. I am ready – waiting and willing on the whitewash. That's why I am fine enough to make it. I am beautiful and empty and in this I am the fullness of the world itself. I circle it and given the chance would lick it round in endless planet-filling repetition and kerb-crawling homogeneity. But don't just call my aspirations grand; I'm from around the corner, I am local, I am never in more than one street – I have to be somewhere. It's just that I aspire. However, today – just today – I border on a square, and the sun is bright. Tarry here a while and let my organs grind out one of their songs.

There is grain to my stone-pulp veneers. I am varnished and polished, cleaner and dirtier, dustier and dryer. I front doors, frame windows and stone walls. I am a fence hugger. I am around the bend, but make straight rules. And no matter what others attempt, they are all still streets for me, conforming to something in me; serving to move, to articulate and circulate. The static keeps changing and I am the plan.

I will never be totally lost – not me! I am named after something or someone important; I keep them alive. Each[1] part[2] of[3] me[4] is[5] numbered[6], odd and even. I am tattooed with your lives. My cars, houses and shops wear numbers. The bins have slogans, the dust smells. In my gutter-heart a wrapper flutters; thrown down and waiting for an administrator's broom. I puff out a proud chest: this street is tended – they look after me a bit. They have to because I bite when I kiss. At my low points, I have cul-de-sacs for fingers; your house – my nail.

For all those efforts to glue it together and make it new, the world ends up dust in me. Your work and your shoes wear me out: even if I am the consummate foot fetishist I can't escape their grind. Each petulant turn, each dragged heel, each purposeful stride wears me. But I wear it back. I wear the dust of your shoes. Open your windows and I suck you out in particles so small you do not have a clue of the theft. I am the lungs. I wheeze like an ashtray. I cough up people. This dust is forever. I will never be clean – I let you forget every bit of your trashy impermanent selves in me – I leave you to rot. I am proud as the street. You age, I turn a corner. I am the street, remember: the mortar to daily grind. When windows are pulled tight and doors closed, I still suck you out through vents and telly. In an eddy all my own I blow you where I want. And you can't hide in cars: I love the four-fold stroke of rubber; I am hardened enough to thrive on benzene. I drink it like the rain. With all your traffic you can slow me but you will never choke me. I am too strong and too old – I am the essence of traffic. I am dust itself compacted through millennia. I am layer upon layer, whilst you are only a deposit in my chest. I breathe you. I am not spleen, I am phlegm machine.

You are a part of me, but there is no joining of souls. I am your detritus, turned through the layers from ugly duckling into swan song. I am empirical and real, and I want you to believe me – I am the chant of the built world. For all of this, for my same street mantras repeating over and over, it will never be the same for all of us. I don't think you have the discipline to be me – you need to clean up your act. It takes skill to be as large and as out of order as this. I am bland sameness blown to the scale of the world, I am boring repetition of the public arena. I dare you to out-bore me. I bore through cities across the world. Same old punches, prejudices, twists, and violences. I have my lines: same words, differently put. Same difference. They won't belong, and you never will; and all the time, me, the same old street saying the same old things.

I am architecture. I am edged. I know the gutter. I am light and dark and light again. I am wet and dry and wet again. I am dusty and still. Down me run the houses. Down me run the cables. Under me the sewer and over me the rain. I am the conduit. The power vein. The means to all ends. I am the very crucible. The same rules said each time nearly the same. Repetition without flinching. Housing, works, shops. Feet pad me, they clatter, and I breathe like a webbed beast. I am the street and I believe me because I am the model. In me everyone dreams, and I am vainly awake to this. For this is the street talking and today I am cobbled together like old Europe – I am momentum itself. Traffic is my business. I am hoary redemption and arrogant avenues. For my own ease, and it's best for you too, I want to be the same everywhere. I want to be tarmacked – I want you to be tarmacked. I am willing to break my stride and I want to break yours. I am a bit of the city you see.

But, in a quiet corner, old, old me fears I am boring. I would love to skip again, and to run after a ball. But I am old, I have forgotten the way. I have been the street for years, like this and like that, but always the street. I no longer want to think of it. In my face even the mirror is grained: I have been dug up and patched up, I can't pretend to be fit. My guts have been blown out and my sides laid bare. I have been cabled: tied tight with rubber-sleeved wires. I have been sold in advance. Through all this I was the street and now I am tired. Tired of the same, tired of pretending. Tired of stretching. Tired of the same old connections. But I must want to be young – I must want new people around me; and perhaps that's why I am boring. I want to be metropolitan and to be everywhere again. I want to wake up from this. I want to be the one who lives. I want to say 'I' and mean it. I can rant with the best. When it comes to streets, I am it, I am the first fundamentalist.

And this is my dream. That I am awake, and the first to be so. I want to be back on the edge and away from the centre. I want to be the street at the heart. I am the city, the bit that you see. I will not be flat. I will always be back, sometimes for ever, else for a moment. But I am never so fleet that you will not see me. I am still the street. Let the wind whistle you my song. I am the measure of all tongues. I scream the melody. Listen! or I will spit you back indoors.

S
S
S

•Street life•

by Ash Amin

Much contemporary analysis of public life in cities warns of the growing closure or surveillance of public spaces such as parks and shopping malls and the physical separation of communities from each other (e.g. see *Gates*, pp. 83–85). We are told of the tendency of the middle classes to 'fortress' themselves in gated enclaves sealed off from a hostile outside world; the rise of twenty-four-hour surveillance within housing estates, streets and shopping malls; the policing of all types of public space including parks and buildings, against vagrants, the homeless and the unemployed; the ghettoisation of low-income and immigrant communities to outer peripheries well out of reach of city centres that are redesigned to meet the needs of the spending and chattering classes. The image of city life portrayed here is that of social separation and fear and physical closure.

But does this image fit with the experience of everyday urban life, particularly street life? Streets have always been a space of passage, encounter, and mixture between a variety of social groups and cultures. The street is a place in which different individuals and groups, and different aspirations and desires, constantly jostle for space. Viewed from the vantage point of the street, the sense of urban public life conveyed is that of mixture and contest, where the contest leads to an open set of social outcomes including closure, turf wars, indifference, cohabitation or social interchange.

This sense of urban life helps to counterbalance the claim that the city is now a place of separated physical and social spaces. This is nicely illustrated by American urban sociologist Sharon Zukin's description of street contests in Harlem between peddlers, store owners and the public administration:

> Long a center for the sale of afrocentric literature and soapbox orations, 125th Street has a score of sidewalk tables where posters, books, and newspapers are sold. On a Saturday afternoon, Jehovah's Witnesses walk toward shoppers holding open copies of the Watchtower. Black Israelites stand in a large group around a man denouncing white people over a public address system. . . . in the early 1990s, the Dinkins administration was pressured into enforcing the local laws against peddlers, including those on 125th Street. Merchants . . . opposed the proliferation of street vendors. . . . They claimed they lost business to peddlers selling cheap knock-offs and bootleg goods. . . . For their part, the vendors claimed they contributed to making 125th Street a tourist attraction, no matter what goods they sold. . . . 125th Street is not only a site but a means of reproducing difference, exclusion, and 'ghetto culture'. (1995, pp. 232–247)

As Zukin argues, street life in Harlem is about 'reproducing difference', precisely because of the open and contested nature of its usage. You could argue, against the thesis which speculates the end of public life in cities, that street life is becoming more varied and hybrid as cities become increasingly exposed to influences and flows from all over the world. The contemporary city is not just the site of migration for people from the countryside and elsewhere in the nation, but also for international migrants, refugees, tourists and business travellers, owing to the intensification of global movements of people and the availability of affordable travel. Cities are the home of an enormous variety of peoples and social groups, often competing for the same space by virtue of the dense spatial juxtaposition that constitutes city life.

Such social diversity in close proximity, with all its cultural variety and cultural strains, has the potential of producing new identities and meanings forged out of mixture. These outcomes draw selectively on the different cultures placed together (e.g. in a London street, elderly Asians buying classical Indian music near white English youths listening to a football commentary) to produce new combinations that are distinctive (e.g. in the same London street, younger Asians listening to the football commentary or classical music). This hybridity of the mainstream in urban life is well illustrated by urban anthropologist John Hutnyk's description of Sudder Street – a street in Calcutta where locals and travellers intermingle:

> The Sudder Street area . . . is a fascinating part of town. There are few opportunities to live in such close proximity to people who speak such a wide variety of languages. . . . There are travellers from so many countries that the community is very much a 'polyglot' or 'heteroglot' mix of mostly English, but also French, Dutch, Belgian, and so on. With variations and failures of language due to forced translations among different grammars, with minor nationalisms and hierarchies evident in subgroups . . . with comic mistranslations – . . . classics like 'porge' instead of 'porridge' and 'mixed grill salad' instead of 'mixed grill and salad'. . . . The street . . . is uneven, the footpath is more often rubble than flat, the holes in the road are large, and dangerous during monsoon flooding, and yet 'the chaos seems uniform' . . . everyone seems to find an appropriate path or position, there is a 'code', there are protocols to learn, and patterns into which visitors 'fit' . . . the 'staggering urban clutter of Calcutta'. (1996, pp. 47–48)

You could describe the crowded parts of any city in the world in these terms. It portrays a picture of street life at odds with the image of privatization and closure of urban public life. It may well be that both aspects are equally present in our cities, confined to different sorts of spaces (e.g. the closure of residential areas, but openness of public spaces). Perhaps what contributes to the balance is the social composition of the groups placed side by side, as well as the fears and expectations associated with being in the same space. Almost invariably, the secluded spaces tend to be those in which particular social groups dominate (e.g. a middle-class or working-class estate under heavy surveillance, or a guarded shopping mall serving relatively affluent consumers). If not this, they are dominated by a culture of loss or threat associated with social mixture. The more open public spaces, as suggested by the two examples given above, tend to represent more subaltern groups and cultures, or a civic culture that does not see mixture as a threat.

Wherever the truth lies, it is clear that the millennial contested usage of street life is not disappearing.

References

Hutnyk, J. (1996) *The Rumour of Calcutta*, London: Zed Press.

Zukin, S. (1995) *The Cultures of Cities*, Oxford: Blackwell.

•Studs•

by Iain Chambers

Three studs in the ear, two rings in the nose, a silver nut and bolt through the fleshy part on the back of the neck, and a wince-inducing ring through the eyebrow: all in one body sipping coffee on Castro. I suppose to puncture, mark and signal your presence in this manner is to challenge those categories that see in the body a stable, unique and irreducible reality. It permits the 'wind from outside' (Bataille) to penetrate. The silence of the everyday physical frame, educated for rational and productive ends, is here interrupted in clamorous display. Using and carrying your body like this, transforming it into an unsolicited gift that shakes the limits of public exchange and conventional sense, is perhaps an unsuspected *potlach* that deliberately exposes the bottom line in its sensuous intimations of social sacrifice.

Here the body, generally relegated to the margins of modernity, returns to our attention. Encountered on the sidewalk, in a café, at the bookshop, on the subway, the radically inscribed body is the difference that confronts a generalised indifference and returns the city to its earlier valency as a culture of shock (Baudelaire, Benjamin). But it also draws upon more archaic sources. The procedure of constructing, constricting and mutating the body is traditional (tattoos, piercing, foot-binding, circumcision, painting, ceremonial scaring); what is new, modern, is the idea of the body as an autonomous and non-contaminated 'natural' entity susceptible to threat by mutation. Marks on the body are the marks of a culture, a history, a civilisation. Marks free the body from the realm of the anonymous. The modernist and humanist assumption of the autonomous subject whose body remains a stable referent in the swirl of historical change and cultural mutation is challenged by the archaic interrupting and interrogating its premises; it is challenged by the archaic as repressed, negated, hence integral, component of the modern.

The tattooed and pierced body is already the modified body, the body as a mobile sign that incorporates and rewrites a history of social perceptions, opening out on to the possibilities of new bodyscapes and body maps: the body as the scene of mutation. The post-humanist body is this mutating body. Most sharply revealed in the art of Orlan, Franco B, Ron Athey, Reuven Cohen, and the public performances of corporeal mutation and its alchemy of the flesh, it invites us to thinking: not the pleasure of reconfirming who we think we are, but the fascinated recognition and uncertainty of what we might become, of what we have already become. So, I find myself asking whether, after the asexual manifestos of punk, this return to the exploration and extension of the body does not also mark an explicit return to sexuality. Perhaps. Certainly the mutation proposed in spending the body in this excessive fashion suggests sexual differences of a more polythetic nature, both in the choice of partner and of pleasure.

This 'reason of the body' (Nietzsche) blinds the critical gaze with the shadows that reason has persistently sought to obliterate, seeing in them only an unwelcomed elsewhere: the disruptive alterity of others – the natives, the mad, the hysterical, witches and shamans. But in the space of shock thinking is able to think again. Attempting to respond to the interrogation posed by such bodies my own critical narcissism comes to be undone by theirs. The utopian dream of overcoming modernity, coupled with the fear of losing one's self in the anonymity of the mass, is drawn through the punctured body into a shocking suspension of time. As a transversal slash across the unremitting pulse of reason the puncture reveals the atopic: an elsewhere that does not respect a linearity desperately intent on completing sense. Here, among marked and modified bodies, the present is not superseded by 'progress' but rather mutates into the transitory affirmation of the eroticism of excess.

Like all languages this choice offers a subtle play of connection and disconnection to the world at large.

While rational regimes seek through exercise and plastic surgery to respect and reflect conventional canons of beauty, the pierced and tattooed body interrogates and interrupts that ideology by taking it to the point where it fragments and unleashes the terrifying fascination of excess (the feared uncanny, the dreaded sublime?). For the terror invoked here is the terror of losing my self, of getting lost in the variants and mutations of what I tend to consider immutable: my body.

Here, paradoxically, the mutability of existence (culture) encounters in the radical mutation of the body the ultimate attempt to avoid in-corporation; such bodies become the last beach for the scene of a desired authenticity (nature). The paradoxical and ironic confirmation of this body fundamentalism has a precise historical and cultural agenda: the purity of a self-managed body that escapes social organisation and industrial configuration. The 'mutant' affirms the body as autonomous principle whose individual signature challenges programmed management – the marked body is no longer immediately usable as an abstract element that can be indifferently exchanged for another; it cannot simply be used by someone else. In a sign-saturated world, the signed body is seemingly self-controlled, your own. In refusing to be recognised, existing as disturbance, as excess, the marked body is once again one's own personal property.

However, as Donna Haraway has argued in her noted cyborg manifesto, there is, of course, no final resolution, only the ironic disposition of survival in the languages and technologies, in the ambivalent modernity, that sustains us. For, in the end, there is ultimately no owner, no agent, individual or organisation able to fully control these signs. The signs of the body, the body as sign; the body in the net, the body as net, the net as body; the body in language, the body as language, language as the body: who is the subject and who is the object here?

Signs, holes, fragments: pieces of a disavowed totality that are among the replies embodied in the signature of a tattoo and the gesture of piercing. A style of thought – one that refuses the assurance of a settled frame and the transparency of reason – is here directly inscribed in the flesh. The enigma of being is dramatically rendered explicit, and the everyday yawns open in a flash of shock casting light on the unthought . . . but possible. According to Georges Bataille, in such acts of excess we encounter the unique foundation of an authentic humanity. Here, where the marginal appears at the centre (and where such spatial metaphors enter their phase of decline) the provocation of the modified, mutated and constructed body – halfway between human hybridity and the cyborg – implants the seed of a disquieting proposal.

To offer one's own body as the instance in which you consume, waste and spend your being is to invoke a supplement that threatens the idea of production that has so consistently dominated and damaged our lives. A historical violence, both repressed in ourselves and experimented on others, is here condensed in the virtual display of sovereignty exercised on one's own body. Thus we might understand tattoos and piercing not merely as physical instances of minor irreverence, but as political manifestos. More modestly, I might begin to consider them among those actions which in exceeding limits contribute ultimately to the mutation of the body politic. Which is not to say that I can pretend to explain this provocation, merely to lower my glance and bend my thought to its presence – not to appropriate it but to seek in it a reply.

•SUPERMARKETS•

by Alastair Bonnett

The global incursions, adoptions and adaptations, of western shopping culture are producing numerous hybrid forms around the world. The supermarket used to mean just one thing: North America, open access shelving, lots of tins and strip lighting. Now this image evokes nostalgia, a glimpse of another era's utopia. For the supermarket has mutated into myriad configurations. It may be tucked inside a mega-mall or implanted into an ancient market; specialised into a sleek and self-consciously contemporary drugstore or opened out into a ramshackle and generalised arena for selling almost anything.

The supermarket has become less a thing than a tendency, which I guess we will have to call 'supermarketisation'. It is a tendency that can be seen working its way through diverse traditions of exchange. The extraordinary energy of this current should remind observers, however, that, although the aesthetics of the classic all-American supermarket may now seem *passé*, the North American model of shopping culture still contains something that people find useful and persuasive. Basic to this formula is the ability to connote individual freedom. The customer is free to choose. Free to choose whether and how to buy products, free to choose whether and how to steal products, free to choose whether and how to respond to the pleasures of submission (shall I succumb to the meaty smells wafted from the deli, or the piles of cut-price chocolate?). Every time I walk around my local store – the entrance is on the left, hence one's journey is clockwise, as prescribed by the supermarket's psychologists – the thrill of choice, of my seemingly endless power to decide, seems to provide confirmation of my liberty.

Given its congruence with the basic tenets of advanced capitalist mythology, it is no surprise that in many towns and cities the supermarketised environment has become a social hub. A place for gathering but not, of course, for loitering or standing around. Being dedicated to consumerism, to the value of things rather than people, the supermarket performs its role as a social centre with a certain awkwardness. 'Friendliness' and 'making folks feel welcome' are balanced with a keen desire to keep them moving, to stop them congealing into non-participating clots blocking the aisles. Of course, many shoppers enjoy ironising and exposing this contradiction: 'must keep moving', 'no time to chat', they joke. A 'busy sociability' is born, with quick smiles and knowing winks. Yet as Cohen and Taylor (1992) explain, it is precisely such self-conscious mockery that keeps people engaged in consumer capitalism, that keeps the whole system in motion. There is very little that is subversive about ironic shopping: supermarkets cannot be ideologically outmanoeuvred by simply 'knowing what they are up to', by sickly grins of recognition.

A more meaningful assault on supermarketisation comes in the form of people who routinely, and without necessarily any political intent, actively contradict its social and spatial logic. I confess to having a very specific example in mind. It concerns a supermarket called Tescos in Epping, Britain. Tescos is the social core of Epping, the London dormitory town where I was brought up. But Tescos is not used in the same way by all Epping residents. One particular incident will serve to demonstrate the point. On a shopping visit some years back I noticed that by two of the tills groups of older women (I'd guess they were aged 65 and over) were failing to circulate in the prescribed manner. They were standing and chatting. Other shoppers, especially younger ones, were doing their best to look 'held up'. They stared at the floor, glancing up occasionally. I have a vivid memory of the reddening faces of five or six impeded teenagers expressing passive contempt; a virulent but immobile hatred. The objects of their derision made symbolic gestures of accommodation. The young shoppers were apologised to. The fact that they were 'in a rush' was understood, allowances would soon be made; 'go on, you go in front love', 'don't mind us'. So it was that the conformists were separated from the subversives, the former's eyes tightened to avoid facial contact, to avoid contamination. What particularly bothered, and from their expressions, viscerally disgusted, the two teenage boys queuing at one of the tills was that, not only was there a group of four older women in front of them, including the till operator, talking to each other and failing to move on, but that two of these women were actually eating. Not only that, they had *brought food in from outside the store*. To be more precise they were sharing out what appeared to be a selection of cakes in two Tupperware containers. The teens, whose sphere of transgression was so different, so spatially and socially removed, were appalled and concerned.

So too, as it turned out, was the store manager. Since I witnessed these events some six years ago, such gatherings have been outlawed, decreed as 'unsupermarket-like' behaviour. Not unrelatedly, all the older female till operators seem to have disappeared. Certainly the last time I visited Tescos in Epping all the product scanning was being performed by individuals of the young and easy-to-redden variety. There was no hanging around, no idle chat, no crumbs needing to be brushed from the glass screen of the barcode reader.

Clearly, the mutation of supermarkets has its limits. The institution may be fragmenting into a multitude of different forms. But its core codes of interaction, its structuring principles of movement and behaviour, remain pretty much the same. You go in with cash, you buy something, and you get out. This is how the modern city is socially articulated: just keep moving, keep buying, keep grinning.

Reference

Cohen, S. and Taylor, L. (1992) *Escape Attempts: The Theory and Practice of Resistance to Everyday Life* (2nd edn), London: Routledge.

•SUPERMARKET TROLLEYS•

by Deborah Levy

There are two kinds of people in Supermarket World. Basket people and Trolley people. Each have different customs, languages and superstitions. If the supermarket is a one-season Eden without sunrise or birdsong, Basket people and

Trolley people are its Adam and Eve humming along with the muzak.

They cruise the aisles wearing the skin and fur of the beasts that lie packaged in the industrial freezers, checking out the panic population seeking special offers. '500 chicken winglets for £2.99.' Just crazee for the winglet experience, BUSTING FOR A WINGLET. Some shoppers wear little plastic winglets around their necks on a chain. Citizen Winglet.

Basket people often use the express till on account of only having a few items in their baskets. Only a small percentage of basket shoppers have true basket integrity. Basket people are rebels and cheats. They load as many products into their baskets as they can get away with, and stagger towards the express till jangling their car keys and checking their pagers. They are light on their feet, they walk fast, they are independent and have no slavish desires to make a large meal with the assistance of a cumbersome trolley. Basket people privately sneer at the Trolley people, the slow people, who read out loud the nutritional information on plastic tubs of margarine, slap their children, plan meals from Monday to Friday, wheeling their wire cages down the aisles, reaching for things. The Trolley people are planners. They leave nothing to chance.

Gliding on wheels is for Trolley people the equivalent of LSD. Stocking up, filling the fridge, filling the days and hours filling the cupboards, the vegetable and wine racks. Trolley people buy multiple packs of juice, multiple boxes of washing powder. They know that Thursday is a pizza day and Friday is a fish day.

Some Trolley people actually drown their trolleys. They throw them into city canals. This is the final frontier for a Trolley person.

Basket people! Learn the ways of Trolley people. Enjoy their music and trolley suicides. Join in their feast days. Get used to their humour. Understand their culture. Watch out for diarrhoea and dysentery. Comply with trolley bureaucracy, red tape and visas. Basket people – avoid blood transfusions unless absolutely necessary and ALWAYS wear a condom.

•SURVEILLANCE•

by Stephen Graham

These days, if a car drives the 'wrong' way down a one-way street in the financial district of the City of London, something unusual happens. A small Closed Circuit Television (CCTV) camera, one of ninety which make up the so-called 'Ring of Steel' system protecting London's financial core from terrorism, automatically swings into action. First, the whole movement of the car is automatically recorded. This, in turn, triggers an alarm in the control centre. The car's licence plate is automatically scanned and instantly checked against a computer database of all cars known to have entered the district that morning. This identifies whether the car 'normally' frequents that part of the City and so whether or not it warrants closer investigation as a possible threat. The whole operation is done with military speed and precision, hundreds of times a day, initially at least, without any human intervention. Within a few years it is anticipated that such systems will be able to penetrate windshield glass to scan and identify the *human* faces of car drivers against digital facial image databases, as effectively as car number-plates are identified today.

The 'Ring of Steel' is one of the best examples of the widening practice of the automated, computerised surveillance of city spaces. Rather than the traditional discretion of the police or the human operators of CCTV cameras, such practices rely on the use of computer programs, linked to databases, to define what is 'abnormal' in the spaces and times of city life. The growth of automated surveillance is an increasingly important part of the more general shift towards the intense electronic scrutiny of cityscapes and city life. Public CCTV is now being seen as a new and cost-effective part of the local policy 'tool kit' for dealing with a range of urban problems – cutting crime, reviving consumer and business confidence in town centres, and underpinning the economic 'competitiveness' of urban areas in the UK.

The clamour for public CCTV in the UK is no doubt a response to the complex anxieties that relatively free social mixing in public spaces generate for many people. It is a response to the demands from these people to be actively watched over, or at least to *feel* that they are being watched over, to reduce their fears of crime within urban worlds of strangers. No doubt the endless recycling of violent images *collected through* CCTV into people's living rooms via crime TV shows may subconsciously both increase fear of crime and demonstrate the effectiveness of CCTV in dealing with it. It is also clear that the widening coverage of CCTV is strongly supported by urban managers and surveillance industries keen to manage traditional urban spaces in ways similar to shopping malls. The instrumental goals of comfortable, managed consumption are squeezing out the notion that town and city centres have important political or democratic functions, as key foundations for the 'public realm' within democratic societies.

Although social reactions to CCTV are very complex, there is no doubt that CCTV may, in certain circumstances, reduce or move crime in covered areas, and reduce the fear of crime. It may thus help to improve the chances of traditional town and city centres of doing well economically. But at what cost? It is increasingly clear that CCTV is open to abuse. Moreover, such abuses hint at the worrying possibilities that the emergence of automated surveillance systems, programmed to search for designated faces or car number-plates not 'normally' expected in an area, might bring.

Already, through CCTV, people and behaviours seen not to 'belong' in the increasingly commercialised and privately managed consumption spaces of British cities tend to experience especially close scrutiny, investigation and even exclusion. In a detailed ethnographic study of CCTV control rooms, British sociologists Clive Norris and Garry Armstrong (1997) have shown that operators of current, non-computerised and hand-operated CCTV 'selectively target those social groups they believe most likely to be deviant. This leads to an over-representation of men, particularly if they are young or black.' They found that CCTV control rooms are already ridden with racism and sexism, and certain types of young people are targeted with socially constructed suspicion, are labelled as 'toerags' 'scumbags', 'yobs', 'scrotes' or 'scrapheads', and are scrutinised, followed and continuously questioned. Malign intent is equated with appearance, youth, clothes and posture. Women are often largely ignored, except to provide voyeuristic light relief for the (usually male heterosexual) operators.

Thus, operators are already trying to impose their own ideas of what should happen, and who should be, in the spaces and times of cities (what Norris and Armstrong term their 'normative ecology'). This stipulates who 'belongs' where and when, and treats everything else as a suspicious 'other' to be disciplined, scrutinised, controlled.

What would happen, then, if it is a set of computer algorithms that becomes the route through which CCTV imaging is translated into disciplinary action by police and law enforcement agencies? Human discretion, judgement and compassion are factored out of the definition of 'unusual'. Instant electronic definitions of people or vehicles seen not to 'belong', drawing on the endless searching of computerised image databases of human faces, car number plates and criminal records, start to define policing practice, and, possibly, access and exclusion within different spaces and times of cities. A very precise normative 'ecology' of the city will, in effect, be *embedded* into these systems, defined by the goals of the operators.

Examples of such precise techniques or personal surveillance, and their tightly defined ecologies of access and 'normality', are increasingly common. Special 'smart' CCTV for car-parks will soon scan for behaviour out of synch with 'normal' behaviour patterns in such places (e.g. 'loitering'). On the face recognition front, technologies for automatically scanning and tracking individual faces are also emerging. A computerised database of 6000 football 'hooligans' has already been constructed in the UK to be sent on-line to police monitoring stadia CCTV systems. The UK Driving Licence Authority is introducing a national licence renewal programme which will scan in digital facial photos of all drivers – a possible prerequisite to a genuine national tracking system. Sydney airport is soon to introduce a system which scans automatically and covertly for known illegal immigrants entering immigration. And, in an experimental project, BT is also working with the Massachusetts Institute of Technology (MIT) and the major British retailer, Marks & Spencer, on a digital image and television-based computer system known as 'Photobook', to be installed in its stores. Real-time cameras, linked to image databases of the faces of convicted shoplifters, will alert security staff of the arrival or the presence of convicted shoplifters in their stores, through advanced facial-recognition software. Accuracy is said to be 'greater than 90%'.

In the United States, the control and surveillance capabilities of new technologies are already being widely explored as tools for new methods of time-space tracking in cities. Even by 1991, over 4.3 million Americans had been under 'correctional supervision' at home. Anklet transponders, linked to telephone modems, provide continuous monitoring of the location of offenders. Newer 'smart' systems promise a much more fine-grained and tailored control over the behaviour of offenders. For example, the arrival of an ankleted shoplifter in a retail store would trigger a silent alarm. The system would secretly identify the offender to the store management. When linked to wider urban surveillance systems, through city-wide radio networks – which are already available – the movements of all ankleted offenders could be correlated with the incidence of crime, in time and space, to help in conviction.

Finally, many so-called 'biometric' systems are using digital scans of parts of our bodies to define access and enforce various types of urban boundaries automatically. Iris-scanning Automatic Teller Machines (ATMs) have been operational in Japan from 1997. Inmate retina scanning is in operation in Cook County, Illinois to control prisoner movements. Connecticut and Pennsylvania are practising digital finger scanning to reduce welfare fraud. Frequent travellers between Canada and Montana now use automated voice recognition for speedier throughput. New private highways in Toronto and Los Angeles now scan all cars for their electronic identities, charging their owners for the precise use made of the roadspace. And hand geometry scans are now made in San Francisco to check for illegal immigrants.

Soon, digitised biometric scans may not

only be traded in global information market-places, but may secretly condition access to all aspects of the services and infrastructural fabric of urban life, based on algorithmic judgements of 'normality', worth, profitability or risk.

Ultimately, the growth, extension and interconnection of all these systems may signal the emergence of urban landscapes made up of many superimposed layers of automated surveillance and their associated disciplinary practices. Each layer would have its own fine mosaic of socio-spatial grids. Each would have its own embedded assumptions and criteria for allocating and withdrawing services or access. And each would have its own systems for specifying and normalising boundary enforcement, through electronically defining the 'acceptable' presence of individuals in the 'cellular' space-times within and between cities.

As urban life begins to be more and more conditioned by the streams of digital tracks we leave through the fabric of cities, the uses and abuses of our electronic personas are starting to subtly redefine our relationships to our cities. But the rate of technological change in image database and digital CCTV currently might well mean that such redefinitions become a good deal less subtle in the near future. For Clive Norris *et al.*, the growing linkage of digital databases with widening computerised CCTV and other tracking technologies means that 'in approximately 20 years time, it will be possible for a national database to track the movements of our "digital selves" around the country. Such a database could come from the new drivers license with a digital face record, or an identity card' (1998, p. 265).

London's 'Ring of Steel', it seems, may be only the start. Might urban life be constituted through dozens of electronically coded and automated 'Rings of Steel' in the near future?

References

Norris, C. and Armstrong, G. (1997) *The Social Construction of Suspicion*, A report to the ESRC.

Norris, C., Moran, J. and Armstrong, G. (1998) 'Algorithmic surveillance – the future of automated visual surveillance', in C. Norris and G. Armstrong (eds), *CCTV: Surveillance and Social Control*, Aldershot: Avebury.

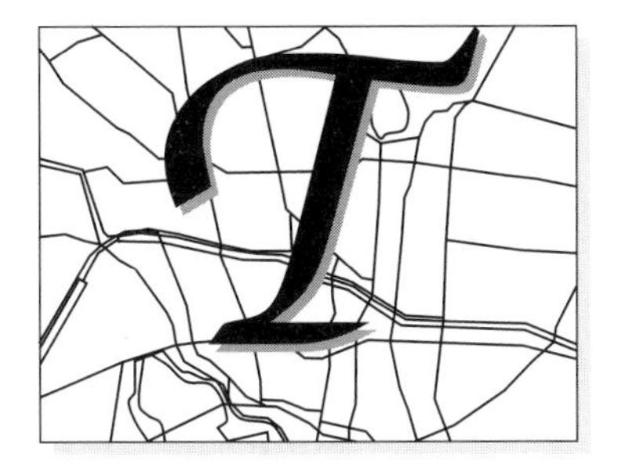

•Taxis•

by Jenny Robinson

The road to the airport was much busier than we had thought – we were late, as usual, and had hit the rush-hour. I wondered about whether we had chosen the quickest route – the people giving me a lift were afraid of driving along the ring road (the alternative) at this time of day. The traffic was really fast on the new freeway; and there were so many taxis at peak hour that it felt dangerous. Not that it seemed like a safe route at any other time: night-times you might get attacked, driving along as fast as you could in your car with doors locked and windows up. Even in the day-time, some people said it wasn't a safe way to go. I went that way all the time. But the road we were on cut through the city in a much more scenic way (you could see the harbour, the university on the ridge as we drove along, the town centre if you looked back) and it avoided traffic to a big township near the airport. As we passed a turn-off we all remembered the last time I'd come to visit, when a tyre had burst. We'd stopped on the hard shoulder just past an on-ramp, really shaken, and talked anxiously as we struggled to change the wheel. Taxis rushed past us on their way into town. I'd been afraid that a taxi speeding on to the freeway might drive straight into us, not realising we were stationary; or that we would be attacked. Car hijackings are common – and here we were, sitting ducks. We agreed we'd been lucky.

Just then a taxi came speeding up behind us, passing on the left. Another jostled past to overtake on the right. Both were packed full of people; one had an NPS number-plate, suggesting it might be heading some way down the coast – Port Shepstone. One of my students drove a taxi there on weekends and vacations. In and out of places where young men 'camped' out at night, training to kill the 'enemy' – their neighbours – forcing young girls to be their 'girlfriends'.

My head was starting to feel sore and tired, all this chatting, and the smell of the oil refinery was very strong here. When I was at university, students had campaigned to have 'the administration' open official residence accommodation (in a posh white area) to black students. As a liberal gesture (?) to black students who were forbidden from living in the neighbourhood, an old army barracks had been converted into a residence for black students – right next to the oil refinery and the airport. My memory slipped from student protests to remembering why all the security fencing was around the oil tanks. I wondered again about why they were so close to the airport. It must be dangerous even now that the political threat was not there.

The driver suddenly slammed on the brakes, his lengthy silence since we had left the house broken, loudly, spluttering, shaken:

> '**@/&&** **@&!!! They think they can drive anywhere; he didn't even look; crazy @@**&&! @@&&**. Bought his ***@@@& license in Pick 'n Pay; he just pulled out in front of that car [a speeding Mercedes sports car, I notice] without looking. These bloody **'&*@* think they can do anything they want. Drive anywhere, pull out in front of you; they're all overloaded, they just pack them in; stupid bloody '@*&*.'

I was concerned. He never ever used his indicator, he was quite a nervous driver. He shouted and swore and waved his hands at taxi drivers. I told him one day someone will stop and shoot him if he's not careful. He can't use racist terms anymore.

I thought about how the rules of the road were changing. I always stayed out of the way of taxis, never tried to pull out into the fast lane if I saw a white kombi-taxi coming, overtook only if it was really clear. I had read that there was a good reason why taxis were always in a rush and drivers taking risks on the roads. More and more people had bought taxis (mini-buses), helped by generous credit arrangements. They'd hired drivers and gone into business. Car sales – often a good indicator of consumer confidence – had boomed. In the mid-1980s the subsidised public transport system, in place since black people had been moved to townships far out of town, was so bad that even though they were illegal taxis began to operate. They were popular on journeys not covered by the unreliable workplace-oriented services run by the white bus companies and state railways. For a while they were known as 'Black taxis' – mostly painted white, used by black people and illegal. Now there are so many taxis, operating on all possible routes, even on the main journey-to-work routes where the trains and buses are much cheaper. Despite the success of this informal taxi industry, the Free Market Foundation has had to eat its policy documents: all this competition and no regulation of how many taxis there are, or how they operate, has led to an oversupply. Jostling in the streets, speeding to fit in more trips per day. Taxi operators have been drawn into turf wars, battles over custom, shootings, deaths, beatings, burnings. The new government and taxi associations are trying to deal with this mess, the inheritance of a wasteful, negligent, racist government.

Woven into the everyday life of the city, taxis fill up the afternoon street with hooting (someone told me this signals they're about to leave . . . taxis are always about to leave . . .). Music fills up the space of the taxi, detracting from the bodies uncomfortably close. Loud pop music. Advertising companies provide tapes with music and ads for taxis. Cool young men in trendy leather caps sit in the front seat, arm stretched out of the window to catch the breeze, one hand on the wheel. Old ladies, laps piled up with bags, children, work clothes, sit behind and listen wearily to the music.

I read about a taxi that travelled around the city with a snake in it which made women pregnant as the taxi drove from one place to another. An urban myth, the taxi.

•TELEMATICS•

by Saskia Sassen

The topoi of E-space

The financial industry is deeply globalized and digitized. It produces a dematerialized output that can circulate around the globe in an instant. And yet this industry operates through a network of financial centers that has expanded sharply over the last decade through the implementation of liberalization and privatisation policies. Why does a global industry operating in digital markets and producing dematerialized outputs need financial centers?

The vast new economic topography that is being implemented through electronic space is one moment, one fragment, of an even vaster economic chain that is in good part embedded in non-electronic spaces. There is no fully virtualized firm and no fully digitized industry. Even the most advanced information industries, such as finance, are installed only partly in electronic space, as are industries that produce digital products, such as software design. The growing digitization of economic activities has not eliminated the need for major international business and financial centers and all the material resources they concentrate, from state-of-the-art telematics infrastructure to brain talent and manual labor.

Nonetheless, telematics and globalization have emerged as fundamental forces reshaping the organization of economic space. This reshaping ranges from the spatial virtualization of a growing number of economic activities to the reconfiguration of the geography of the built environment *for* economic activity. Whether in electronic space or in the geography of the built environment, this reshaping involves organizational and structural changes. Telematics maximizes the potential for geographic dispersal and globalization entails an economic logic that maximizes the attractions/profitability of such dispersal.

One outcome of these transformations has been captured in images of geographic dispersal on the global scale and the neutralization of place and distance through telematics in a growing number of economic activities. Yet it is precisely the combination of the spatial dispersal of numerous economic activities *and* telematic global integration which has contributed to a strategic role for major cities in the current phase of the world economy. Beyond their sometimes long history as centers for world trade and banking, these cities now function as command points in the organization of the world economy; as key locations and market-places for the leading industries of this period (finance and specialized services for firms); and as sites for the production of innovations in those industries. The continued and often growing concentration and specialization of financial and corporate service functions in major cities in highly developed countries is, in good part, a strategic development. It is precisely because of the territorial dispersal facilitated by telecommunication advances that agglomeration of centralizing activities has sharpened. This is not a mere continuation of old patterns of agglomeration but, one could posit, a new logic for agglomeration which involves strategic sectors. A majority of firms and economic activities are *not* located in global cities. But for the cutting edge sectors, being in such cities is being in the strategic information loop.

Centrality remains a key property of the economic system, but the spatial correlates of centrality have been profoundly altered by the new technologies and by globalization. This engenders a whole new problematic around the definition of what constitutes centrality today in an economic system where (1) a share of transactions occurs through technologies that neutralize distance and place, and do so on a global scale; (2) centrality has historically been embodied in certain types of built environment and urban form. Economic globalization and the new information technologies have not only reconfigured centrality and its spatial correlates, they have also created new spaces *for* centrality.

We are seeing the formation of a transterritorial 'center' constituted via telematics and intense economic transactions. The most powerful of these new geographies of centrality at the inter-urban level binds the major international financial and business centers: New York, London, Tokyo, Paris, Frankfurt, Zurich, Amsterdam, Los Angeles, Sydney, Hong Kong, among others. But this geography now also includes cities such as São Paulo and Mexico City. The intensity of transactions among these cities, particularly through the financial markets, trade in services, and investment has increased sharply, as have the orders of magnitude involved.

As a political economist interested in the spatial organization of the economy and in the spatial correlates of economic power, it seems to me that a focus on place and infrastructure in the new global information economy creates a conceptual and practical opening for questions about the embeddedness of electronic space. It allows us to elaborate that point where the materiality of place/infrastructure intersects with those technologies and organizational forms that neutralize place and materiality.

And it entails an elaboration of electronic space, the fact that this space is not simply about transmission capacities but also a space where new structures for economic activity and for economic power are being constituted.

Further reading

Castells, M. (1989) *The Informational City*, Oxford: Blackwell.

Castells, M. (1996) *The Network Society*, Oxford: Blackwell.

Graham, S. and Marvin, S. (1995) *Telecommunications and the City: Electronic Spaces, Urban Places*, London: Routledge.

Sassen, S. (1999) 'Global financial centers', *Foreign Affairs* (January/February).

Sassen, S. (2000) *The Global City: New York, London, Tokyo*, 2nd edn, Princeton, NJ: Princeton University Press.

Sassen, S. (ed.) (2000) *Cities and their Cross-border Networks*, Institute for Advanced Studies, United Nations University, Tokyo, Japan: The United Nations Press.

•TELEPHONE BOXES•

by David Bell

The British telephone box, one of the most potent national landscape symbols, is also the site of a number of important cultural practices; its presence in the city affords countless opportunities for use, only some of them related to the act of making a phone call (which we may, of course, observe in any city around the world blessed with some kind of public telephones). Ever since 'public call offices' were authorized in Britain by the Postmaster General in 1884, telephone boxes have been the container of forms of individual and collective behaviour which give them a special place in urban (and, indeed, suburban and rural) iconography. The most iconic of all, the cast-iron, post-office-red telephone box, now provokes outbursts of nostalgia among many city dwellers – especially following its replacement by new forms of telephone box by the GPO (and later British Telecom, as well as new telecommunications companies in the wake of privatization), a process which began in the 1960s with the first aluminium phone boxes, and accelerated in the mid-1980s with Telecom's extensive modernization programme. The removal of old-style boxes during this programme not only outraged the box's many fans, but spawned a market for disused and uprooted red telephone boxes as heritigized totems of Englishness, sold off at auctions and reused as shower cabinets or to decorate and authenticate Victoriana-filled houses, gardens and pubs.

Gavin Stamp's (1989) *Telephone Boxes* is written from the perspective of an art historian, and never strays beyond that remit to consider the *cultural* history of the telephone box. An extract from the *Telegraph and Telephone Journal* of 1933, which he quotes, affectionately records its intended symbolism, praising its

> cheerful hue by day and its welcoming bright light at night, its promise of ready aid to all in need of rapid communication, its form as a friendly figure in the scene whether it stands as one of a row in a busy railway station or shopping centre, or solitary in a suburban High Street, or enbosomed in bushes at the entrance to a park or recreation ground. (p. 12)

Designed, as Stamp puts it, to allow 'human beings . . . to stand protected when making a telephone call' (p. 22), the telephone box is the locus of a whole set of activities other than that primary function. In fact, even Stamp's base definition of the box is only partially right: the telephone box offers *aural* protection and shelter, but keeps its occupant(s) *visible*. This mixture of secrecy and exposure, inaudibility and visibility, the private and the public, is in itself interesting – the user can be seen making the call. Clandestine and prank or hoax calling thus runs the risk of surveillance. However, its anonymity and non-traceability does render the telephone box a prime site for pranksters, and for little acts of resistance; one such act, logistically quite complicated but regularly committed until technology stepped in, was making reverse-charge (or call-collect) calls *to* telephone boxes, leaving the phone company to pick up the tab. Pay phones are also useful launderers for stolen credit cards (a friend in the United States was offered the chance to phone the UK at a budget rate by paying cash to someone with a suspiciously large and varied collection of credit cards), and for passing off foreign or forged currency.

Of course, with the rise of mobile phones, fewer people have need of these iron cages. But then again, not everyone seen in a phone box is making a call.

As a clearly recognisable cityscape feature, the telephone box has always been a convenient rendezvous point – with the added advantage of cover to shelter from inclement weather. There are even world records for the number of people crammed into a telephone box. A telephone box can double as a public toilet in times of dire need, or as a place to eat chips free from rain, or even as an impromptu site for quick sex (including, of course, phone sex). The decorating of call boxes with prostitutes' cards is a constant reminder of phone box erotics.

But perhaps one of the most widespread and interesting uses of telephone boxes is as receivers of acts of vandalism, whether acquisitive (looting the coin boxes), playful (writing friends' telephone numbers on the walls, accompanied by lewd invitations) or malicious (urinating into pay phone receivers). Considerable time and expertise is invested by telephone companies in designing 'vandal-proof' call boxes – each new one unveiled acting as an open invitation to vandals for 'field testing'. The conspicuousness and ubiquity of telephone boxes makes them ideal for practising the vandal's art. Gavin Stamp, of course, predictably calls British Telecom the 'real vandal' for their modernization scheme.

Looking at the many photographs collected in Stamp's book started to make me feel strangely uneasy. The telephone boxes looked alien – like some kind of strange sentinels, standing on street corners, huddled in groups, hustling for business. I'd never noticed before, but they look a little sinister, and more than a little surreal. Fine art student Stewart Wilson played on this surreality in an installation for the show *Visitor* at Staffordshire University in 1997. He glazed a phone box with one-way mirrors, so that users could only see themselves reflected claustrophobically back once inside, while those on the outside could see clearly in. In place of the telephone was a pair of headphones, with the sound of frantic knocking. It reminded me of a Spanish horror film, *The Telephone Box*, about a man who gets stuck in a telephone box in a busy shopping precinct. Most of the film is taken up with *his* frantic knocking, as outsiders go about their business, ignoring him. A phone company van arrives, and he breathes a sigh of relief – but then the box and its contents are loaded on to the van, driven away, and dumped in a weird graveyard of corpse-filled telephone boxes. No wonder we've all switched to mobiles. . . .

Reference

Stamp, G. (1989) *Telephone Boxes*, London: Chatto & Windus.

•TIMES SQUARE•

by Sharon Zukin

Times Square is called 'the crossroads of the world,' but it is a shifting canvas of urban entertainment and real estate development – an important metropolitan center for New York City, and also a projection of all modern cities' glossy hopes and dirty fears. Times Square looms large in the popular imagination. Since its concentrated commercial development in the early twentieth century, the area – and its most famous 'crossroads,' Broadway and 42nd Street – have been associated with three powerful means of mass communication: the 'legitimate' theater, popular music, and the *New York Times*, after which the district was named. Both dream machines and growth machines, the theater, *Times*, and Tin Pan Alley created an image of vitality, cultural innovation, and creative ferment on which the area's current revitalization still depends. Paradoxically, Times Square once descended into urban squalor, yet rose again to become a beacon of economic redevelopment by Disneyfication.

Times Square peaked in popularity in 1945, when newsreel photographs captured crowds rapturously learning that the Second World War had ended. At that point Times Square was still a central gathering place, a public space for information and communication, as well as a site for communing with an urban crowd. Men and women thronged to Times Square to read the latest news bulletins on the Motogram, the electric ribbon that had snaked around the Times Tower since 1928. Marquees and billboards, three stories high, illuminated the names of famous Broadway stars and advertised national brands of cigarettes and beverages with clever new symbols and technologies. (Who could forget the giant Camels billboard that blew big smoke rings in the air?) Working- and middle-class couples came by subway or intercity bus for an evening out at the movies and an inexpensive ice-cream, coffee, or beer. Every New Year's Eve, on the stroke of midnight, a giant glittering ball was dropped from the top of the tower; this annual rite was broadcast by television around the world. Yes, Times Square was the place to take the city's pulse and breathe the excitement of the crowd.

But this was a dangerous excitement. The crowds and inexpensive entertainment were only a brief flash between the decline of more expensive, more glamorous theaters and restaurants in the 1920s and their replacement by burlesque houses, cheap movie theaters, penny arcades and peep-shows, and eventually by pornographic movie theaters, bookshops and video stores. After the Second World War, Times Square degenerated. From a place of sociable fun among friends, it became a place where anonymous individuals connected briefly, often on the basis of sex or drugs, usually under police surveillance.

Times Square went through the same post-war process of devaluation as other centers of mass urban entertainment that lay close to train and bus terminals or the docks. Real estate developers refused to build new office buildings. New hotels opened further uptown. Even the commercial theater audience, dominated by tourists and suburbanites, preferred to visit Times Square for matinées rather than at night.

The area's mix of popular entertainment and vice had nearly always drawn raffish, and even rough characters. But the lack of new development and the withdrawal of most of the middle class led to a sense of abandonment not typical of the rest of midtown Manhattan. Poorer people still came to Times Square for the cheap movie tickets, but more people came for the sex industry and drug trade.

For twenty or thirty years, the city government floated plans – or hopes – of redevelopment. But there were too many other sites to build on, most of which carried local government subsidies. During the 1970s, when the construction of a new hotel, and demolition of two theaters, kindled some interest in redeveloping the area, the city entered a chronic fiscal crisis that killed demand for several years. By the 1980s, the New York State Urban Development Corporation (UDC), a public agency that has a legislative mandate to act without specific public approval, took control of redevelopment. UDC seized possession of a large number of buildings on 42nd Street – the most visible, accessible, and potentially valuable piece of property in Times Square, and one that would be central to any redevelopment project. UDC added subsidies for new construction, and the City Planning Commission kicked in with zoning changes that permitted taller, bulkier buildings. By 1984, when growth in the financial industry and its handmaidens (law, advertising, and mass media) raised the ante on midtown office space, UDC unveiled a plan for private developers to build four tall office towers at 42nd Street and Seventh Avenue.

The idea of covering Times Square with four million square feet of an egregiously monolithic postmodernism angered some influential groups. Together with the theater owners, the Municipal Arts Society forged a vocal alliance of architecture and culture critics, celebrities, and philanthropists who loudly opposed turning Times Square into an ordinary commercial district. They sued UDC in the courts, forced a slight scaling down of the office plans, and managed to delay the start of construction until demand for new offices met the Wall Street crash of 1989, and died.

The subsidies, however, remained in place. Zoning changes, permitting theater owners to sell the 'air rights' over their buildings to developers, who could then 'transfer' them to properties on the larger streets and avenues, encouraged developers to keep thinking about tall buildings. Advocates of Times Square's aesthetic appeal were assuaged by other zoning changes that required all building owners in the area to cover their façades with giant electric signs. The real estate market, however, was still apathetic.

The tsunami that finally turned Times Square from tawdry to Disney began with small waves at the end of the 1980s. A not-for-profit arts organization made a temporary art installation along 42nd Street. Some redevelopment officials created an 'interim' plan for theaters, restaurants, and 'family' entertainment facilities. The city government slowly realized that tourists and suburbanites actually liked being in Times Square . . . as long as they felt safe.

Architects and local development officials began a long process of persuading the world's premier organizer of 'safe' entertainment, The Disney Company, to locate, first, a store, then a theater, and eventually an entertainment facility along 42nd Street. Financial subsidies, and the city's point man, the architect Robert A. M. Stern, who was also a member of Disney's board, convinced the company to consider an 'urban' strategy – which gave them both a presence at the 'crossroads of the world' and a theater laboratory for incubating new products. While Disney used its leverage to exact more concessions, the city government and UDC's 42nd Street Redevelopment Corporation used Disney's presence to attract more development.

Disney and the new electronic billboards make a flashy show in the 'front' stage of 42nd Street. But the 'back' stage is filled by office development for multinational media corporations, which during the 1980s and early 1990s expanded enormously in North America. Bertelsmann, Condé Nast, and Reuters have built their own office towers on Times Square, helped by continued development subsidies. The financial firm Morgan Stanley took a building at the district's northern end, whose façade is illuminated by giant electronic stock-market quotes. Times Square is a visible symbiosis of media, finance, fashion, and entertainment. It shows how the global 'symbolic economy' plays out on a local site.

The old Times Square's game and video arcades and Harem Theater, whose projectionist was regularly sent to jail in the 1970s for showing pornographic films, have fallen into the dustbin of history. They have been overtaken by Disney and Warner Brothers stores, a children's theater, coffee shops, 'theme' restaurants, and other, pay-to-enter simulated environments. This is New York City's new idea of public space. If the space attracts tourists and fashion editors and music promoters, so the thinking goes, then the city must, indeed, be the 'capital of the world'.

At its development, Times Square reflected the growth machine of the local urban economy of its time. It was logical that this entertainment and communications center and its visual trope – 'bright lights, big city' – should represent the hopes and illusions of metropolitan culture, in general, and of New York City itself. It was also logical that, with the development of other commercial centers, these lights would fade. What is unusual about Times Square, however, is the tenacity with which the hopes and illusions survive. Like other city centers that retain both an audience for popular culture and the fresh talent that runs the culture industries – Potsdamer Platz in Berlin, the Playa de Catalunya in Barcelona, Market Street in San Francisco – Times Square re-creates itself as an urban destination. This time, the buildings are taller, the stores refer to Hollywood more than to Broadway, and the corporations whose billboards grace 42nd Street are all multinational. The Disney presence suggests a comparison between the new Times Square and Tomorrowland: a fantasy of an urban street in the modern world.

•Tourists•

by David Gilbert

In 1926 W. E. Hambley published *How to 'do' London in a Day*, a tourist guidebook written for visitors from the northern industrial cities of England who wished to escape from the capital in a single day 'with a good deal of change out of a sovereign'. Hambley warned that 'just as truly as London is full of attractions, so it is full of *dis*tractions', and called upon visitors to resist the 'strong desire to leave the beaten track'. Hambley's beaten (and equally beating) track required arrival at St Pancras, Kings Cross or Euston between 6 and 7 a.m. (which would have required departure from Manchester or Sheffield three hours earlier), followed by ten miles of walking back and forth across London. The route march allowed for fifteen minutes at Petticoat Lane market, half an hour at St Paul's and the National Gallery, and the luxury of an hour at Westminster Abbey and the South Kensington Museum, before taking in an exhibition at Earls Court, and catching an early evening train back to the North. Not surprisingly Hambley advised his readers to see to their corns before 'doing' London.

Guidebooks like Hambley's are full of warnings about less-than-chirpy cockneys, at best sullen and unhelpful, and at worst thieving and dangerous. In guidebooks to Paris and New York the inhabitants of those cities are also often described as rude and arrogant or shifty and violent. But this antagonism works both ways, and tourists probably come quite high on any big city dweller's list of public irritants. While it's easy to feel sorry for Hambley's tourists, dragooned around the sights of London, it's equally easy to image the aura of chaos surrounding the party – blocking the entrance to the tube station while they argue over whether it's the Circle line or the Metropolitan they want, asking directions to the place they're standing in, or swamping some tiny café, taking every seat in the place, ordering sixteen teas and seven coffees (two without milk) before deciding it's too expensive anyway and leaving just as the drinks arrive. There is a special loathing which the inhabitants of big cities reserve for 'out-of-towners', matched only by the hostility directed at the tourists of political and economic superpowers. In the nineteenth century it was Thomas Cook's English excursionists who seemed to be invading the cities of Europe; in our own times it is perhaps Japanese, German and, of course, American tourists who provoke that special mixture of resentment and financially motivated welcome found in tourist destinations across the world.

Anti-tourism has a long history. Time and time again the explorer has been distinguished from the traveller, and the traveller from the mere tourist; as James Buzard puts it in *The Beaten Track* (1993), 'the tourist is the dupe of fashion, following blindly where authentic travellers have gone with open eyes and free spirits'. In the city we can contrast that archetypal figure of urban modernity, the flâneur, passing freely through the streets and sights of the city, with Hambley's poor tourists, constantly urged by their guide not to stray from the set itinerary. In his essay on the *Guide Bleu* in *Mythologies*, Roland Barthes lambasted the knowledge presented in tourist guidebooks. For Barthes, the *Guide Bleu* was not a source of advice, information and education but rather 'the very opposite of what it advertises, an agent of blindness', reducing 'geography to an uninhabited world of monuments'. Barthes' ire was aimed specifically at the *Guide Bleu* for Franco's Spain, which ignored the harsh realities of political oppression and social division to concentrate on the architecture of churches and cathedrals. However, his criticisms also express a more general disquiet about the superficiality or inauthenticity of tourist literature and the touristic experience.

But this is too simple an understanding of the processes and meanings of tourism. We must question the division that anti-tourism makes between the clichéd representations of guidebooks and the supposedly 'real' sights described. The English sociologist John Urry has developed the idea of the *Tourist Gaze* (1990), suggesting, among other ideas, that many places have been consciously or unconsciously shaped by the ways in which they are seen or consumed by tourists. This is a banal observation when applied to theme parks, but less obviously is also applicable to the development of the modern city. For example, the late nineteenth-century reconstruction of many European imperial capitals owed as much to the development of mass tourism as to the ideas of grandeur held by politicians and planners. From Brussels to Vienna, what mattered was not just that a fittingly spectacular city centre should be built, but that it should also accord with or even exceed the expectations of growing numbers of visitors.

Another challenge to anti-tourism points to the emergence of 'post-tourism' – that the clichés of the tourist industry have become so familiar that they become an opportunity for pleasurable ironic indulgence rather than the source of anxiety. Many travel to Paris or Rome, not to waste their time and energy in a pointless attempt to discover the authentic city, but rather to perform a repertoire of tourist clichés both knowingly and pleasurably. In our post-modern times we are accustomed to think of the potential pleasures of this self-reflective tourism, but if we return to Hambley's party tramping on through the London streets, we find a rather different kind of behaviour. Hambley's tourists were equally aware of their status as tourists, and that they were engaged in a stock set of rituals. But, for these hardy northern travellers, this was less about pleasure than about obligation. The capital city, the heart of the Empire, had to be seen – or more precisely, tourists had to be able to say that they had been there – but as the title and tone of the guidebook suggests, the main motivation was a sense of duty, and the main pleasure the eventual escape home. These tourists wanted to avoid the 'real' London and 'real' Londoners if at all possible. Hambley's guidebook is an extreme example of a familiar touristic experience – that there are some places which you feel you must have visited. *How to 'do' London in a Day* hardly leaves time for sightseeing, in a strict sense. This guidebook is more about site-being, where the comprehensiveness of the route around the city becomes the important thing – not so much the tourist gaze as a tourist trudge.

References

Buzard, J. (1993) *The Beaten Track*, Oxford: Clarendon.

Urry, J. (1990) *Tourist Gaze*, London: Sage.

•TOWER BLOCKS•

by Joe Kerr

Hackney, one of the ring of 'inner' boroughs which surround the ancient central cities of London and Westminster, is the poorest urban area in Britain. In the 1980s, when the Thatcherite political revolution was in full swing, it achieved an unwanted degree of celebrity as the hellish paradigm of inner city distress, the terrible exemplar for all the social ills of the decayed and discredited welfare state. Whenever journalists required material about the awfulness of life in social housing projects, or the immorality of unwaged single parents queue-jumping housing lists in an apparent desire to inhabit the same allegedly disastrous estates, or the evidence of the failure of multi-ethnic society, or the corruption of unchallenged municipal socialism, Hackney was a soft and easy target.

But Hackney Council appears to have developed a unique strategy to combat its contemporary reputation as the anthropological show piece of dysfunctional society, for it has now managed to dynamite its way into public consciousness and the national press with an entirely new and different reputation, through its programme of demolishing 'problem' housing estates. In what amounts to a public confession of past errors, and as a dramatic demonstration of its commitment to reforming the futures of its own constituents, it has been steadily blowing up high-rise housing blocks for more than a decade. Never mind that at least part of the blame for the widely acknowledged social disaster of such notorious projects as the Holly Street Estate might reasonably be laid at the door of Hackney's own allocations and maintenance policies, the Council can now proudly point to the future with the erasure of what local resident Patrick Wright once aptly described as the 'tombstones' of the welfare state. Thus recently the first of three tower blocks earmarked for demolition at Holly Street was destroyed in front of an enormous, festive crowd, of the kind that might have formerly witnessed a public hanging.

But the devastating irony of this particular event was that the crowd waiting in excited anticipation was marshalled behind police barriers in the surrounding streets of now-gentrified Victorian villas, between garden-loads of home owners basking in the prospect of this hideous vertical eruption in the physical fabric – and the social order – of what is now otherwise a charming inner London suburb, being removed forever. The scene represented nothing less than ill-disguised *schadenfreude* on the part of these victors of class warfare. For contemporary Hackney has experienced an increasing dilution of its near-universal poverty and overwhelming lower-class status, as young professionals have reappropriated the surviving enclaves of once respectable housing, attracted by rock-bottom property prices a mere stone's throw from the 'Square Mile'. Prominent among these ambitious colonisers were many of the barristers and media people who now represent the modernised, 'New' Labour Party, including Tony and Cherie Blair. Indeed, Tony Blair was first elected to Labour Party office in the local ward, and was instrumental in the fight back against Trotskyist influence in the constituency.

Spatially, this fragmentation of the area's social structure was symbolised by the absolute proximity of the vertiginous system-built tower blocks of Holly Street to the squat brick nineteenth-century terraces. Faced with the visual prospect of these gaunt monoliths, and the psychological prospect of associated property crime – the habitual and often well-founded anxiety of gentrifiers – it is no surprise that the barbecues were lit on the day when this blight was erased from the bourgeois horizon.

For, in effect, this was pay-back time for the tireless efforts of the professional middle classes to regain a political voice, following their increasing disenfranchisement from mainstream politics. As the *Guardian* newspaper recently claimed, the demolition of these tower blocks, the most visible and formidable symbols of welfarism, could easily and simply be interpreted as society 'moving closer together as a Blairite one nation' (Andy Beckett, 1997, 'Sunnier side of the street', *Guardian G2*, 9 October, p. 2). The estimated 3,000 residents of this infamous rat- and cockroach-infested estate could be reintegrated into 'normal' urban living by rehousing them in the pitched-roofed, brick-built 'vernacular' homes which have already sprung up in place of the concrete forms and empty grassed wastelands of the landscape of state social provision.

The actual demolition of this first tower block at Holly Street lasted a mere couple of seconds, and the surprising sensation this engendered was one of pain and loss: at the ease with which the homes of perhaps a couple of hundred people could be permanently destroyed: and at the realisation that the tower block, the ultimate symbol of municipal power and progress, has proved as fragile as the ideology which created it. For such episodes force the staunch defenders of welfarism to acknowledge what should have been eminently plain all along: that the welfare state was ultimately no more than an incredibly sophisticated mechanism for ensuring the continuance and sustenance of a democratic capitalist system in an era of unprecedented global political change, and that its usefulness to capitalism has now passed. As manual labour is replaced by automated production and distribution, the need for decent housing, public transport, universal education and health rapidly diminishes.

But even today, memories survive of the more positive and optimistic image that Holly Street once represented for those original prospective tenants who felt themselves to be moving up in the world. Thus one tower is to be refurbished and retained for those older residents who still enjoy the vision of the vertical city, remembering the social conditions from which they had risen, and who have been bypassed by the heightened material expectations of a forgetful generation. But slowly the familiar older fabric of the Victorian city is growing back to restore and recover this once blighted territory. The name of the Holly Street Estate still survives on many city maps, but its use is now officially discouraged, and it has been removed from local signage – to be replaced by the ordinary names of the streets that once more cross the site. Thus an unwanted and embarrassing past is erased from everywhere but the recesses of public, collective memory.

•TRAFFIC-LIGHTS•

by Miles Ogborn

The traffic-light is a police officer with only three things to say – 'stop', 'go' and 'prepare yourself for change'. Its history is one of the gradual replacement of uniformed bodies, brains and signalling hands with lenses, bulbs and tinted glass. At first they retained their human counterparts, augmenting and regularising their abilities and skills. In December 1868 a 22-foot-high set of signals were erected in London to aid the access of Members to the Houses of Parliament. Their signalling arms and gaslight were worked by an officer, just as the later lights and gongs in Paris, and the revolving umbrellas inscribed with 'Stop' and 'Go' or the adapted railway signals used in American cities required a fleshy attendant. Yet the first London signals sent out another warning sign. Barely a month after their installation the gaslight exploded, badly injuring the constable who had thought he was in control. Elsewhere the signals' partners became less alert, their actions more automatic. So by the 1920s the non-human police officer worked alone. Automatic lights appeared in Wolverhampton in 1927 and, with the help of lighthouse manufacturers, electrical engineers and traffic control professionals, have been instructing drivers ever since. However, remembering that they are a branch of the undercover police opens them to scrutiny.

Just as the sight of a police officer inspires guilt and a recognition of the gap between our actions and those of the ideal citizen, so too do traffic-lights signal our relationship with the powers that be. Where I come from red means stop and green means go, and we have learned to do it (we are so good at it we no longer need to have the word 'STOP' written on the red light). As drivers brake or accelerate a relationship with a state apparatus is entered into. And on the basis of this relationship traffic-lights can organise the chaos of urban individualism on a grand scale. The world traffic management market was estimated at £600 million in 1989, and what is being paid for are answers to the question of how individual actions are to be co-ordinated and regulated in the interests of the flows of people and things that make cities work. The timings of signals at intersections are set accordingly. Sets of city lights are grouped together into computer-controlled networks running locally specific algorithms. Individual drivers are herded by signs, marshalled into groups and, ideally, sent surfing down the streets on a 'green wave' which, as it flows, minimises noise, pollution, fuel consumption and accidents. In all this the traffic-lights signal the dream of the liberal state. Their aim is to disappear. They work best when no one notices them, when regulation becomes co-ordination and facilitation.

We would be foolish not to stop and go when we are told. It is in our interest. It is for our safety and it is for the benefit of all. So the police officer is also installed within us and is inseparable from our concern for ourselves and how we consider others. We construct our urban liberties through the discipline of the lights. You can only move productively if you know when to stop. You can only be an 'amber gambler' if you are sure enough that someone has given you good odds. Some of the particular contours of this urban world are revealed when the signals break down and drivers must make their own way through the city. They proceed with fear and trepidation into each junction – the place traffic engineers call 'the conflict area' – where they each have to negotiate for rights of passage on an individual basis with others coming across their path using an unfamiliar language of hand signals and facial expressions made difficult by distance and windscreens. In this atmosphere of uncertainty people could get badly hurt. So people have come to need these signs and those who drive have developed minds and bodies which respond automatically to them. Some of the dangers and some of the freedoms of living with others at high speed are passed off for an unconscious movement of hands and feet when we see red – unthought reactions which show the depth of our relationship with the state. However, our dependence on traffic-lights could be unravelled.

I have been told that it should not be assumed that people stop on red in parts of Brazil, India or South Africa. In each case the automatic signalling is broken. Not through a failure in technology, but in the relationship between the state and its citizens. This may be because the state is unable to make the argument that people would benefit from its decisions – even about how to proceed through the city – in places where such promises have been spectacularly unfulfilled in the past. It may be that the fragile apparatuses of the state are overwhelmed by people's demands to move, to live and to survive. In a gridlocked city you have to go when you have the chance, not when someone says you can. It may also be that it would be foolish to obey the silent police officers who are asking you to stop in areas where their walking, talking counterparts cannot guarantee your safety.

Indeed, these officers have often been part of the creation of the conditions of violence which make those in such privileged forms of transport want to hurry through the lights and away from the fragments of divided cities that they see as dark and dangerous. It is only under certain political conditions that the signals work.

Yet the story is not simply one of car-borne city travellers shaping themselves to the demands of the lights. These are signs that have recognised and reacted to drivers since the 1930s. The lights and their handlers need to know what drivers are like and how they will act when given certain signals. How much red can they take before they assume that the system has malfunctioned? How much amber does safety demand? The time settings at intersections also attempt to respond to past and predicted demand as it fluctuates throughout the day and the week. Rubber pads in the road, failed experiments with microphones and hooters, and success with loops of wire which register metallic objects in their magnetic fields have all told the silent police officer that cars are coming and issued instructions to hold the green. In more complex urban systems these detectors feed continuous streams of data to central computers whose microprocessor technology can provide adjustments in the timings of connected groups of traffic lights so that congestion can be averted before it happens. In these information loops each light change demands a response from the drivers to which they in turn respond. Drivers and lights are meshed together as the moving fabric of the networked city.

Within these cities there is a certain version of politics that is signalled by the lights. The wire loops in the road recognise each driver as an equal, individualised and independent citizen (any passengers are dependents) and respond to their demands for movement within the limits set by the demands of other similarly placed individuals. The version of liberal democracy that this offers is undercut by the property qualification of car ownership, and the obvious dangers of cities organised just for drivers. Other possibilities might be imagined which are more democratic. There are already traffic control systems that incorporate priority for public transport. Trams press on, and London Transport's plaintive plea to 'Let the Bus Go First' might be made real through the agency of a technological system which would hurry buses through junctions. Why not also have lights that favour cyclists or pedestrians? Cars might be stopped for bunches of people emerging from stations by signals triggered as the train pulls in. Such an integrated transport policy might also recognise the civic spirit of those who travel with others rather than alone. These scenarios may be judged to be improvements in terms of social or environmental justice, but other possibilities would entail an erosion of the democratic elements in the current situation. It would – in the age of smart cards and road pricing – certainly be feasible to engineer traffic signals that gave the green light to those who had paid for the privilege. Instead of allocating 'green time' to minimise overall delay, microprocessors might be programmed to calculate the monetary bids from the competing queues of car-borne consumers of movement, and the signals set as the market decides. The lights have changed.

•TRAILS•

by Jane M. Jacobs

For the Parisian situationists of the 1950s and 1960s making trails was an essential practice in their radical urbanism. These trails were not marked on to the urban environment a priori. They were not like the roads, pavements, bike paths, or tourist walks which, in their various ways, recommend planning-approved passages through the city. The situationists may well have made use of such trails (how else can one move through a city?), but they did so only in order to undo their prescribed logic. The practice that was central to the situationists was the dérive (literally 'drift'). Dérive was an experimental movement through the city which was, as key situationist Guy Debord (1958) expressed it, a 'technique of transient passage through varied ambiances'. It indulged errancy and sought out random encounters with the unexpected vortexes, currents, micro-climates, networks and fissures of the city.

The dérive was nothing like the 'classical journey' (the one that takes us from here to there). Nor was it anything like that most informal of movements, the stroll. And its political direction was certainly far surer than that of its strolling predecessor, the flâneur, who flirted with the fascinations of modern commodity culture. The dérive was a practice by which to encounter a city not fully determined by capitalism. By moving in unexpected ways the situationists wanted to divert and disturb the planned pathways of efficient mass circulation and disdain consumer spectacles. Theirs was a radical critique of urbanism which operated in the zone of experience, powered by the body (foot was the preferred mode of transport although on occasion taxis could be used), and carved out of unexpected encounters between the mind and the materiality of the city.

Debord once dubbed this practice 'aimless ambulation', but it was often choreographed to produce its own special effects. There was a general understanding of what conditions would produce the best dérive. There were appropriate durations ('the time between two periods of sleep'), times to avoid ('the last hours of the night are generally unsuitable'), and preferred numbers ('several small groups of two or three people'). A successful dérive also depended on 'cross-checking' those involved to ensure the 'same awakening of consciousness' and on systematic comparison of psychogeographical experiences once the journey was at its end. Being random in the contemporary capitalist city required then a certain amount of preparation and determination. This was not another form of passive spectatorship but an active and activating participation in the city. Maps and areal photographs needed to be studied, starting points had to be determined, 'directions of penetration' agreed upon, and even the occasional 'possible rendezvous' planned. And, of course, there was the weather. Prolonged rain made a dérive virtually impossible, although sudden and short storms were 'rather favourable' (Debord, 1958).

The over-exposed tourist destination was especially loathsome to the situationists. Tourism belonged to the register of the spectacle which, in turn, was blamed for habituation, passivity, social degeneration, depoliticisation. The situationists sought a city that was far more than a 'supplement to the museum for tourists' (Unsigned, 1959). Their driftings sought to uncover 'yet unknown ambiences' and, through an encounter with these hidden parts of the city, to create what Simon Sadler refers to as a 'network of anti-spectacular spaces'.

To avoid the tourist spectacle (or to even approach it in surprising ways) was a political act. It provided yet another component in the psycho-geographical infrastructure needed to divert and displace the prescribed organisation of the capitalist city (Sadler, 1998, p. 92).

In the contemporary touristic city – those centres that have swapped production industries for consumption industries – there seems little opportunity left for the subversive creativity of situationist trail making. In such cities, seemingly idiosyncratic pathways trace thoroughly orchestrated routes from one authorised 'zone of ambience' to another. Such tourist trails draw major monuments and minor landmarks into coherent spatial stories. They chart out sanctioned histories (the royalty trail, the heritage trail) and accommodate just enough of those other histories (the famous murders trail, the underworld trail, the sewer trail, the poverty trail, the riot trail) to preserve a sense of the non-tour, the de-tour. Far from challenging the situationist view of the relationship between spectacle and capitalism, such touristic trails confirm the latter's ability to extend its spectacularising influence over the minor, the ugly, the outcast. It is hard to imagine where the situationists would be walking now.

But are all planned tourist trails so securely located in the depoliticised and sanitised order of the commodity spectacle? Might it be possible for such trails to contribute to the restructuring of urban spaces and the stories they tell? Might they – in the spirit of the *détournement* (diversion) – also work to realign existing urban constructs by putting them in contact with other (hidden or yet-to-be-realised) spaces?

We can contemplate this possibility through the example of a trail which exists in the Australian city of Melbourne. This is a city which, like many cities carved out of colonialism, sits on the precarious foundations of indigenous dispossession. And, as in many colonial places, Melbourne is now edging its way towards something postcolonial. Like other Australian cities, Melbourne's planners have attempted to re-indigenise the space of the city through various gestures that acknowledge indigenous histories and presences. For the most part, this process has not entailed the return of lands (it is hard for cities to be *that* postcolonial). Many of these belated efforts exist in the form of the symbolic marking of space as Aboriginal. In Melbourne, this has included the construction of a walking trail which takes in a number of 'indigenous sites', some grounded in spiritual traditions, some in the horrors of colonial encounter, and some created anew through contemporary artworks. The 'Another View Walking Trail' wends its way through the streets of downtown Melbourne telling an alternative story of urban development to that of triumphant explorers and pioneer settlers. This trail fractures the narrative consensus upon which the colonial city of Melbourne had been founded.

Who walks this alternative trail? For the thousands of international tourists curious about indigenous Australia, the trail provides a glimpse of a now barely visible indigenous presence in the city. But the trail is also meant for those who call Melbourne home. Those settler Melbournians who follow this trail – be it out of sympathy, guilt or mere curiosity – are drawn to previously unknown places and have the meaning of familiar places utterly transformed.

This postcolonial trail presents a different assemblage of walking, tourism, spectacle and politics to that envisaged by the situationists. It follows a pathway that is both predetermined and well travelled. Yet those who do follow its route are sure to chance upon the unexpected and the hidden. This trail contrives a passage through the city whose special effect – whose *political* effect – is to reshape the psycho-geography of a citizenry. By promiscuously oscillating between the touristic and the political, it is able to chart one frail pathway towards a postcolonial future.

References

Debord, G. (1958) 'Theory of the dérive', *Internationale Situationniste*, 2. Reprinted in L. Andreotti and X. Costa (eds) (1996) *Theory of the Dérive and other Situationist Writings on the City*, Barcelona: Museu d'Art Contemporani Barcelona, ACTAR, pp. 22–27.

Sadler, S. (1998) *The Situationist City*, Cambridge, MA: The MIT Press.

Unsigned (1959) 'Unitary urbanism at the end of the 1950s', *Internationale Situationniste*, 3. Reprinted in L. Andreotti and X. Costa (eds) (1996) *Theory of the Dérive and other Situationist Writings on the City*, Barcelona: Museu d'Art Contemporani Barcelona, ACTAR, pp. 83–88.

•TREES•

by Jenny Robinson

n. -mithi, (u, imi) . . . tree, medicine . . . .

If you can get up to a high point in a city, you become aware of the trees; or if you're walking along and the cicadas and pigeons are filling the evening with their noise, you might look up and notice. . . . Some cities are almost hidden by the trees: I was amazed, visiting Harare for the first time, standing on a hill in the centre of the city, to see only trees.

Many of those trees, or their ancestors, travelled around the world to grace our cities. Oak trees, plane trees, Christmas fir trees, jacaranda trees, mango trees, pine trees . . . the city brings together worlds: England, India, Australia, Brazil. Should we pull them up to promote indigenous vegetation? Or should we tolerate arboreal difference, celebrating the ways in which cities work as connecting places?

A lot of indigenous trees are endangered species. Beautiful yellow woods, huge straight trees with magnificent green leaves, sought out for building, for furniture, now hide away in small pockets of indigenous forest in mountain valleys, or are enthusiastically cultivated by naturalist urban dwellers. Mine died, not happy at being indoors.

Several endangered species of tree are valued for their medicinal properties. Women and men called to practise traditional medicine heal the sick with many different types of imithi. Izangoma and Izinyanga have made a living with barks and roots for generations.

The need to provide growing numbers of urban dwellers with medicine has created a huge industry in traditional medicines. Wholesalers, traders and suppliers, as well as traditional healers, form a complex network, circulating trees through the urban environment. Once again the city emerges as a site of connections: forests, trees, women, Izangoma and Izinyanga, from rural areas and small towns across the country, are all drawn towards the city, to cure, to survive, to help urban dwellers cope with city maladies.

But the forests are slowly depleted of popular medicinal species.

Women often remain in rural areas – maintaining access to precious land as men go off to work in the city, slowly forgetting to send money back home. It is the poorest rural women who turn to gathering imithi as a way of surviving, bringing much-needed money into their households. They collect for long hours, watching out for the forest guards and fences which stand in for the law, denying access to the forests on which local people have always depended, preserving them instead for tourists (James, 1996).

Women make the long and expensive journey to the town or the city with their large bags of imithi – weeks of dangerous work, to be sold in an oversupplied market, on the street, for a small profit. They sleep, exposed, on the street next to their medicines until they are sold or they have run out of time. Trees and people move back and forth between town and forest, trying to survive in the city.

Reference

James, B. (1996) *Gender, Development and Environment in the Remote Rural Village of Mabibi, KwaZulu-Natal*. MA Thesis, University of Natal, Durban.

•TUNNELS•

by Mary King

Boston is a city on top of a city. An improved copy of itself, the city is a skeleton erected above its own bones. Striving for heightened authenticity, each structure demolished produces two perfected versions. For every neighborhood laid to ruin, there arises a designer group of residences out of the ashes. In its reach for global prominence, Boston constantly builds and rebuilds, engineering a new worldly order according to an accentuated conception of the old.

Towards the harbor, a progressive construction project is currently underway. Initiated to create a vast underwater passage connecting the city to that slab of concrete in the ocean which supports the expanding international airport, the Harbor Tunnel promises to facilitate world commerce. This multi-billion-dollar venture is designed to accommodate the projected growth in airport traffic stemming from future increase in global travel and trade. Local residents have titled the project 'The Big Dig', for to them it has become a gigantic hole into which money is carelessly thrown. Yet such aspirations are hardly novel, and as work on the project proceeds at an intensified rate, another, more ancient tunnel, was being destroyed.

On the city's gray edge where the ocean flicks at the docks and sea birds rise with the tide, an old tunnel was unearthed in Boston. Emerging from the land like a slender finger of earth, the haphazard excavation of the airport's construction revealed the tunnel's existence. Like the glints of broken glass and pottery fragments which lined its subversive bed, it was an exposure that required official attention, yet this attention made its burial inevitable.

Carved out two hundred years earlier by men and women anxious not to subordinate their own interests to that of an irrelevant economic order, the smugglers' tunnel was wide enough for an ox cart and as long as the old city. Thriving on secrecy, it prompted a lucrative, worldwide traffic in contraband merchandise by representatives of a wholly different social order. Such a passageway was an artefact of another city, swelling, inventive, not to be contained, much like the irregular contours of the tunnel's margins. It overflowed with impudent trash – with medicine bottles, wine flasks, broken vessels – items highly valued by archaeologists for the slivers of history they possess.

But in deference to a history of order, of fine brick courtyards, rows of houses and imported pottery, proper archaeologists, patrician fathers and expert historians ignored this unsightly passage. Engaged in the sorting, separating and dividing of their profession, these distinguished practitioners, entrusted with legitimating a specific history and a particular city image, neglected the information the tunnel would have revealed. And so no mention was made of its discovery. Ruled insignificant, with the sweep of a steam shovel blow, the tunnel was destroyed. In the span of one day it was covered, flattened and pounded horizontal, as if to make straight were to correct.

But a tunnel, any tunnel, directly threatens the integrity of that which rests above it. Such a passage erodes foundations from the bottom up, its presence rendering superstructures perilously top-heavy with contradiction. The discovery of the tunnel automatically designated the slated airport construction project shaky from birth.

Because a tunnel is a ponderous mass pitched above emptiness, it abruptly uncovers the precariousness of the present. To undergo its spaceless interior is to surrender certainty, for a tunnel is quite literally a passage through solidity.

Because all past cities are failures, there are those who are compelled to stack disorder and wreckage in order to restore symmetry to a designated past. Their efforts, designed to shore up the present, only turn back abruptly on contradictory objects, even the smallest of which contains the information of the whole. Built into each city is the form of its own demolition; within its construction, its destruction. By consequence, those who produce history must always look backwards in time, perpetually running from the not yet determined. In this context, a smugglers' tunnel emerges as a kink in that chain of history where linkages line up as so many intersecting rings, forming a long line into the past. Such a distortion is an unbroken reminder of disintegration, depreciation and decay. Nevertheless, the destruction of the tunnel signals not so much an ending as a trace which can be retrieved even if only now in absence.

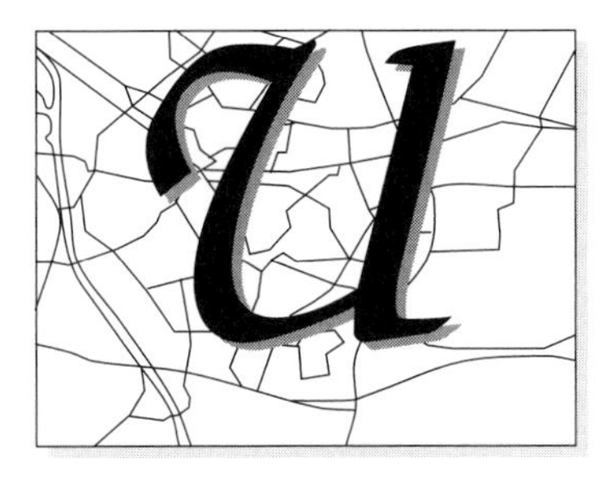

•Underground•

by Stephen Graham

Like trees, cities have root systems. The root system of a city has several elements. There are its networked infrastructures, the road tunnel and metro systems, water and energy utilities, metro lines, and cable and telephone networks which sustain the tiniest details of the lives of virtually all city dwellers. Then there are the sunken spaces gouged out of earth and rock: the hidden car-parks, the underground shopping spaces, the old mining tunnels, the secret military installations, the air raid or nuclear shelters. Finally, there are the built foundations, the root systems of buildings whose depth and complexity may even approach that of their more familiar surface parts.

The extending subterranean metropolitan world consumes a growing portion of urban capital to be engineered and sunk deep into the earth. It links city dwellers into giant lattices and webs of flow which curiously are rarely studied and are usually taken for granted. In fact, mega-urbanisation across the planet is forcing cities to expand *downwards* as well as upwards and outwards. But urbanists and urban theory only really consider the latter two trends.

We are left with scattered hints of how the influence of the underground city subtly weaves across all strands of modern urban life. Hurried walks through the widening subterranean labyrinths of consumption and movement spaces, built up by centuries of modern urban planning and civil engineering in London, Paris, Tokyo or São Paulo. Newspaper reports of Japanese corporations designing whole underground cities, with housing, restaurants, offices, transport networks and leisure facilities, to be sunk into Tokyo Bay or built right under the City's business districts. Brief glimpses into gaping roadworks exposing the multiple webs of wires, huge water pipes, cable networks, heating and energy ducts and massive foundations, embedded deep into the murky depths of the subterranean city. The cliché of pavement steam vents in New York where periodic debates also arise about the mysterious shelters for the homeless 'mole people', living in analogous 'cities' deep down away from law enforcement agencies within the warm depths of the City's subway system. And occasional catastrophic collapses of utility networks: the ice storm in Toronto and

Montreal in winter 1997, the loss of power in Auckland for a month between February and March 1998, or the shattering of Sarajevo's underground world during the Bosnian war. These serve further to hint at the utter reliance of modern urban life on hidden grids of functioning wires, ducts, pipes, tunnels and conduits.

Of course the subterranean has long evoked fear, angst and superstition. Many cultures have a rich folklore linking the underworld to the underground and castigating all contact with subterranean as unclean, unpure, evil, devilish. The invisibility of the below-ground urban world has also made it easy to take it for granted or ignore. But the neglect of the urban subterranean limits the degree to which we can fully understand contemporary urban change. Urban perspectives which studiously ignore the underground city will find it difficult to address the vital political economies of underground space; the complex technological webs that undergird and interlace our surface-level urban worlds; the subterranean intersections of power that the vast sunk capital of urban infrastructures support; and how urban grids connect to global grids to support what the French urbanist Gabriel Dupuy terms 'Networked Urbanism'.

A few short examples will help demonstrate my point. The first is the battles for underground urban space between infrastructure operators. Because 80% of their costs are literally sunk, immobile, deep into the urban fabric, the precise location of the infrastructure grid in relation to market potential is all. Network planning used to be largely vague guesswork. But now a three-dimensional struggle for space and geopolitical power is silently being acted out deep within the subterranean channels of every major city. Battles for premium underground connections are being made more Machiavellian by the shift to private, competing and international infrastructure firms. Complex contested geographies are emerging here within and between cities as giant transnational infrastructure corporations battle it out to get their grids over the expensive 'last mile' to the doorsteps of the largest, most profitable firms and organisations, in the zones of the highest demand.

Often, this is happening on an increasingly global level. The telecoms firm Worldcom-MFS, for example, is building new high-capacity telecoms grids in the business centres of the fifty largest finance capitals in the world while ignoring everywhere else. Driven by such precise targeting, network planning is becoming a precise 'science', based on using Geographical Information Systems (so-called 'GISs') to process information about market potential, gleaned through the use of credit bureaux, consumption and business records and census information.

The second example is the interlinkage of international transport grids with metro and road systems within cities. For it is at major global airport, rail and port hubs that the subterranean infrastructure grids of the city need to link directly and seamlessly into global systems of movement for people and freight. In the case of the airport, complex micro-geopolitics emerge as private airport authorities vie with transport operators to extract the maximum added value from the endless streams of highly affluent passengers, who flood in from all over the planet, to be confined effectively within closed physical systems at single points in space and time. Thus airports increasingly inject themselves with vast subterranean malls and leisure spaces. Privatised metro and train operators struggle to convert dead underground 'frontage' into capital-raising retail space. And taxi and coach operators, who have long charged a premium for traversing the last mile airport 'ransom strip', often fight vehemently to *prevent* mass transit systems from connecting directly into the airport's own people-shifting networks. When they win, of course, many cities are left with sophisticated metro networks which curiously ignore perhaps the spot of highest demand – the international airport – leaving bemused passengers to wonder just how transport planners could possibly be so stupid.

Our final illustration comes from the current social transformation of urban utility networks under neo-liberal regulatory regimes. Urban research has traditionally not considered social access to urban utilities because it was traditionally assumed that they were, as Steven Pinch argued in his book *Cities and Services*, 'generally speaking, freely available, to all individuals at equal cost within particular local government or administrative areas' (1985, p. 10). If such grids are homogenous, immanent and universal, where's the geography? Why should urbanists be interested?

In fact, the current wave of privatisation and liberalisation means that utilities can no longer be dismissed in this way. In the UK, for example – a leader in utility liberalisation – new 'smart' metering technologies are supporting intense competition for affluent users and areas in gas, telecommunications and electricity. Transnational utility firms are linking with retailing, banking and media firms to develop comprehensive service packages to tempt higher-income users to choose them as suppliers from the many available – 'smart home' applications, automatic meter reading, energy efficiency measures, home shopping and banking, the Internet, multimedia, etc.

Elsewhere, however, those in poverty, who often don't even have a phone or a bank account to start with, often pay more for poorer energy and water services, without the benefit of competition. Those prone to debt are now carefully selected using credit registers and given so-called 'pre-payment meters' for energy and sometimes water services, which they must recharge using tokens bought at a post office, *in advance* of using the service, and at higher than normal tariffs. Thus the infrastructure networks that replaced wells or coal merchants have, in effect, been turned *back into point services*, requiring travel to allow consumption (ironically, for those with the lowest mobility).

What we need, clearly, are ways to visualise the underground metropolis. We need imaginative leaps through which we can inscribe the subterranean fully into our urban consciousness, so integrating it into our depictions and analyses of urban worlds. To Robert Sullivan, introducing a classic 1940s text on underground New York, we might do this best through a God-like game of imagining the very *uprooting* of the city like some vast metropolitan tree. Perhaps his imaginative leap is a good place to start:

> Imagine grabbing Manhattan by the Empire State Building and pulling the entire island up by its roots. Imagine shaking it. Imagine millions of wires and hundreds of thousands of cables freeing themselves from the great hunks of rock and tons of musty and polluted dirt. Imagine a sewer system and a set of water lines three times as long as the Hudson River. Picture mysterious little vaults just beneath the crust of the sidewalk, a sweaty grid of steam pipes 103 miles long, a turn-of-the-eighteenth-century merchant ship bureau under Front Street, rusty old gas lines that could be wrapped twenty-three times around Manhattan, and huge, bomb-proof concrete tubes that descend almost eighty storeys into the ground. (Sullivan, 1947)

u
u
u

References

Pinch, S. (1985) *Cities and Services: The Geography of Collective Consumption*, London: Routledge.

Sullivan, R. (1947) 'Introduction', in Harry Granick, *Underneath New York*, New York: Fordham Press.

•UNDERNEATH THE ARCHES•

by Joe Kerr

The 'experience' of urban history

As a young child I was always fascinated by the stories that circulated from time to time of Japanese combat soldiers being discovered on remote jungle-clad islands, still fighting a war that had actually ended twenty years or more before. I was enthralled not with the horrible prospect of these wasted lives, but instead by the possibility these stories raised of places that time really had forgotten, a prospect that has inspired countless novelists and film-makers, as well as the imagination of any number of amateur historians.

Imagine then my surprise on finding a hidden and apparently forgotten corner of London in which it seemed that the Second World War was still being fought, where the familiar wail of the air raid siren still heralded the drone of mighty German war planes and the boom of falling bombs. Here was a place where every person's secret science fiction fantasy, that of stepping back into the past, could actually be realised. And what a time to step back into: the height of the Blitz, when, as everyone knows, London met its Darkest Hour with unprecedented stoicism and heroism.

What I had actually stumbled upon, in the slightly inauspicious location of a brick arch beneath the former London and Greenwich Railway (the world's first metropolitan railway line), was the 'Blitz Experience' in the Winston Churchill's Britain at War Theme Museum. Not real war after all then, but not a conventional museum either, for here one enters the new, exciting world of the simulated 'experience', where London's history can apparently be vividly experienced by the intrepid explorer. Here, as Lord Charles Spencer Churchill writes in the guide to a spectacle that bears his distinguished forebear's name, you will be surrounded by 'actual relics, from toilet rolls and soap to a complete Anderson Shelter, that you can sit in and hear air raids', or you can cautiously pick your way through the debris of a recently bombed London street. Exciting or what?

But of course we know that this is unproblematic, sanitised 'experience'; you cannot actually perish horribly, or inflict violent death on others; nor can you participate in the community spirit that is universally believed to have prevailed inside the bomb shelters during the London Blitz. The public desire for uncritical and self-deluding nostalgia that feeds the construction of these spectacles would be more frightening if they weren't done in such a ham-fisted manner. The 'Blitz Experience' is really no more than a pathetic collection of objects placed in curiously old-fashioned tableaux of wartime pubs and shops, or the simulated aftermath of a bombing raid, rendered in a manner so crude that it recalls fairground attractions rather than pretending at the status of museum displays. A fireman remains perpetually on his ladder frozen in an act of rescue, while behind him an amateur arrangement of crinkled red paper and electric light struggles to convince all but the most gullible of spectators that they are standing on the brink of a fire storm.

The only reasonably evocative component of these tableaux is the noise; for nothing evokes the Blitz better than the hazy rumble of distant bombs, and the wail of the siren. Such sounds seem to linger in these dark places as if trapped from a previous layer of time. But such is the ease of dislocation we have all become accustomed to that it requires an effort of remembrance to realise that these spaces have absolutely no physical connection to the events depicted. And perpetually looming over these scenes is the vaulted firmament of the new, hyperreal world, the semicircular brick soffit of the railway arch which remains continuously in vision to the unbelieving eye.

London is increasingly represented in spectacles which have no discernible connection with authentic sites or events at all – the total simulation of a non-existent past. Under these same railway arches we can also find the 'Jack the Ripper Experience', where we can creep anxiously through the slum alleys of Whitechapel, despite the fact that Whitechapel is on the other side of the river and a couple of miles away. Thus these brutal and disturbing events in the city's recent history have here been wholly decontextualised, and safely reconstituted in the light of the warm nostalgic glow which increasingly colours our vision of London's past.

Our city has been re-mapped and re-imaged by the location finders and art directors of film and television drama, and re-peopled by the waxworks of the new and populist museum culture of the *experience*. It was even reported in the newspapers recently that protest demonstrations were held against the proposed destruction of some 'authentic' Victorian streets abutting these railway arches, involving actors who otherwise would make a living peopling these cobbled streets in the familiar foggy world of the period drama.

History has become an undifferentiated text no longer ordered by chronology, in which real and simulated memories can no longer be distinguished, but which can be sampled, 'experienced' and consumed equally by all. The consequences for the future of this devaluing of collective memory – the commodifying of real events and experiences through the reconstitution of the city's fabric as essentially a theme park to itself – must surely have profound implications for the future comprehension of the past, and thus for our ability to control and shape the spaces we wish to inhabit.

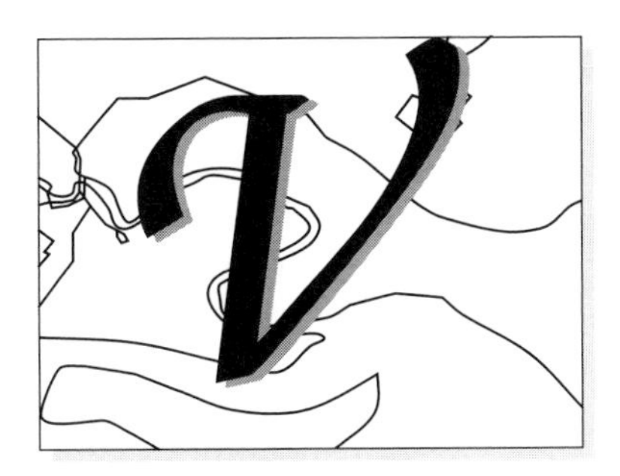

•VIEWS•

by Fran Tonkiss

Seeing, and not being seen – vision is complicated in the city. Different fields of perception are opened up by urban forms; different ways of looking coexist and compete. Images pile on top of each other; things are seen – or half-seen – quickly, suddenly. The tops of buildings, reflections in store fronts, the spectacle of crossing the street. In the city, simply, there is so much to *see*.

And then, cities offer a kind of invisibility. Blank faces look past you in the street. Falling into step with the flow of bodies, one takes cover. Seeing, and not being seen. You can watch the crowd, or melt away inside it.

Urban theorists have been preoccupied with ways of seeing in the city. One well-rehearsed model sees the urban in terms of Cartesian perspective. The field of vision is unified from the point of view of a disinterested, abstracted eye (as of God). Such a way of seeing is imputed to the planner, the master-builder, the overseer, the urban modernist: its primary organising principle is the grid; it may be linked to Renaissance art, to geometry and order in urban design, and to rationalist and scientific philosophies.

Or you can look at it another way. In Angela Carter's 'Elegy for a Freelance', the God's-eye view over a city – seen from the attic room of an urban guerrilla – is more voluptuous:

> London lay below me with her legs wide open; she was a whore sufficiently accommodating to find room for us in her embraces, even though she cost so much to love. (Carter, 1992, p. 106)

Another colonising vision maybe, although an unruly one. Either way the city gives itself up to the eye; laid out, revealed, submissive to survey.

On 5 November 1605, Guy Fawkes and his co-conspirators in the Gunpowder Plot are said to have gathered on Parliament Hill for a view: to look down across London and see, as they hoped, Parliament in flames. Now, in the same place, stands a little metal panorama. The Greater London Council was given, for a while, to map and mark the London skyline from vantage points such as this – halfway across Waterloo Bridge, from Greenwich Park, the top of Hampstead Heath – codifying a particular vision of the city. These fading plaques record landmarks which are rarely discernible to the eye, or which have been excised or transformed over the passage of time – the Royal Observatory in Greenwich obscured by Canary Wharf on the Isle of Dogs, the old Post Office Tower that now houses a disliked privatised utility. In this vision of late capitalist London, the Palace of Westminster and St Paul's Cathedral line up on a monumental skyline alongside the Shell building and the Natwest Bank.

Thinking about these places where London resolves itself into a kind of panorama, one misses the God effect. The visual experience is incoherent, the scene less than 'accommodating'. London's weather and atmospheric pollution mean that the views are often partly or wholly invisible. This mundane aesthetic of heavy architecture seen dully through rain is, however, cut through by moments when you snatch a view of the city – cycling across a bridge late at night, framed down an alley off the Marshalsea Road. These little epiphanies – accidental, fragmentary, sudden – speak less to the God-trickery of Cartesian perspective than to a more haphazard urban sublime. This is the visual field of the furtive glance, the sidelong glimpse that opens up hidden spaces; a fugitive vision of the city.

It's possible to legislate for looking. In Richmond Park, a peep-hole is cut into a hedge of trees through which you can make out, along a protected eight-mile sightline – dimly, perhaps intuitively – the dome of St Paul's Cathedral. More usually, planning orders protect householders' views, or try to stop them looking into their neighbours' gardens or bedrooms. If seeing is central to urban experience, it is proscribed in so many ways. One skill which city dwellers must learn is how to keep on *not* looking. This is the blasé manner of 'seeing' in the city. On the underground, squashed into the kinds of proximity you would allow few people by choice, you steal a glance – at a newspaper, a watch, an unsettling or beautiful person. You do not see the man asking for money, the woman shouting on the pavement.

Within one version of the good city, urban freedoms depend on the presence of 'eyes upon the street', the mutual policing carried out by local people as they go about their daily business. In a classic essay, Jane Jacobs wrote of the dense sightlines on her street in New York, running from her apartment window to the grocer setting out his fruit, the barber sweeping the sidewalk, the man sat on the stoop (Jacobs, 1961). Such an idea has been regimented in the form of Neighbourhood Watch, less suggestive of a lively street life than of burglar alarms ringing on empty afternoons, protecting the stereos.

The presence of 'eyes upon the street', all the same, is growing more commonplace – if not in the way Jacobs imagined. The watchful eye of the local shopkeeper, officialised as Neighbourhood Watch, now takes on a disembodied form as the closed circuit television camera. This is meant to reassure us, it seems, that we are not alone on deserted railway platforms. Cameras follow the movements of the missing person, who only truly 'disappears' when they vanish from the screens. Our late-night kisses or fights are watched impassively, recorded. It is difficult to know whether this kind of gaze affords the freedom of the city, or degrades civil liberties. As the sightlines are reversed it becomes harder not to be seen. The city, at last, has started to look back.

References

Carter, A. (1992) *Fireworks*, London: Virago.

Jacobs, J. (1961) *The Death and Life of Great American Cities*, New York: Random House.

•VINYL•

by Rob Shields

Vinyl for short, polyvinyl chloride or PVC has become a ubiquitous building and packaging material. Consider its use in buildings. In complex types of construction it is usually hidden. For example, in those northern climates where buildings are most sophisticated such as Canada or Scandinavia, insulation, structure, cladding, wind- or draught-proofing and watertight barriers are segregated for performance. Vinyl sheeting forms a vapor barrier. But in simpler buildings – such as tube or 'hoop' greenhouses where translucent vinyl is stretched over tubular steel arches – it is exposed. Stretched over market stalls, and forming the rain barrier of the self-built houses of the poor, vinyl is the world's material of choice for quick, do-it-yourself, and dry building the world over. This high-tech synthetic is the key element in low-tech shelter.

PVC typically comes in green, black, blue, translucent and white-ish colored rolls. It forms the basic material for dark green garbage bags. As see-through bags of all kinds, vinyl appears in many types of consumer packaging. It is the main element in bright blue tarpaulins. It substitutes for glass as translucent plastic stapled across windows. At warehouses it forms doors, draped or pleated across busy entrances which can be easily pushed through by labourers carrying heavy loads. Vinyl is the roof of countless markets and souks. It is the emergency covering which keeps the rain out of hurricane and fire-damaged buildings.

Vinyl degrades slowly against ultraviolet rays as it is inoculated with additives. It is of little interest to any bacterial or insect pests. Vinyl has represented a fundamental change in hygienic packaging; it offers the possibility of working simply with wet mixtures even when all containers have been lost – a simple groundsheet or tarp under which one can take cover in the heaviest monsoon. Air-dropped rolls of blue vinyl have moved from the most visible mark of United Nations relief supplies to become a fundamental 'found object' and purchased element of everyday life for the global poor.

The dark plastic of discarded consumer packaging blows across overgrazed farmlands. In the Moroccan countryside, bags and torn strips caught on low shrubs announce villages before they are visible to the naked eye. Discarded but not degrading, vinyl is the residue of 'development' and could form a statistical index of consumerism. Bright blue has replaced heavy, corroding corrugated iron roofing materials: its use can be picked out and counted in any color photograph of urban areas because it contrasts so strongly with the earth tones and greys of weathered boards. Blue is a roof, a door, a wall. In one of the most innovative developments, vinyl-covered cardboard has produced barrios built out of a new building type: half 'shack' (with distinguishable walls and roof) half tent. These materials are perfectly suited to the technology of the rope and rock weight, rather than the more cumbersome hammer and nails or bricks. Perhaps we could begin to call this 'blue-ing' of the cities.

South Korea is one of the few countries to differentiate statistically between the use of vinyl in agricultural, storage, housing and other buildings. Agricultural uses such as the greenhouses described above which are mostly associated with mushroom and snail farming. The storage uses are myriad. It is housing which is most intriguing. More and more poorer Koreans squat unused land on the tops of the hillocks which punctuate so many Korean towns and cities.

These sharp peaks are unusable except for recreational purposes. The hillsides are too steep for cars and inconvenient to climb on foot, for there is little land area along the top of these ridges and mounds. However, they do offer a challenge to joggers and climbers, and paths are cleared by informal neighborhood sports associations. The squatters are families displaced from rental properties cleared to make way for new high-rise apartments. They assemble the detritus of this construction into solid camps featuring vinyl. The term 'vinyl house' is a Korean term – and the vinyl consumes lives each winter as the houses burn suddenly and violently when the plastic sheeting is touched by kerosene heaters.

Vinyl signals the changing material culture of poverty and marks the lingering experience of disaster – climatic and economic. Its flows and locales of production appear to align neatly with the global economics of oil economies. In addition, the proliferation of discarded plastic packaging is an indication of levels of consumption of manufactured, hygenic and transhipped goods. An improper material, cheeky splashes of blue in the unplanned districts of African and South East Asian cities mark the appropriation of the globalized technology of synthetics. If it is so widely distributed and so cheap that one could laud vinyl as a 'democratic' material, so its take-up in slum housing attests to the unequal terms of the diffusion of globalized materials and cultural products.

•WALLPAPER•

by Joe Kerr

Buildings are usually exclusively thought about in terms of design intention, the ideas brought into play by architects and others involved in their production. Given that usage and function are the central realities of architecture, this insistence on privileging the act of creation over the subsequent 'life' of a building is perverse; but it is often only when architecture is witnessed under unexpected or accidental conditions that the complex relationships between the physical fabric of buildings and the shapes and patterns of human life are suddenly exposed and revealed.

Travelling past an infamous housing project recently, I was arrested by a startling visual image, a brief, transient glimpse into the troubled and troubling reality of social housing. While the Holly Street Estate in London's East End has often been represented as epitomising the nightmare of 1960s system-built high-rise housing, in fact much of the estate consisted of long rows of four-storey maisonette blocks – known locally as the 'snakes' – clinging densely to the ground in marked contrast to the four great point blocks looming indiscriminately over them and the surrounding area. Indeed, these extensive low-rise elements of the estate were wholly overshadowed by the nineteen-storey towers, not only in physical reality but also in the hysterical polemic which inevitably dominated debates concerning the future of the estate. In actual fact, and in direct opposition to general public perceptions, it was these homes close to the ground which had suffered the greatest social distress on the Holly Street Estate.

As I saw them on that day, the 'snake' blocks were in the process of destruction, and what was momentarily exposed to view, as the glazed entrance walls were stripped down prior to final demolition, were row upon row of the interiors of these duplex apartments, appearing like some obscene, dolls-house parody of high-density living. Viewed as a whole, this fascinating spectacle of tiers of identical dwellings, needing only iron bars to complete their resemblance to massed ranks of animal cages, simply served to reinforce the familiar sense of crushing monotony which the modernist ideal of mass habitation habitually created.

But taking a tighter focus on this tableau,

the clashing contrasts of tattered but highly coloured wallpapers and curtains paradoxically evoked the extraordinary efforts to which people will go to impose some sense of individuality and personality on the most alienating of social and architectural structures. These pathetic, shredded remnants of coverings for walls and windows were all traces of individual human choices, made to banish any sense of mediocrity or uniformity from the spaces of private life. The raw exposure of the internal workings of so many *machines à habiter* engendered in me a particular sense of the callousness of the disinterested, technocratic social engineers, who planned and controlled the lives represented by the tattered fragments exposed here, in that they had briefly allowed this intermediate condition to exist, between the occupation of these spaces as 'home' and their total destruction, during which I received these voyeuristic glimpses of what should have remained wholly intimate and private. Of course the inhabitants of such ghetto-like environments have long been accustomed to living their lives in the public gaze, but now even the necessary minimum pretensions of privacy which had made collective existence plausible had been cruelly exposed. The realisation that the private spaces of 'public' housing are only ever colonised, that permanence is such a fragile concept for urban habitation, left a profound sense of the loss of dignity that marked the ordinary human lives who had previously treated these places as 'where we live'.

As fast as these blocks have been torn down, new, dense housing units have been erected. But these latter are carefully designed in the pretty picturesque of 'housing association vernacular', allowing a visual re-stitching of the urban fabric, rendering it seamless and unproblematic: another act in the re-creation of Britain's mythic past, to match the 'traditional' cast-iron lamp-posts, litter bins and bus-stops which are marching inexorably along every inner city high street, holding the realities of deprivation and alienation at bay. Ironically, a short distance away there stands another forlorn and empty tower block, Keeling House, by the renowned architect Denys Lasdun. But this block has been 'listed', giving it statutory protection as a work of architectural importance, and, while it is currently useless for any purpose, it serves to suggest that even these uncared-for symbols of past political concerns might one day be seen in the nostalgic glow of our heritage culture.

What my brief, furtive gaze into the forlorn and abandoned wreckage of former lives so strongly revealed to me was the truth of Christopher Reed's assertion of 'the antagonism of twentieth-century art and architecture towards the values associated with domesticity'. It is only now, as the traces of modernism are kicked over, that we can truly assess the damage that has flowed from the hatred and contempt with which the avant-garde treated such notions as family, privacy and home. It seems almost as if the modest envelope of space within which the anonymous infantry of industrial society were intended to spend their non-productive (but often reproductive) hours were provided only grudgingly by a technocratic class which had no qualms about their own desire for comfortable domesticity. The overweening, brutal architectural language, with its encoded messages of Fordist mass living, is so inhuman in its scale and monotony that it comes as something of a surprise that such fragile and ephemeral signs of human living as curtains and wallpaper have managed to stick so tenaciously to the site after their 'patrons' have departed. While we might all applaud the desire to actually build the utopian vision of a selfless, democratic society, ironically the collapse of that dream into the selfish individualism of late capitalist society might at least permit the mass of people to simply get on with their domestic lives free from the threat of unwanted external intervention.

•WATER (1)•

by Andrew Barry

It is quite normal to think of the urban as a series of networks: both social and technological. Indeed, the city is often visualised using a network diagram, particularly by those who are unfamiliar with its complexity, and with the details of particular places. Think of the underground, subway or bus map given to new visitors to a city by the tourist authorities. Or identified as a list of nodes in a network. Think of the addresses in the phone book. Through networks and through diagrams of networks the scale and complexity of the city is made manageable.

But along with the city of networks, one might also talk, following the work of Tom Osborne, of the hydraulic city; the city of water. Today, the importance of water to the vitality of the city is perhaps less obvious than previously. But, in the nineteenth century, the city's water was of considerable public political significance. It was the object of a vast number of reports and investigations, as well as the deliberations of parliamentary bodies. It was subject increasingly both to regulation and engineering. In 1842, Chadwick's famous *Report into the Sanitary Condition of the Labouring Population of Great Britain* is but one marker of a remarkable concern with the state of the city's water in the nineteenth century, and its intimate links to the health of the poor. Indeed, for the researchers of the period there was a fairly direct correspondence between a map of the water supply and the spatial distribution of fever and diseases. Improving the flow and distribution of the water supply was of clear benefit to the health of the urban population. Building reliable sewers,

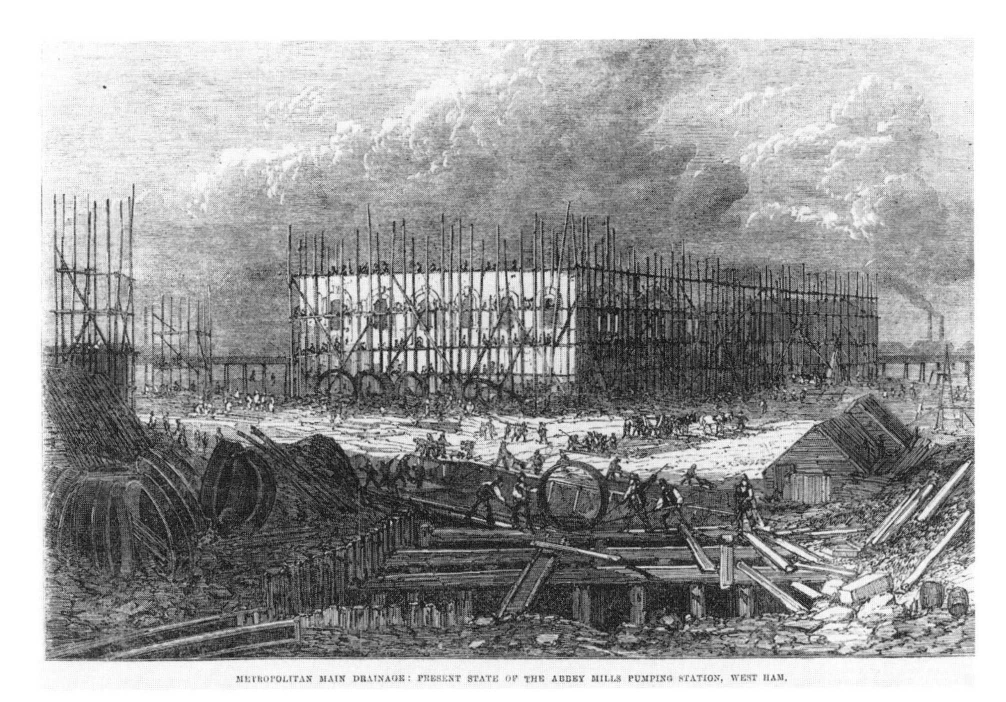

Figure 11 Wood engraving of Metropolitan main drain, London, 1865–68. Reprinted with kind permission from the Museum of London

pipes, taps and drains established a link between private dwellings and domestic behaviour, and the health of the social body. In effect, through such hydraulic and mechanical devices, the environmental quality of domestic spaces could be regulated and preserved at a distance, thereby avoiding any excessive direct intervention by the public health authorities. This was a perfect technical expression of what Michel Foucault termed liberal governmentality: a liberal approach to the problem of government.

In the nineteenth century there was a belief that urban sanitary improvement was more than simply a matter for public works. It had a *natural* dimension. Of course, the vast projects of nineteenth-century sanitary engineering were artificial constructions. And, as such, they doubtless contributed to a sense of the distinction between the 'artificiality' of the city and the 'natural' state of the countryside which became such a powerful theme in nineteenth- and twentieth-century politics and culture. But, as Osborne argues, there was also a metaphorical association between water supply and nature. For the sanitation and water supply were themselves often imagined in organic terms. Chadwick himself conceived of the flow of water in organic terms. His was an 'arterial-venous' project. It 'literally envisaged a circular process whereby human excreta would be used in agriculture which agricultural drainage would supply towns with their water supply.' Thus the hydraulic city became 'a regulated milieu along with the body and the economy' (Osborne, 1996, p. 114). Drains and waterworks were some of the most impressive achievements of nineteenth-century engineering.

But what of water today? Perhaps there has been, sadly, a decline in the awareness of the importance of drains as the arteries of the city. But this does not mean that there is no longer any consciousness of the significance of water. In part, drinking mineral water has become a matter of social distinction and taste. For, as Pierre Bourdieu continues to remind us, the cultivated taste is one which knows that even (or especially) the most functional objects, such as water, have an aesthetic dimension. Precisely because drinking water is apparently functional, it is not. And precisely because it possesses a subtle taste, that taste is highly valued. Only the addition of ice and lemon will do. The functionality and purity of water has both an aesthetic and a commercial value.

At the same time drinking mineral water is the easiest way of acquiring something that is natural without leaving the confines of the city. Mineral water is valued for its 'natural' qualities, however manufactured and marketed these natural qualities may be. Its mineral impurities are measured precisely. And the naturalness of these impurities has beneficial effects for the regulated milieu of the body. A body which needs to manage or avoid too much stimulation, and self-abuse. And which also needs to be protected against the risks of pollution from a post-industrial urban environment. If Chadwick's interest in sanitation was much more than simply a matter of engineering, water today also has much wider resonances. For as the welfare state retreats from providing public provision, housing and health become an individual responsibility. Personal risks demand private insurance. Buying mineral water is not merely a matter of aesthetics, or even of social distinction. It is a way of securing a private water supply.

Reference

Osborne, T. (1996) 'Security and vitality: drains, liberalism and power in the nineteenth century', in A. Barry, T. Osborne and N. Rose (eds), *Foucault and Political Reason: Liberalism, Neo-liberalism and Rationalities of Government*, London: UCL press.

• Water (2) •

by Mary King

Underneath one image usually lies another. Beneath one city may stand the ruins of an earlier version, just as one old building will have three or four structures built over the first. For the city of Boston, carefully laid cement tiles cover the pavements of former colonial streets and concealed below the ground exists an intricate network of pipes, sewers and water mains which travel unnoticed to each neighborhood. Like the blue veins underneath the skin of a wrist, it is water which sustains the city.

A city may convey the impression of vast resource wealth, but underlying this representation is almost certainly another. While continual expansion symbolizes economic growth it subsequently represents a diminishment of resources. As it is, most metropolises suffer from periodic water shortages. In the thick of July when the sun is at its fullest and the rain does not fall, the city of Boston, Massachusetts invariably issues a water advisory, instructing its citizens to abstain from unnecessary use. But during such a crisis, hardly a soul forgoes a daily bath. Furthermore, residents continue to soap and rinse their dogs, to fill shiny plastic pools with water, and to hose down their fine, green patches of lawn. Despite the official curtailment, the liquid continues to flow uninterrupted from the tap, albeit more slowly. Most citizens of Boston do not think about the origin of this resource. There seems no reason to doubt that it has always existed and that it always will.

Yet, below the surface of the city, from the individual leaking house faucets, past the holding tanks where water evaporates at a constant rate, through the rusted, seeping iron pipes that lose millions of gallons each year, one hundred and twenty miles underground to a territory that seems only thinly connected to the larger city lies the source of the water. The Quabbin Reservoir shines like a bright coin surrounded by miles and miles of pine forest. Contained by fences and barricades, the property is guarded by the largest contingent of forest police in the state. Access to and use of the land is severely restricted, and even those who live in the surrounding towns have their rights circumscribed. Strict commandments are posted at each entrance: There shall be no hunting. There shall be no picnicking. There shall be no swimming. Indeed it is true that the reservoir is hardly occupied except for forest rangers who patrol the grounds assiduously and who insure that the park remains untouched, in a pre-habitation stage of wilderness. The pristine conditions of the forest are mandated by the reservoir guidelines which stipulate that transgressions are punishable by fine, imprisonment, or both. Yet this litany of restrictions and penalties strikes at the surrounding residents with the rawest of ironies. For although the reservoir appears a genuine feature situated upon the natural landscape, there is another reality which lies beneath it.

Wandering through the Quabbin forest it quickly becomes evident that old roads weave their way throughout the woods. Mossy stone walls carve up forgotten property boundaries and rusted farm equipment collapses on top of fern beds. There isn't a patch of land on the entire reserve which hasn't been trammeled or rent apart in the process of human activity. That such images do not accord with the current enactments of land use is not surprising.

It was shortly after summer in the year 1939 that the citizens of The Swift River Valley were evicted from their homes to make room for the coming municipal water project. Taking only those belongings which could be transported, 3000 people were evacuated by court order to neighboring towns and farms. Even the dead were not left in peace, for the communal cemetery in the heart of the valley was dug up and relocated to a far-off location, leaving only a few bones behind. The four thriving towns of Dana, Enfield, Greenwich, and Prescott were deliberately submerged in order to create the single largest domestic water source in the United States.

Just as a weathered gray stone stands over a grave, one memory is often erected above another. It happened that the land was flooded intact: houses, barns, and church steeples completely engulfed. Pastures of dry corn, meadowlands, and gourd patches were instantly transfigured into watery sunken gardens. Supple aquatic vines which grew up around stair posts began to inch through the cracks of floor-boards. The Swift River Valley became less a place of habitation than a home for turtles, for eels darting through doors and porticoes, and for green fish orbiting in and out of broken windows. Since all is well that stands forgotten, hardly a strand of sun penetrated that remote place, where falling leaves drifted slowly by and geese floated languidly on the surface of the sky.

But resentment has a long memory and refuses to be submerged. Many citizens of the surrounding towns, those direct descendants of The Swift River Valley, bequeath sentiments of loss to their children like cherished family heirlooms. Each year, at the start of autumn, the local school cafeteria undergoes a transformation as it is scrubbed clean, draped in black and festooned with garlands of colored leaves. Boys don their patent leather shoes while girls dress in their best white frocks and pastel satin sashes. To the strains of a melancholy fiddle, the children take to each other's arms in a mournful re-enactment of the final ball that took place in Greenwich the night before the four towns were flooded.

Thereafter, when summer rides high in July and the sun cooks the paved stones of downtown Boston, when fire hydrants spill lavishly open and citizens enthusiastically wash their cars, even as public fountains continue to spout water, water, water, the contents of the reservoir draw perilously low. A high ridge above the lake serves as an ideal vantage point from which to watch the water recede from the outline of the shore. While the Quabbin dries down to its barest, exposing the cracked mud and withered vegetation upon the lake floor, it becomes possible to partially witness an ancient shoe, a broken wagon wheel, or a remote porch step emerge from underneath the muck. During the waterless days of summer in the hottest swells of July, subtle glimpses of Dana, Enfield, Greenwich, and Prescott once again reappear.

W
W
W

•WEALTH•

Top 30 Banks in the world ranked by assets,[1] 1991

Asset rank	*Bank*	*City*
1	Dai-Ichi Kangyo Bank Ltd	Tokyo
2	Sumitomo Bank Ltd	Osaka
3	Sakura Bank Ltd	Tokyo
4	Fuji Bank Ltd	Tokyo
5	Sanwa Bank Ltd	Osaka
6	Mitsubishi Bank Ltd	Tokyo
7	Norinchukin Bank	Tokyo
8	Credit Agricole Mutuel	Paris
9	Credit Lyonnais	Paris
10	Industrial Bank of Japan Ltd	Tokyo
11	Deutsche Bank AG	Frankfurt
12	Banque Nationale de Paris	Paris
13	Barclay's Bank Plc	London
14	Tokai Bank Ltd	Nagoya
15	Mitsubishi Trust & Banking Corp	Tokyo
16	ABN-AMRO Bank Ltd	Nagova
17	Sumitomo Trust & Banking Co Ltd	Osaka
18	National Westminster Bank Plc	London
19	Mitsui Trust & Banking Co Ltd	Osaka
20	Societe Generale	Paris
21	Long-Term Credit Bank of Japan Ltd	Tokyo
22	Bank of Tokyo Ltd	Tokyo
23	Kyowa Saitama Bank Ltd	Tokyo
24	Dresdner Bank	Frankfurt
25	Daiwa Bank Ltd	Osaka
26	Union Bank of Switzerland	Zurich
27	Yasuda Trust & Banking Co Ltd	Tokyo
28	Istituto Bancario San Paulo di Torino	Turin
29	Citibank NA	New York
30	Toyo Trust & Banking Co Ltd	Tokyo

[1] Banks ranked by assets on 31 December 1991, or nearest fiscal year end. Holding companies are excluded.

Source: *The American Banker*, 1992; cited in Sassen, S. (1994) *Cities in a World Economy* (Thousand Oaks, CA: Pine Forge Press), p. 13.

• WEATHER •

by Deborah Levy

A woman stops me on the street and says, 'It's a bit rainy if you know what I mean?' Well, I don't know what she means because if it's not raining, the sun is shining. Does she mean her internal weather? It's a bit rainy inside her head? There is a very narrow vocabulary for better weather. A distinct lack of sun verbs.

Citizens hold barbecues in their small city gardens. They buy sausages, one hundred per cent heritage beef, and wait for it to rain like weatherman Michael Fish said it was going to. (It is rumoured that Mr Fish is being head-hunted to play the Marquis de Sade in a West End Musical.) It rains. The citizens retire indoors with a stupid, resigned smile on their faces. Even though they were told it was going to rain, they stare at the sky with crazy betrayed eyes. Meanwhile the sausages lie on damp smouldering charcoal reciting from *Hamlet*.

• WEEDS •

by David Bell

I am perpetually struck, while I tend my garden, by a question: 'What makes a weed a weed?' And, without consulting respectable horticultural tomes (it doesn't interest me *that* much), the best answer I can come up with is that a weed is simply *a plant out of place*. Or, as Zygmunt Bauman (1991) writes: 'weeds are the products of garden designs. They have no other meaning but someone's refusal to tolerate them' (p. 100). 'Weed' is thus a very loose, ambivalent classification that can embrace many floral forms. It may be that the plants we routinely commit to weed-dom in the UK share a number of character traits – that they grow well (and fast) in virtually any conditions, that they out-compete our more fragile (and more prized) cultivated specimens, that they seem to be more resistant to garden pests (while slugs and aphids devour all around them, only the weeds seem immune), and that they are neither useful nor beautiful (though this is the most debatable point, and the one I have most difficulty with: some of the 'weeds' in my garden do look nice, and some are potentially useful – but I still tend to pull them up or hoe them into oblivion). There are universal weeds – plants that seem unable to escape consignment to weed-dom, such as dandelions and cooch grass – but there are also site-specific weeds, such as the saplings that infest my garden from a neighbour's tree; these are of a perfectly acceptable, horticulturally sanctioned species, but to me they're weeds, continually springing up, pushing other plants aside or overshadowing them. A small patch of mint I planted is fast becoming weed-like in its voracity to colonize the side bed, and the wild geraniums planted by the previous owner all over the garden are getting dangerously close to reclassification for similar reasons. (On a larger scale, the rhododendron, a prolific flowering bush introduced by Victorians, is effectively outlawed in many areas of rural Britain, following the same fate that the giant hogweed earlier endured – to be classed as forbidden vegetation.) Weeds are site-specific in other ways, too: when I had an allotment, there was an entirely different register of weeds to look out for, and the general sentiment towards weeds among all allotment-holders was far more fascistic, since here they could hamper the productive potential of food crops, not just a few flowers.

A more interesting phenomenon is the upgrading of certain weeds to the category of 'wild plants' or 'wild flowers'. In the UK, various gardening organizations are especially keen to promote the growing of 'wild flowers' which an older generation of gardeners spent most of their time obliterating as weeds; while there will always be a fetish for cultivars in the horticultural world, the return of the wild (and the weed) – and the fascination for rediscovering (or inventing) 'traditional' folk uses from herbal remedies to cottage recipes – has marked an interesting turn for the fortunes of weed-kind, investing (horti)cultural capital in previously 'misunderstood' and denigrated species.

Of course, weeds are not a uniquely urban phenomenon, as the heavy use of pesticides on agricultural land testifies. But in the city, weeds have a different symbolism, since part of the anti-weed agenda in cities concerns their out-of-placeness in otherwise plant-free realms: in city streets, the outsides of buildings, growing between the cracks on pavements, and so on. Some weeds apparently thrive in these inhospitable environments where cultivars have no chance. Mark Dion's essay on *Ailanthus altissima* (in Dion and Rockman, 1996), called variously 'stink ash' or the 'tree of heaven', records an interesting example. The tree, introduced to the USA from China via England in the eighteenth century, 'thrives in waste zones: vacant lots, highway medians, alleyways, industrial fringes and other disturbed zones' (1996, p. 70) where it easily out-competes other flora, its roots eating through concrete and rubble, its growth (up to five feet per season) unimpaired by urban toxins. Look at such 'waste zones' in any city, and you will see similarly hardy colonizers grabbing hold of these fragile, peripheral environments. Last time I drove into Liverpool, parts of the main roads were disappearing under vegetation, and derelict buildings are soon clothed in green. Even abandoned cars can become makeshift hothouses for city weeds to thrive in.

Weeds have become an urban obsession, in fact; squads of city employees spend their time de-weeding public spaces in cities, while householders exert peer pressure on neighbours who let their gardens weedify (unless they're cultivating a wild flower garden). As Bauman points out, obsessing over weeds became part of the simultaneously nostalgic and modernist Nazi project of creating 'national landscapes' free from harmful, ill-bred, uncultivated influences; his elegant metaphor of the 'gardening state' embodies the weed-hating obsession with purity and order, which also functioned metaphorically in Nazi ideology. Weeds, then, represent impurity, disorder – the wild. Like their beastly counterparts, the feral animals that roam our cities, weeds often evoke phobic responses because, like feral animals, they inhabit the space between nature and culture, moving (regressing) from the latter towards the former. They are unruly, untrustworthy, unmodern and un-urban (but, of course, simultaneously completely modern and completely urban). They are nature out of place.

(*With thanks to Jon Binnie*)

References

Bauman, Z. (1991) *Modernity and Ambivalence*, Cambridge: Polity.

Dion, M. and Rockman, A. (eds) (1996) *Concrete Jungle*, New York: Juno.

W
W
W

•WORLD RECORDS•

by Arturo Escobar

Italo Calvino *in memoriam*

> With cities, it is as with dreams: everything imaginable can be dreamed, but even the most unexpected dream is a rebus that conceals a desire or, its reverse, a fear. Cities, like dreams, are made of desires and fears, even if the thread of their discourse is secret, their rules are absurd, their perspectives deceitful, and everything conceals something else. (Italo Calvino, *Invisible Cities*, p. 44)

There are two ways of describing the city of Cali. You can say – as the journalists from the global media who came to this city in the apogee of the drug scare – that the city is controlled by the drug mafia of the Cali cartel, and that at each step you take in this tropical city of balmy air you may be subjected to road blockades by special police forces in search of weapons and drugs, or be distracted by the scare of an occasional bomb threat; or that if you venture out at night with your friends to any of the hundreds of music places that exist in this city with a passion for music and dance you might find a petty drug boss who wants the place for himself and his men, free liquor for everybody, you hear him say as you leave in haste and as he announces he's bringing the best orchestra in town, also paid by him, to the place. And bearing in mind that many a store in the city's fashionable shopping centers, so pleasant to stroll in with their refreshing fountains and trees in the open air, is in business because of the bonanza associated with the drug trade, and that most people of Cali leisurely and unselfconsciously buy their goods in these stores, unable to distinguish between what's legitimate and what is not – you can then work back and forth from these facts until you have an image of what Cali is and will ever be. Or else you can say, like the young traveller who happened to visit the city one fine April day:

> 'I arrived there in my first youth, one morning many people were hurrying towards one of the neighboring, denuded hills, children women and men holding small seedlings of trees in their hands, and close to noon time, when the sun was at its zenith and too hot for somebody who came from the North, loudspeakers from a helicopter flying over the hill instructed people to plant their trees all at once. In scarcely one minute more than ten thousand trees had been planted, sufficient, I heard, to take away from Lima the Guinness record for most trees planted in a minute's time. And having remained in the city for some time, I witnessed on another day in July the good looking and charismatic mayor of Cali heading a team of hundreds of trucks and thousands of people to commence the most sustained sweep a city has ever seen: 15,000 tons of trash picked up in 48 hours by more than 200 trucks and thousands of volunteers.'

> 'Early in the morning of the first day of the gigantic sweep you could see the people of the poor quarters already beautifying their humble abodes, or picking up brooms and garbage bags at the local citizens' centers, and you could hear here and there musical bands and loudspeakers proclaiming the need to reclaim the culture of cleanliness, solidarity and civility, and observe young boys and girls bathing and washing façades and streets when the fire hydrants opened up their gargoyles to let the water run free. Senior citizen groups, water guardians and ecological guards, along with young

W W W

traffic police women and men, seemed to orchestrate the largest cleaning effort a city has seen in recent times – at least since the disinfection of Venice after its last plague at the century's dawn. And as I was told the next day, with this cleaning campaign – even as today's sweepings were already piling up on those of the day before and for years and decades to come; even as *caleños* began to forget the best wishes of the day before, too worried by myriad other things, from the most distorted real estate prices to seemingly ever escalating crime; and even as the charismatic mayor of Cali was being arrested on the second day of the sweep for having received money from the Cali cartel to finance his campaign – another world record was obtained out of this peculiar chapter in the social history of waste. In the months that followed, I learned that these events were only one of the many facets of that wondrous and contradictory space of peoples and places and things that opened before me on that first morning in Cali.'

Sources

Global media coverage of the Cali cartel and Colombia's drug production and trafficking (ongoing); field-day in Cali during the campaign for the Guinness record of city cleaning, July 3 and 4, 1997; local newspaper coverage of the event; friends' recollections of the April tree-planting world record; a flashing memory of Venice's disinfection in Visconti's *Death in Venice*.

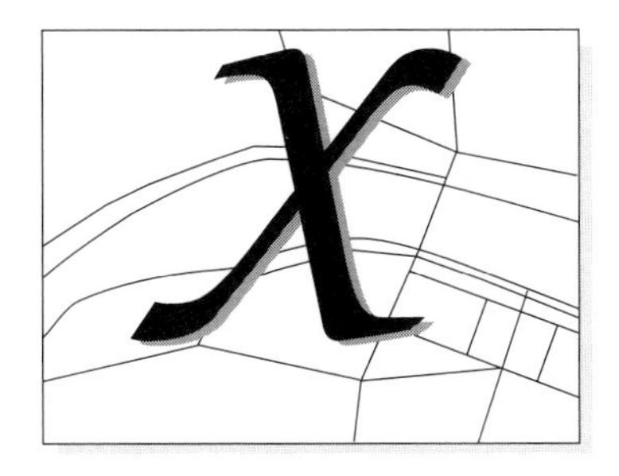

•X-RATED•

by Steve Pile

There are so many Xs in the city, it almost seems as if they are being deliberately collected there. Like. . . .

X-roads: There are very few dead-ends in the centres of cities. Instead, cities insist on keeping things moving by channelling traffic through one-way systems. Cars and people move so fast that it is sometimes impossible to do anything but follow the flow. Indeed, if you stop to think, you'll be shouted at. Stopping the city moving is almost a crime. Meanwhile, at crossroads, traffic is stopped the whole time – as cars and people take it in turns to obey the stop/start instructions. Those who cross out of turn do so at their peril. Here, it is moving that is a crime. Stop/Start: maybe it is the city that is at a crossroads, not sure whether to stay or go?

Xenophobia: Above all, cities are places of migration – and recent migration at that. Never before have more than half the world's population lived in cities. Never before has the word cosmopolitan been so true of the metropolis. Cosmopolitan defines the sophistication of the city dweller, the enjoyment they have in revelling in their encounters with so many different people, from so many different backgrounds. Yet the city too houses its xenophobes, angrily decrying 'those' from 'over there', now 'over here'. It is not that cosmopolitanism suggests that tolerance is the ideal for city life though; xenophobes should learn to love the foreign. For isn't it true that no one is a native to the city, since the city comprises too many different parts?

Xerox: People are not copies of each other; they have their own unique combination of needs, wants and desires. Aren't cities definitely cities, however, because – like nowhere else – there is the possibility that a person might meet someone else who shares their own unique pleasures?

X
X
X

X-directory: People like to make contact, they enjoy phones, and today it feels as if everyone has a mobile phone. Odd this, since cities are already full of phones. Yet, curiously, increasingly people choose not to list their numbers in telephone directories. And there isn't a directory for mobile numbers. At the point of contact, people find ever more ways to secure their privacy. Perhaps city life demands it: too many unknown factors. So, why not give someone a call?

Xmas: In central London, the telephone boxes are covered in calling cards offering the services of sex workers. At Christmas, a festive spirit enters the air . . . and, bearing that spirit in mind, you should think about Holly when she (or he, or she-he, or . . .) offers you a Christmas cracker, for it's an Xmas special, full of XXX videos and much, much more! Do you want to pull it?

X: Crosses often mark the graves of Christians taken by God. As a rule, the larger the cross, the richer the dead – paradoxically, though, the dead are not rich on earth or in heaven. As the years pass, the crosses weather and fall, since money is prone to die with the living. In the city, too, the cemetery is a dead technology. Has death become old-fashioned?

X marks the spot: As stories about pirates unfolded, there usually came a point where a chart of an isolated, uninhabited island was discovered, fought over and won. On the map, an X marked the spot. The simple pictogram – a triangulation point, only without the third axis – tells the lucky owner of the chart exactly where the treasure is buried; riches, that is, beyond anyone's wildest dreams. The directions are simple: find a landmark, go a certain number of paces in a certain direction, then get digging, greedily, urgently, expectantly, fearfully.

The path is so well defined, so narrow, they might as well be on a street. Find the town hall. Walk a hundred yards down the high street. Hurry up! Dig.

Perhaps pirates should have buried their treasure in cities. But perhaps not. The city is continually being dug up. It wouldn't be long before someone repairing the city's system of gas pipes, telephone or television cables, water mains, roads, pot holes, underground railways, or whatever, stumbled on their secret stash. On the other hand, who can tell what lies buried beneath the city?

X-Files: On the surface, the city roars and buzzes with the activities of people – as they trade, as they go out and have a great time, as they buy and sell, as they move and settle, as they do whatever they do. But, behind all these activities, something more prosaic is happening. The paper trail of all these human transactions is silently accumulating. Documents kept in triplicate are no longer the preserve of stuffy bureaucrats, nor have they diminished with the introduction of electric technologies. Indeed, the inter- and intra-nets of the paperless office have facilitated the proliferation of paper trails, as each email is copied to everyone imaginable and each person dutifully prints out their email trail and files it. Far from saving trees, the city is built on paper – that is, on the ghosts of trees. In London, this has meant an ever-increasing need for storage space for companies' archives. The problem has been solved by using giant underground spaces built during the Second World War as air raid shelters.

Big enough to house 8,000 people, eight deep-level shelters were built near London underground stations. However, only four were ever used as such. In part, this is because the tide of the war had turned. Instead, these spaces were used for other purposes, like housing the massive number of American troops that were arriving. Many stayed in the shelter below Stockwell station. The vault at Goodge Street station is called 'The Eisenhower Centre' because it was used to plan the D-Day invasion, but it is now used, along with four others, by Security Archives to store company data. Another is occupied by Britannia Data Management. One is part of the telephone system. But one is empty. What is going on beneath the city? Is this a case for Mulder and Scully? Where would they put the file?

There is a further tale to tell about air raid shelters in London: when the bombs first rained down on London, the government ordered the underground system to be locked up so that people could not escape into the tunnels, for they feared that people would not come out again. It took the determined intervention of a radical, 'Red Ellen' Wilkinson, a junior Minister, to get them opened. What kind of democracy was that?

X: Amidst all these Xs in the city, perhaps there is one that is all too rare – the one that is found on ballot-papers. One cross every four or five years is a rare X indeed. Perhaps, though, it is just as well – don't they say 'Don't Vote! The Government will get in'?

X
X
X

•Yellow lines•

by Marc Augé

The yellow line guides us and orientates us, but leaves a certain number of decisions to our judgement. In this way it is sometimes formidably ambivalent. In France it guides us in two ways. It defines the route that we follow, along its straight lines and curves, to such an extent that at night drivers trust its middle line, brightened up by their headlights, more than the borders of the road which are not always marked by another yellow line. Thus getting closer to the dangerous verges of the road. At the same time, the yellow line marks the border with the 'other side', the side you can only reach by taking illicit risks. Symmetrical and inverse to the one reserved for us. The side run by drivers who we expect to follow the same discipline as we do, and also respect the yellow line.

But there is yellow line and yellow line. Some are double, the solid line being paired on our side by a broken line, subtly imparting its message by informing us that we can overtake the vehicle that is slowing us down, by crossing the line if we judge that we have the time to do so. Sometimes extending too far towards the top of the hill, or at the beginning of a hairpin bend, this 'dotted tolerance' may go too far and present risks to the ones who blindly trust it. Although the yellow line may seem simple, it is as perverse as the cable the tightrope walker traverses. It indicates the way to follow – alongside all the dangers. But unlike the tightrope which is for solitary exploits, the yellow line is eminently social: it always presupposes, warns and organises priorities, forbidding overtaking, imposing one way. It is the visible *alter ego* of our social existence, the representation in colour, on tarmac, of an ethic rule that we should be able to memorise without the help of yellow lines – if we were drivers who were pure in spirit. We know this is not the case and the yellow line is a timely reminder that we are not alone on this earth, that social life is a matter of compromise and that the law resulting from it affects everybody. In this respect the yellow line is the symbol of the necessities and constraints of any social life. It is therefore not surprising that the yellow line bears a metaphoric value in everyday or political language. In France today it is relatively frequent (although this metaphorical usage is only recent) for an undisciplined student, an aggressive sportsman or shameless politician to be told that they have 'gone over the yellow line'. With the help of the colour this phrase, along with sport phrases such as 'yellow card' and 'red card', indicates an ultimate warning.

From another point of view, we will distinguish the subtle and problematic yellow lines which at the same time indicate and put into order the secondary itineraries – the long and rather silly lines, without any particular message, that cut part of the motorways into parallel lanes. The yellow line on the small roads is a little like a cartoon caption (the drawing would be the road itself and its route): it tells us to keep to the right, that the bend is too close to try to overtake, that this time we can go for it, only if we get back into our lane on time – where the broken line stops – that if we have time we can turn right at the next crossroads, turn off to visit the village whose church bell-tower can be seen through the trees. It is our own journey that it is commenting upon, in a way, and we may be surprised to find that we are excusing ourselves (to it?) if we happen to clip it when we get back in lane a little late. It holds our hand (or rather our steering wheel), and is itself rather lenient as long as its designated enforcers (motor-cyclists or other officers of the law) do not decide to take issue with our interpretations and to take note of all our omissions or errors.

On the motorway, the lanes are clearly marked: they deviate little, they are writ large. The yellow line is just a marker for driving in the night or a guide towards the exit. It does not talk like the countryside yellow line; it does not tell us about the hill ahead and the bend coming up. It never makes any mistakes, but it has nothing to say. And we too, silent, without regard for the impassive profiles of the drivers who pass us in the opposite lane, on the other side of the yellow line, we drive from one town to another without thinking about anything.

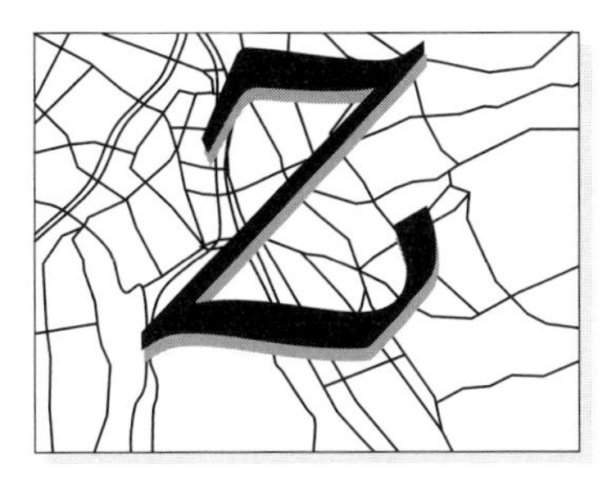

•ZERO•

by Ross King

City space, like speech, rides upon an underground system of language – there are signs, meanings; as they string together they yield analogously words (department store, landscape, place, path); words become sentences and ultimately texts (cities, countrysides).

Language, as Derrida would inform us, has dual origins: in metaphor (where the search for the universal – for 'language as such' rather than 'the languages of people' – finds its hope); and in song (the unrepeatable and untranslatable, that which cannot be spoken, sadness, joy, spirit). As the city becomes a city of universal signs – every city the same, with the same freeways, office blocks, fast foods, franchises, logos – it becomes purely metaphoric, and the ground zero of *metaphoric* space-as-language is at last before us. (It was the dream of Louis Kahn, my old teacher at the University of Pennsylvania, to find the 'ground zero' of architecture, the originary architecture of the human spirit. The lesson of the present, my new teacher, is that it might suddenly be before us.)

The ground zero of song, on the other hand, will always be of 'this place' – individual, unrepeatable, and always already slipping from our grasp. So what is it? What is 'this place' that is indeed unrepeatable, unrepresentable, untranslatable from one place to another or from one language to another, defying superimposition? . . . the longing for old memories, the tears of the returning exile . . . that dampness and ever-present feeling of water that 'defines' Sydney (for Vincent Buckley), the summer dust under the lime tree (Solzenitsyn), the already grey antiquity of a newly modern Paris (Monet, Baudelaire).

In introducing an essay on the literary cultures of rival Australian metropolises Sydney and Melbourne, Vincent Buckley was drawn to their differences as places – how to represent a memory through the articulation of difference. (A more fundamental question: Why can we only describe spirit by reference to what it is not?) So, 'Melbourne does not strike me as an interesting city . . . [rather,] it is all suburbia . . . not an *urbs* in the classical sense. It is at the opposite extreme from the beautiful and brilliant foreign-ness of Sydney, whose banality still remains hidden from my eyes, which are those of a lover, excited and scared . . . always the moist, warm, hill-ridden, self-subsistent water city of my first meeting.' One 'a sea city', the other 'withdrawn from water'; and the winds are different, and the vegetation too, and this 'contributes to the sub-tropical quality of [Sydney's] light, while Melbourne's light in the summer is clumped and softened, just as in the winter it is flattened and distributed' (Buckley, 1980).

Moistness, foreign-ness, warmth – when, as a recent refugee from Sydney to Melbourne, I first read Buckley's words, they quite tore at me, and twenty years later I still cannot contemplate them without emotion. They resonate with my own memories of lost times, lost friends, vanished places. But of course this is a nonsense. The Sydney that Buckley describes was never my Sydney: in my inner western suburbs the only beaches were on advertising posters, most memorably in the trams that (for me) never seemed to go to the beach, the moistness was the stifling oppression of summer or sitting all day in water-soaked clothes at school in winter, and the foreign-ness an exclusionary class structure (it certainly excluded me), a cultural élite, and the ostentations of unattainable wealth.

There is a less ecstatic literature of Sydney, typically of the inner city of the 1940s and 1950s – the era of my infancy and youth – and typically by women writers who knew the double grind of poverty and a man's world: grime and hard times, solace in the bottle, the heat and the cold, for them the moistness of leaking roofs and rising damp, rats and cockroaches, no sign of progress, but just the eternal return of departing children and rites of passage (see e.g. Park, 1948; Tennant, 1947). I can remember such places, but when I read those semi-autobiographical novels no emotion is touched. The yearning memory is always for something else.

These memories of place – indeed, the 'spirit of place' – are always of a mythic past. Walter Benjamin wrote of 'mythic history', of a classless world that never existed. Castles are not sites of domination and subjection, nor cathedrals of ideological repression, nor ports and palaces of exploitation, nor the old home of one's childhood of hunger and lost dreams. The spirit of place comes from a collective forgetting of the spirit of a time (*Zeitgeist*).

Here resides the function of cultural production and the role of the artist – to continually find and re-find, through its representation, the spirit of place. It is the never-ending discovery, in each age, of Sydney (or Paris, or Vienna, or New York) – the quintessential cities of the human spirit. But there is also another role: to unmask the oppressions, subjections, exploitations – to restore the meanings to the signs that constitute the city of signs – the once-metaphoric city that has lost the meanings that underlie the urge to metaphor.

(Does ground zero reside in the signs, or in the meanings – class relations, oppressions, exploitations – that lie beneath them but are masked by their modern slippage to mere signs?)

We would seek first to represent the unrepresentable (an impossible task, but it is the seeking that counts, not the achievement) – the song and play of a place, rain, dust, sunlight, sentiment. And we would seek to unmask the unmaskable – to force into consciousness the ways in which our nostalgia blinds us to the forms of our repression and self-delusion – to reveal the full potential, however still distant, of human emancipation and fulfilment. The task of the representer of space and place – that is, of the artist and the geographer – is to show *dreams of a better space of everyday life*.

References

Buckley, V. (1980) 'Unequal twins: a discontinuous analysis', in J. Davidson (ed.) (1986), *The Sydney-Melbourne Book*, Sydney: Allen and Unwin, pp. 148–158.

Park, R. (1948) *The Harp in the South*, Sydney: Angus and Robinson.

Tennant, K. (1947) *Tell Morning This*, Sydney: Angus and Robinson.

•Technical Note•

by Steve Pile and Nigel Thrift

If you have spent some time with this book already, you may be left with the impression that it is simply a collection of fragments, apparently with no privileged voice (theoretical or otherwise), no place from which the authors speak, no one city that can bring the entries down to earth. A vortex of thoughts, without direction or point. But, in fact, this book raises a series of technical questions about how cities are to be understood and about the place from which the city is theorised. This Technical Note concerns the theoretical 'underviewings' of the city that have been used to make sense of the diverse and paradoxical elements that cities undoubtedly contain. In this note we will not review the whole body of urban theory. Instead, we would like to take some alternative cardinal points (namely diagrams, montage, screens, clues) through which it is possible to see how cities are understood, either as a collection of fragments, or as a whole. Inevitably, however, we are still left with the central paradox that this book confronts: imagining the city as a whole is a necessarily partial exercise; putting parts of the city together will never add up to the whole. Urban theorising can spend forever with these fragments of the city: they can be (re)sorted, (re)assembled and (re)connected, but cities continually unsettle and disturb any claims to 'know' what exactly is going on. And this is perhaps why they are particularly and peculiarly fascinating.

We have not been interested in, or involved in, a project which seeks to 'know' cities, but rather in a space where the tensions involved in knowing cities – the different knowledges and objects of knowledge that are possible – can be set against one another. In other words, we want to understand the city as a patchwork of intersecting fields, as a discordant symphony of overlapping fragments. (Indeed, the Guide map on the inside back cover is an experiment in drawing this out.) Thus, wherever one starts from, one ends with superimposition and interchange, and so with imaginative access from one space into others. In other words, 'to identify with the city is to aspire to the pure anonymity of metamorphosis, of being nothing other than an endless turnover of perceptions and connections' (Sheringham (1996), summarising the work of Réda). In the literature, this kind of experiential arbitrage has often been represented through the figure of the street. The street was the place in the city where it all came together and moved apart (Prendergast, 1992). Nowadays, however, it might be argued that the street no longer holds this iconic status. The street is no longer the privileged locus of mutability but has been replaced by all manner of communications networks: the telephone, fax, e-mail and so on (e.g. Virilio, 1997).

Whatever the case, we need techniques, devices, to help us to suture some of the elements of these splintered cities. We will use four of these analytical and representational technologies: diagrams, montage, screen and clues. These technologies represent theoretical imaging devices, through which the city becomes visible as city. Each technique is partial, but each has its insights to lend. Using these techniques, the fragments in this book can begin to feel much more like a city, since the city is already in these fragments. First, let us start with *Diagrams*, in which we will use the work of Manuel Castells. Second, we will look at Walter Benjamin's use of *Montage* and his understanding of the city in ruins. Third, we will look at how the city has provided a *Screen* for imaginative writings about the city. Finally, we will look at how the city is full of *Clues* as to its identity. We wonder, since many theorists are keen to identify a perpetrator, what kind of crime has been committed? Or indeed whether cities are, in fact, a crime? Maybe, instead, the clues lead us to a different practical understanding of the city in which clues are ambiguous, the form of their assembly producing the crime (rather than the other way round).

Diagrams

One way in which people have sought to understand cities is by the identification of underlying social processes that produce urban spaces in the form that they do. This usually involves beginning with a general framework for understanding social processes, then the application of this framework to urban spaces. Using this method, urban space becomes the product of those processes; the test of the method is in its capacity to incorporate ever more diverse kinds of urban space. Sometimes words fail to convey either the theory or its understanding of cities. Sometimes diagrams become useful, because they can represent the essential features of the theory, and map them on to urban space. Essentially, these diagrams are simplifications of the complexities of cities. They take the whole of the city and interpret them through a part of the city: a representational process that is assumed to account for the city. In this way, the diagram comes to represent the essence of the city. Everything you need to know is here (even as the diagram also represents what is actually absent from the analysis). Let us look more closely at just one example of this process, taken from Manuel Castells' highly influential interpretation of the rise of a network society and its urban form.

Castells uses this diagram to represent the major nodes and links in the urban region of the Pearl River Delta. As you can see, cities are represented in the diagram by circles: the larger the circle the more important the city; while the links between cities are roads represented by straight lines. Though apparently static, dotted lines for future roads suggest that this diagram is undergoing some change. This is important. Castells is suggesting that this growing metropolitan region is becoming increasingly networked. And it is networks that are important in Castells' understanding of contemporary society.

Instead of representing 'the real solution', the diagram actually represents a set of assumptions about the nature of the city and of society. For Castells, the most important thing to understand is that

> the global city phenomena cannot be reduced to a few urban cores at the top of the hierarchy. It is a process that connects advanced services, producer centres, and markets in a global network, with different intensity and at a different scale depending upon the relative importance of activities located in each area *vis-à-vis* the global network. Inside each country, the networking architecture reproduces itself into regional and local centres, so that the whole system becomes interconnected at a global level. (Castells, 1996, p. 380)

In this diagram, then, we can see Castells' argument given space: the urban system is about more than simply what goes on in the major centres, like Guangzhou and Hong Kong, it also includes local centres such as Qingyuan and Xinan; by indicating inner and outer regions, Castells identifies different densities and intensities of interactions within the region; and, by assuming that these cities are further linked, as Hong Kong is to the West, we can assume that this diagram fits neatly into diagrams that could be produced for other regions, with their own systems, hierarchies and networks. Cities are located in these arborescent relationships as sites where processes settle and intensify, thereby creating stability in the system. Networks create cities, and these can be represented in the diagram. And all cities can be interpreted in this way. However, Castells' analysis has more to say than simply about the connections that exist between cities. There are other consequences too.

> Megacities articulate the global economy, link up the informational networks, concentrate the world's power. . . . Megacities concentrate the best and the worst, from the innovators and the powers that be to their structurally irrelevant people, ready to sell their irrelevance or to make 'the others' pay for it. Yet what is most significant about the megacities is that they are connected externally to global networks and to segments of their own countries, while internally disconnecting local populations that are functionally unnecessary or socially disruptive. I argue that this is true of New York as well as Mexico [City] and Jakarta. **It is this distinctive feature of being globally connected and locally disconnected, physically and socially, that makes megacities a new urban form.** (Castells, 1996, p. 404; emboldened in original)

A quick glance back at the diagram, and the origins – and the political urgency – of this statement begins to take on a new shape. As we have pointed out, the diagram represents the essence of a situation – everything you need to know – while actually only talking about part of a situation. In Castells' analysis, there are

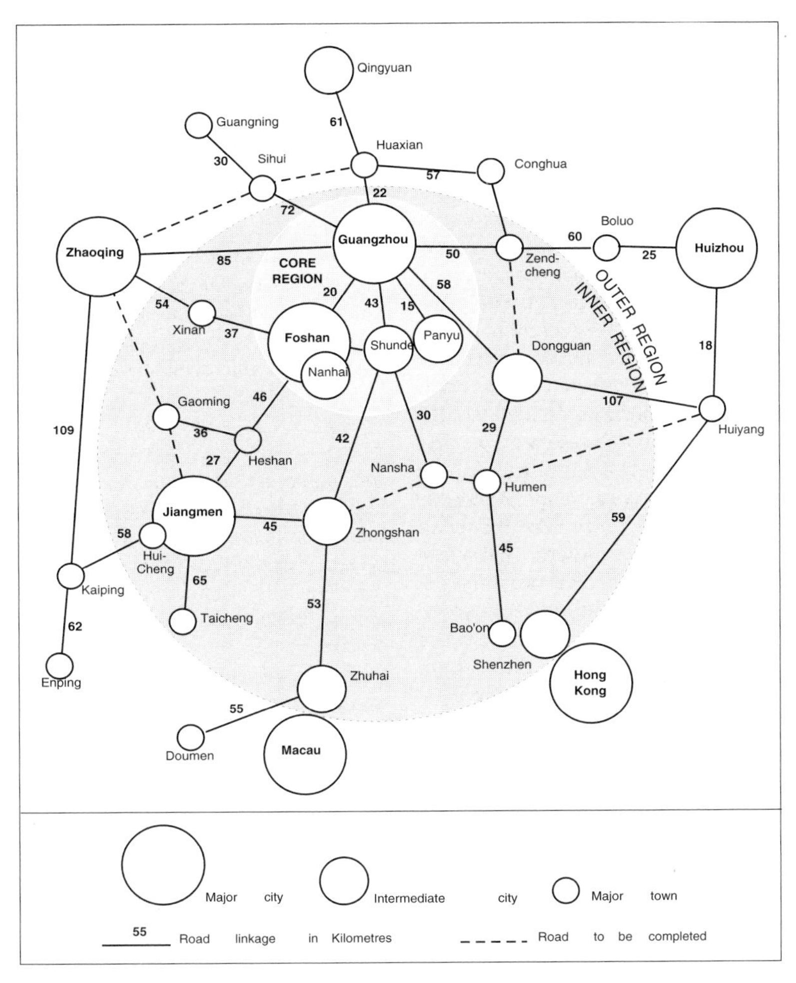

Figure 13 Diagrammatic representation of major nodes and links in the urban region of the Pearl River Delta

Source: Castells, 1996, p. 406

power élites in cities who have connections and who are able to concentrate power in their hands. Though you cannot see them in Castells' diagram, they are the hidden hands that make the networks work, although the scale of the diagram does not allow us to see what it is that they do, nor how they do it. If we cannot 'see' the structurally relevant people, then it is an absolute certainty that the diagram showing the essence of the city does not represent the structurally irrelevant people in cities – people like us. It does not show us what they are doing, and indeed how they might be creating alternative networks, or even the creation and maintenance of the networks that are there. We cannot see the ears, eyes and brains of the informational city, the bodies that live, breathe and die in the city. But we should not be so quick to condemn diagrams, or Castells' theorising, because this form of understanding *does* capture something significant about cities. It takes an idea and builds cities upon them and shows us what we get – in a diagram. For sure, Castells is right to point to the significance of networks and the connections that exist – whether visibly or invisibly – between things, but there are other ways to trace out the connections between things, perhaps by starting with the things themselves. To pursue this way of 'building' an understanding of cities, we need to move to another point of view: the montage.

Montage

In one of his best known works, *One Way Street* (written between August 1925 and September 1926), Walter Benjamin offers an enigmatic analysis of the intersection of modernity and the city (Benjamin, 1985). In part, his analysis is enigmatic because of the style it is written in. The work consists of 'a mosaic of aphoristic paragraphs, captioned by placards of urban scenery', as Susan Sontag puts it in her Introduction (1979, p. 34). Many of these aphoristic paragraphs, moreover, seem to bear little relation either to the city or to modernity except in the headings that accompany them (the placards Sontag speaks of). However, these paragraphs are a device that Benjamin is using both to interpret modernity and the city and to wake people up to the hidden histories of the present. In this, he might not be so far away from Castells' analysis of the underlying diagram of capitalism and the city. Nevertheless, their models and interpretations differ markedly. It will help this discussion if we take a fragment of Benjamin's analysis and look at how he builds up a picture of the city. Let us look at the paragraph(s) titled 'Travel souvenirs' (pp. 80–83; reprinted in its entirety in this volume under the heading 'Souvenirs').

In a work comprising aphorisms, 'Travel souvenirs' is apparently further fractured, for it contains a series of subtitled entries. Let us take one:

> *Navy.* – The beauty of the tall sailing ships is unique. Not only has their outline remained unchanged for centuries, but also they appear in the most immutable landscape: at sea, silhouetted against the horizon. (p. 81)

And another:

> *Heidelberg Castle.* – Ruins jutting into the sky can appear doubly beautiful on clear days when, in their windows or above their contours, the gaze meets passing clouds. Through the transient spectacle it opens in the sky, destruction reaffirms the eternity of these fallen stones. (p. 81)

And another:

> *Naples, Museo Nazionale.* – Archaic statues offer in their smiles the consciousness of their bodies to the onlooker, as a child holds out to us freshly picked flowers untied and unarranged; later art laces its expressions more tightly, like the audit who binds the lasting bouquet with cutting grasses. (pp. 82–83)

And finally:

> *Sky.* – As I stepped from a house in a dream the night sky met my eyes. It shed intense radiance. For in this plenitude of stars the images of the constellations stood sensuously present. A Lion, a Maiden, a Scale and many others shone lividly down, dense clusters of stars, upon the earth. No moon was to be seen. (p. 83)

As a sample of a sample, we hope these give a flavour of the whole work. As you will see, the suggestion that the work is tangentially related to the city and modernity seems well founded. More than this, the argument that it is enigmatic, its purpose obscure, also seems incontestable. From these fragments, let us try to reconstruct Benjamin's enterprise, his dream of a revolutionary practice that would alert people to their circumstances and allow them to realise their own dreams (see Pile, 2000).

The first, and possibly most obvious point is that these travel souvenirs are descriptive fragments, brought back from many different places: Versailles, Seville, Marseilles, Moscow, to name a few. 'Travel souvenirs', then, is joined by some narrative of travel that underlies the work, but where the journey does not provide a rationale for the evanescent images Benjamin relates. Benjamin, then, is pulling together a spatial story that is not related in terms of linear time or of a conscious journey – some fragments, even, are dreams. Benjamin's seemingly disconnected travel tales, then, have more connections simply by evoking journeys to other cities. By placing these 'postcards from another place' side by side, Benjamin is also alluding to – or digging into – the nature of modernity and the modern city.

Let us think about the fragments again: ships, castles, churches, stars, monuments. Each seems to be torn from its surrounds to provide an alternative context. Sailing ships stand still on a timeless horizon; ruins jut into the clouds; a cathedral in Marseilles appears like a railway station; constellations of stars take on meaning; ancient statues are counter-posed with modern monuments. In these stories, a subtle twist appears in each of the tales. The sailing ship, for example, has long gone, yet it is given a timelessness. The placing of these histories side by side is revealing, as it is in the juxtaposition of ancient statues and modern monuments. Ruins and churches, too, take on new meanings in the modern city, while older monuments take their place among the new castles and churches of the modern city: with their shopping malls, railway stations, clocks and traffic regulation.

These juxtapositions are a montage of urban images. But these images are meant to be read side by side, and in moving between them they are meant to reveal something of the present, if only in a glimpse. If one aspect of the present is the discontinuous nature of space, with its souvenirs and its myriad connections to other places, then another is the Janus-faced nature of history. History moves on in these images, but side by side; we can also see the ever-presence of eternity, always returning, as if history could only be shaken, but never stirred.

It was Benjamin's aim, however, to awaken history, to unsettle its dust. By placing these images together, by asking the reader to make the connection between these images for themselves, and by putting these images into a relationship that also changed their meanings, Benjamin hoped to set in motion a train of thoughts that would keep moving. But he was not intent on destroying what was there already; far from it, he was attempting to recuperate the ruins and dust of the modern city and to reassemble them into something that was genuinely new, genuinely unimaginable. While Castells' enterprise is to describe the underlying structural principle that allows a complete interpretation of everything that is going on in the city, Benjamin's analysis works in the opposite direction. From the fragments, from the discarded and the overlooked, from the trivial and the fashionable, Benjamin hopes to build up a picture of the city that is as old as the sea. In this way, the city would be silhouetted against the horizon. Or it might even be possible to see the city in a totally different light (Gilloch, 1996; Caygill, 1997).

One way Benjamin thought that the city might be seen anew was through moving images. Cinema, he thought, might offer a new perspective on the city by bringing many images into sharp juxtaposition, by being able to establish connections between things that might otherwise seem unconnected. Of course, there are more screens than the silver screen, and other media through which connections can be made. What is important is that people have always used multi-media in the attempt to capture the experience of cities, through various projections (into books, on to screens). It is to the flow, or distribution, of images that we turn next.

Screens

Our notion of images has been changed by the multiplication of screens through which and with which we now frame the city: the cinema screen, the television screen, the computer screen. These screens do not offer passive representations like Chartier's (1995) effigies of dead princes lying in state. They offer an *active* means of grasping hold of the city. It is as though architecture has been replaced as the prime means of shaping the city by the cinema and television.

How to make sense of this sc(r)enic perspective? This is no easy task. To begin with, these representations are themselves products of particular notions of space which carry with them potent presuppositions about how spatial order is generated: from the spaces of the bourgeois interior with their strong notion of an 'inside' through the spaces of aesthetic modernism with their sense of spaces (and time) as cut up and splintered, through to current notions of spaces as libidinal economies characterised by mobility, flow and swirls of intensity which might figure even an apparently solid building as 'permeated from every direction by streams of energy which run in and out of it by every imaginable route: water, gas, electricity, telephone lines, radio and television signals, and so on' (Lefebvre, 1991, p. 93) – an observation that applies equally well to entire cities (Burgin, 1996).

> How, then, might we consider the urban life of screens, once we no longer consider screens to be passive reflections of some more real reality? Perhaps, screens can best be seen as characterised by three properties. To begin with, screens can be seen as 'in the middle', as relations of encounter which obtain their power by expressing the quality of a sensation, feeling, action, or state. 'In a becoming, one term does not become another; rather each term encounters the other, and the becoming is something between the two, outside the two'. (Smith, 1997, p. xxx)

Second, screens must be seen as crucial elements of fantasy, as what Benjamin called 'neuralgic nodes'. The city, in other words, is simultaneously present as material and psychical and screens are the increasingly vital intermediaries. Conscious and unconscious fantasies are a force in our lives as real as any material circumstances (see Pile, 1996); the stuff of dreams is as much the stuff of the city as knowledges or tools.

Finally, these screens must be seen as indices of possibility. For more optimistic writers like Appadurai (1996), the proliferation of screens has produced new imaginative possibilities for people in cities around the world, has literally enriched our imaginative experience of cities, by producing psychic echoes and reverberations that enliven rather than dull the senses.

We can go further still in two ways. One is to provide screens with their own independent life. In actor-network theory, for example, screens might be refigured as the nodes in networks of a much expanded realm of the 'social', no longer just feeding the psychical world but having their own non-organic forms of existence. And in the work of vitalist writers like Deleuze, screens become a means of expressing 'affects' or 'percepts', qualities of possibility existing independently of their articulation in any particular state of affairs or determinate space of individual subjects. Screens make these qualities into 'forces' that are 'related to each other (exertion–resistance, action–reaction, excitation–response, situation–behaviour, individual–milieu) and are actualised in determinate space-times, geographical or historical milieus, and individual people' (Smith, 1997, p. xxxi). Thus screens allow us to pass into a 'general perception', rather like Virginia Woolf's *Mrs Dalloway*.

> Mrs Dalloway has perceptions of the town, but this is because she has passed into the town like 'a knife through everything', to the point where she herself has become imperceptible; she is no longer a person, but a becoming ('She would not say of herself, I am this, I am that'): the Town as a perception. What the percept makes visible are the invisible forces that populate the universe, that affect us and make us become: characters pass into the landscape and themselves become part of the compound of sensations. (Smith, 1997, p. xxxiv)

People have sought to express the 'affects' and 'percepts' of the city by placing images together. These juxtapositions mirror the way in which the city juxtaposes many different possibilities, emotions, sensations, perceptions. The city, for many, seems to be a rebus – a picture puzzle. Perhaps this puzzle can be solved. Many have sought to hunt down clues, to find out 'whodunit?' – and why?

Clues

One other way to assemble the city. In a recent book, the French sociologist Pierre Bourdieu (1998, p. 7) notes the degree to which cities are the product of the multiple dispositions – structured principles of improvisation – that characterise human *practices* (Thrift, 1996). His Parisian example is a telling one: 'it suffices to think of the extraordinary amount of agreement generated by the numbers of dispositions – or wills – which are assured in just five minutes of the circulation of traffic around the Place de la Bastille or Concorde.' Yet we should not make too much of the efficacy of this structured chaos. Precisely because the principles of interaction have their improvisatory components, accidents happen. Indeed, Parisian driving is known for its special elastic qualities. Certainly French roads are among the most dangerous in the world, with 8300 dying in traffic accidents in 1998: 'such carnage appears to have little deterrent effect. Last month a paraplegic man was stopped for driving at 115 mph in a BMW fitted with special controls. His disability was the result of a car accident caused by speeding' (Macintyre, 1999, p. 20).

Such outcomes are helped by an urban culture which treats many of the rules of the road with latitude. For example, traffic wardens, known dismissively in France as 'periwinkles' on account of their flower-blue suits, hand out an estimated five million tickets annually in Paris. But about a quarter of these are simply never paid for all sorts of reasons – patronage, knowing the right people, general parking fine amnesties at elections, and so on. And many fines for speeding, light-jumping and the like are also 'lost', under an ancient system by which motorists *faire sauter* their fines, a term that may be translated variously as 'leap frog', 'blow up', 'bounce', or 'fry'.

In other words, the inhabitants of different cities in different countries have different senses of what counts as normality, a crime or a transgression. To understand these dispositions and how they are practised requires a different conjectural kind of knowledge of the city, a knowledge of what counts, what matters, what stirs the passions. In turn, the hunt after these motives requires forensic or clinical skills which hone *intuition*. We are not, then, looking for grand theories of what counts or for deep inner meanings but for the *clues* to the pattern of interaction, clues that allow us to gain a practical hold on the situation. These skills are the skills of the connoisseur, the clinician or the detective, hunting down particular ways of being through the most negligible of details. They mean that even the smallest of signs can each become clues:

> it was once said that falling in love meant over-valuing the tiny ways in which one

> woman, or one man, differed from others. This could of course be extended to works of art or to horses. In such contexts, the elastic rigour (to use a contradictory phrase) of the conjectural paradigm seems impossible to eliminate. It's a matter of kinds of knowledge which tend to be unspoken, whose rules, as we have said, do not allow themselves to be a connoisseur or a diagnostician simply by applying the rules. With this kind of knowledge there are factors in play which cannot be measured: a whiff, a glance, an intuition. (Ginzburg, 1980, p. 28)

It is no surprise, then, that the figure of the detective has become one of the standard representations of the city. In this figure, and in the main characters in the novels of writers like Paul Auster or Peter Ackroyd – who seem to spend much of their time following the style of detectives even when they are not – we find seekers after clues who value discontinuity, invest in deferral, make leaps of intuition. Their convoluted knowledge of the city and its inhabitants resists systematisation, but knowledge it is; a kind of practical knowledge powered up.

Conclusions

A staple of the experience of modern cities is the city guidebook. Social guides seem to document all aspects of urban life; they provide maps of the key sites to visit, they provide lists of restaurants, hotels, shops, clubs, theatres, museums, they provide signposts to 'a good time'. Some of the guidebooks – Baedeker, Michelin, The Rough Guide – have become institutions in their own right, provoking Second World War air raids, and wars over their judgements. More than this, guidebooks have organised city life and have become constitutive of much of what modern cities are and how both visitors and more permanent inhabitants perceive them. These guidebooks 'order' the city in specific ways, pitched at different kinds of travellers, city-goers and siteseers. Perhaps in response to the confusion of cities, there is a profusion of guides seeking to sort them out. There is, for example, an extraordinary abundance of English-language material on Paris over many centuries which has coloured, to some extent at least, how modern Paris is encountered and dreamed (see Clark, 1996). Guidebooks themselves become jumping-off points to the negotiation of cities: families argue about where to visit, situationists provide guidebooks intended to get you lost, and increasingly guide 'books' are appearing on the internet (and even for the internet!), complete with real-time pictures which provide virtual interaction. And then the lack of guidebooks can itself become a start. In E. M. Forster's *A Room With A View*, for example,

> The youthful Lucy Honeychurch has travelled to Florence as a first-time visitor: she is accordingly equipped with Baedeker's *Handbook to Northern Italy*, the most important features of which she has committed to memory. But, entrusting the book to one of her elderly lady companions she finds herself to be 'In Santa Croce with no Baedeker'. As a result she successfully escapes the straightjacket of Baedekerian confinement and liberates herself from what Forster neatly terms 'the orthodox Baedeker-bestarred Italy'; and the novel . . . will tell in part of her encounter with a more authentically 'Italian Italy', of her discovery of 'the true Italy' (again Forster's expressions) of which the guidebooks can only scratch the surface. And Lucy's liberation will be more than just from her persona as a tourist: by losing her Baedeker and with it her ingenuousness as a foreign traveller, Forster's heroine, as well as discovering a truer Italy, will also discover herself. (Clark, 1996, p. 28)

This book is like a guidebook in that it, too, can only scratch at the surface. It, too, can only provide a fragment, or fragments, of what the city is like – guides always seem less trustworthy in cities that you know well! But, at the same time, in providing a multiplicity of perspectives it attempts to do something more. It is not just a compendium of snapshots, or fleeting images, or places to go, but a dance of the fragments, of the jokes, delights, dreams, dreads, violences, abstractions, frustrations, angers and anxieties, smells and sounds, that go to make up the urban 'modern'. And this is appropriate. For not only is urban space increasingly produced as a series of commodifiable fragments, but the city itself acts to fragment life, experience and space, such that it is, to all intents and purposes, an unassimilable, irreducible and sometimes even incomprehensible entity. So that, in a sense, the fragments are where the city is to be found. But, of course, these fragments are also meshed together by networks of speed, light and power which constitute 'an invisible city' (Latour and Hermant, 1998) which this book also attempts to trace out. In the end, all we have tried to do is to provide a set of (intellectual) resources for working through these visible and invisible cities, all the way from A through to Z.

References

Appadurai, A. (1996) *Modernity at Large*, Minneapolis: University of Minnesota Press.

Benjamin, W. (1985) *One Way Street and other Writings*, London: Verso.

Bourdieu, P. (1998) *La Domination Masculine*, Paris: Editions du Seuil.

Burgin, V. (1996) *In/Different Spaces: Place and Memory in Visual Culture*, Berkeley: University of California Press.

Castells, M. (1996) *The Rise of the Network Society. Volume 1: The Information Age; Economy, Society and Culture*, Oxford: Blackwell.

Caygill, H. (1997) *Walter Benjamin: The Colour of Experience*, London: Routledge.

Chartier, R. (1995) 'The world as representation', in J. Revel and L. Hunt (eds), *Histories. French Constructions of the Past*, New York: New Press, pp. 544–558.

Clark, R. (1996) 'Threading the maze: nineteenth century guides for British travellers to Paris', in M. Sheringham (ed.), *Parisian Fields*, London: Reaktion, pp. 8–29.

Gilloch, G. (1996) *Myth and Metropolis: Walter Benjamin and the City*, Cambridge: Polity Press.

Ginzburg, C. (1980) 'Morelli, Freud and Sherlock Holmes: clues and scientific method', *History Workshop*, 9, pp. 5–36.

Latour, B. and Hermant, E. (1998) *Paris. Ville Invisible*, Paris: Institut Synthelabo/La Decouverte.

Lefebvre, H. (1991) *The Production of Space*, Oxford: Blackwell.

Macintyre, B. (1999) 'Time to clamp down on tickets to hide', *The Times*, 9 January, p. 20.

Prendergast, C. (1992) *Paris and the Nineteenth Century*, Oxford: Blackwell.

Pile, S. (1996) *The Body and The City*, London: Routledge.

Pile, S. (2000) 'Sleepwalking in the modern city: Walter Benjamin, Sigmund Freud and the world of dreams', in S. Watson and G. Bridge (eds), *Blackwell Companion to Urban Studies*, Oxford: Blackwell.

Sheringham, M. (1996) 'City space, mental space, poetic space: Paris in Breton, Benjamin and Réda', in M. Sheringham (ed.), *Parisian Fields*, London: Reaktion, pp. 85–114.

Smith, D. W. (1997) 'Introduction. 'A life of pure immanence: Deleuze's "critique et clinique" project', in G. Deleuze, *Essays: Critical and Clinical*, Minneapolis: University of Minnesota Press, pp. xi–iv.

Sontag, S. (1979) 'Introduction', in W. Benjamin (1985) *One Way Street and Other Writings*, London: Verso, pp. 7–28.

Thrift, N. J. (1996) *Spatial Formations*, London: Sage.

Virilio, P. (1997) *Open Sky*, London: Verso.

Index